Fundamentals of Laser Optoelectronics

SERIES IN OPTICS AND PHOTONICS

Editor-in-charge: S. L. Chin

Vol. 1: Fundamentals of Laser Optoelectronics
S. L. Chin

Series in Optics and Photonics — Vol. 1

Fundamentals of Laser Optoelectronics

S. L. Chin

Centre d'Optique, Photonique et Laser
Département de Physique
Faculté des Sciences et de Génie
Université Laval

Published by

World Scientific Publishing Co. Pte. Ltd.,
P O Box 128, Farrer Road, Singapore 9128
USA office: 687 Hartwell Street, Teaneck, NJ 07666
UK office: 73 Lynton Mead, Totteridge, London N20 8DH

FUNDAMENTALS OF LASER OPTOELECTRONICS

ISBN 981-02-0072-2
981-02-0073-0 (pbk)

Printed in Singapore by JBW Printers and Binders Pte. Ltd.

TO MY WIFE, MAY,
and my sons, ERIC and PETER

CONTENTS

PREFACE

This book is based on a course given by the author to third and fourth year undergraduate students of physics, engineering physics and electrical engineering. The purpose is to introduce some of the fundamental principles underlying laser beam control in optoelectronics. The contents of the book are scattered in many sources and there seems to be no single source available at the undergraduate level. That's why the present book is written. Although the aim of the book is high, it is inevitable that there are errors and flaws. The author would appreciate it very much if the reader could drop him a line indicating any weakness and error he or she has found at the following address.

Tel: (418) 656-3418
Fax: (418) 656-2623
Mail: Dept. of Physics
Laval University
Ste-Foy, Quebec
Canada G1K 7P4

(It is the intention of the author to update and/or modify the content every two to three years. Thus, all comments are appreciated.) The following are the major references used by the author. At various places in the text, the names of the authors of these sources will be mentioned and the reader is requested to consult these sources for detail.

(a) M. Born and E. Wolf, "Principles of optics", 6th edition, Pergamon Press (1980).
(b) A. Sommerfeld, "Optics", translated by O. Laporte and P. A. Moldauer, Academic Press (1954).
(c) A. E. Siegman, "An Introduction to Lasers and Masers", McGraw-Hill (1971).
(d) A. Siegman, "Lasers", University Science Books (1986).
(e) E. Hecht, "Optics", Second edition, Addison-Wesley (1987).
(f) A. Yariv and P. Yeh, "Optical waves in crystals", Wiley-Intersciences (1984).
(g) W. Koechner, "Solid-State Laser Engineering", Springer-Verlag (1976).

ACKNOWLEDGEMENTS

The author wishes to thank Mrs. Claudette Nadeau for typing the manuscript patiently. It is indeed a pleasure working with her. Miss Jennifer Decker has read several chapters of the book and has made a number of suggestions for correction and improvement. The author has taken these into account and appreciates very much her help. The following are those who have also helped in preparing the manuscript, especially drawing the figures: Mr. Alain Laverdière, Mr. Simon Lagacé, Mr. Gaston Godin and Mr. Denis Lessard. I thank them all. Much of the thinking in the book is based on my personal experience in research work on intense laser interaction. Since the research work has been consistently supported by le Fonds-FCAR of the Province of Quebéc, the Natural Sciences and Engineering Research Council (NSERC) of Canada and, of course, Laval University, I owe them all for the indirect support for the present work.

INTRODUCTION

The laser is now a tool for many scientists and engineers. The application (or use) of laser is so widespread that in almost every field of science, engineering and technology including medicine and everyday life, we can find laser in use. At the same time, new engineering and technological disciplines are created. Names such as photonics, optoelectronics, optronics etc. are becoming normal vocabulary. As we step into the 21st century, optical technology is certainly going to play a significant role in our life. More and more people are going into the vast field of optics, once only narrowly defined as lens maker's little world. Such a big change (or revolution) is in part due to the advancement in electronics but it is the invention of the laser 30 years ago that made this revolution possible. In fact, laser is the heart of most optical, optoelectronic and photonic applications. There is thus a need among the undergraduate students and new comers to the field to have a book at hand that discusses the fundamental principles of controlling a laser beam. It is the hope of the author that the readers, after reading through the book will have a good background to go into any laser or optoelectronic or photonic laboratory so that they can quickly learn to use the equipment and control the laser beam (and light beam) without difficulty.

Controlling laser (and light) beams means making any of the following types of changes (or manipulations) of a laser beam; several of them are interrelated.

a) Temporal change: temporal laser mode control, switching (ON/OFF), laser pulse slicing etc.

b) Spatial energy (intensity) distribution change: spatial filtering, apodization, spatial laser mode control etc.

c) Directional change: deviation by reflection, refraction and diffraction (grating, active or passive etc.), guiding (fibers and waveguides).

d) Total intensity (power or energy) change: Amplitude modulation, switching, mode control etc.

e) Frequency/phase change: laser mode control, phase (frequency) modulation, nonlinear optics etc.

f) Polarization change: propagation through wave plates, retarders (active and passive), dichroic polarizers etc.

Each manipulation has a certain purpose and application. Knowing the basic principles underlying the techniques of controlling the laser (light) beam will be a great help to all those who work with lasers and optical applications.

Three major subjects underlie most of the above mentioned laser beam control techniques, namely, optics of isotropic media, optics of anisotropic media and laser. This book explains the principles of the physical phenomena underlying the last two subjects at the undergraduate's level, especially the part on optical anisotropy. This is because optical anisotropy is the heart of many electro-optic, magneto-optic, acousto-optic, non-linear optical, Q-switching and mode locking devices. The principles of these latter devices are also treated. Optics of isotropic media is classical optics and is assumed known to a certain extent to the readers. Whenever necessary, there will be a reminder about a particular subject. No attempt is made to deal with fiber optics. The book terminates with a review of the principles of short laser pulse generation.

Problems for students are not explicitly given at the end of each chapter. Instead, they are given along the course of the text and are indicated by the underlined word "exercise".

CHAPTER I

MAXWELL'S EQUATIONS, WAVE EQUATION AND WAVES: A REVIEW

§1.1 A pictorial view of E-M waves

The concept of the electromagnetic (E-M) wave is central to this book. Though we talk about manipulation and control of laser (and light) beams, we need to use the wave concept to interpret propagation and interaction. Such concept of an electromagnetic wave originates from the Maxwell's equations. In general, they are given by:

$$\nabla \cdot \vec{E} = \rho/\epsilon_o \tag{1-1}$$

$$\nabla \cdot \vec{B} = 0 \tag{1-2}$$

$$\nabla \times \vec{E} = -\frac{\partial \vec{B}}{\partial t} \tag{1-3}$$

$$\nabla \times \vec{B} = \mu_o \epsilon_o \frac{\partial \vec{E}}{\partial t} + \mu_o \vec{J} \tag{1-4}$$

Eq. (1-1) is the differential form of Coulomb's law where $\vec{E}$ is the electric field of a system of charges of density ρ. ϵ_o is the dielectric constant in free space, or the permittivity of vacuum.

$$\epsilon_o = 8.854 \times 10^{-12} \text{ F/m} \tag{1-5}$$

Eq. (1-2) is the characteristic equation for a magnetic field $\vec{B}$ which doesn't have a point source; i.e. there is no magnetic monopole.

Eq. (1-3) is the differential form of Faraday's law and eq. (1-4) represents the modified Ampere's law.

$$\mu_o = 4\pi \times 10^{-7} \quad \text{H/m} \tag{1-6}$$

is the permeability of vacuum and $\vec{J}$ is the current density.

In free space, $\rho = 0$, $\vec{J} = 0$ and eq. (1-1) to (1-4) become:

$$\nabla \cdot \vec{E} = 0 \tag{1-7}$$

$$\nabla \cdot \vec{B} = 0 \tag{1-8}$$

$$\nabla \times \vec{E} = -\frac{\partial \vec{B}}{\partial t} \tag{1-9}$$

$$\nabla \times \vec{B} = \mu_o \epsilon_o \frac{\partial \vec{E}}{\partial t} \tag{1-10}$$

Let us first of all interpret eq. (1-7) to (1-10) qualitatively. Mathematically speaking, when the divergence of a vector field is zero,

i.e. $$\nabla \cdot \vec{V} = 0$$

where $\vec{V}$ is the vector field, it means that every field line will form a closed loop and they do not cross each other. From eq. (1-7) and (1-8), we can say that both the electric and magnetic field lines form closed loops in space, as shown schematically in Fig. (1-1a and b). If there were only these two equations, then the $\vec{E}$ and $\vec{B}$ fields would be completely independent. It is eq. (1-9) and (1-10) that link them together.

From eq. (1-9), we understand that the time rate of change of the magnetic field generates a circulation of electric field around it (Faraday's law). This is shown pictorially in (Fig. 1-1c) where around the loop of $\vec{B}(t)$, loops of E(t) are generated at every point of the $\vec{B}$ - loop. Only four points are indicated in (c). Note that the direction of $\vec{E}$ conforms with the negative sign in eq. (1-9). The electric field in turn generates magnetic fields around it according to eq. (1-10) which says that the time rate of change of an electric field produces a circulation of magnetic field around it. This is shown in Fig. (1-1d). We stress that because of the constraints of eq. (1-7) and (1-8), the electric and magnetic fields should form loops as in Fig. (1-1a and b). Also, the circulations in eq. (1-9) and (1-10) require them to form loops. Eq. (1-9) and (1-10) provide the coupling between them.

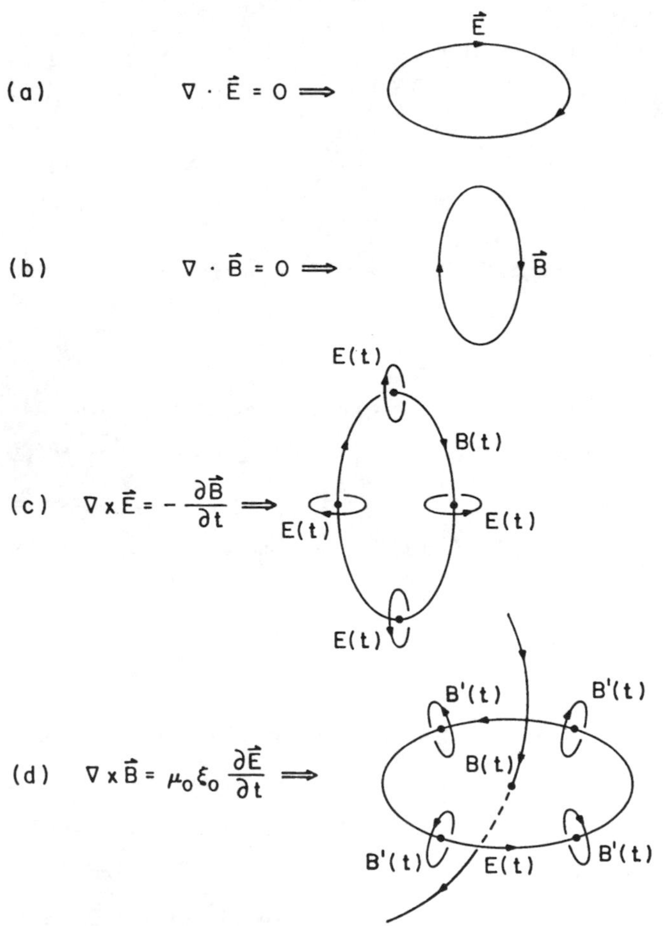

Fig. 1-1. Pictorial interpretation of Maxwell's equations.

Now, the newly generated magnetic field (Fig. 1-1d) will in turn generate a new electric field around it according to eq. (1-9) and the new electric field will generate newer magnetic field according to eq. (1-10) and so on. Along one direction in space, the relationship between these successive generations of $\vec{E}$ and $\vec{B}$ field loops look like a chain as shown in Fig. (1-2). Such "chain reactions" expand in all directions and this constitutes the propagation of the $\vec{E}$ and $\vec{B}$ fields. We have now a feeling of wave propagation. At every point P in space at time t, there is a resultant electric and a resultant magnetic field. Each is a superposition (vector sum) of those expanding field loops that reach this space point P at time t. Such <u>superposition principle</u> is an assumption that if we put a test charge (i.e. detector) at point P and time t, it will experience the resultant electric field mentioned above. Similarly, if we put a test current (i.e. detector) at P and t, we assume that it will experience the resultant $\vec{B}$. Such assumptions are proved to be valid experimentally.

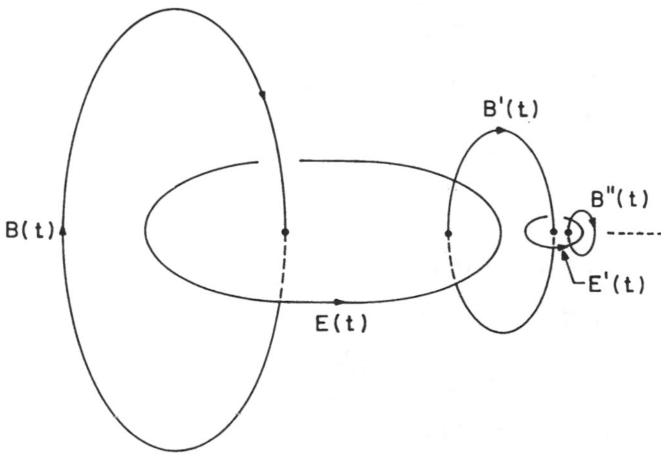

Fig. 1-2. Pictorial visualization of electromagnetic field propagation by combining Maxwell's equations.

§1.2 Wave equation and plane waves

A rigorous description of the fields $\vec{E}$ and $\vec{B}$ results from the solution of Maxwell's equations. We give a quick derivation of the wave equations. The following vector identity will be used.

$$\nabla \times \nabla \times \vec{V} = \nabla (\nabla \cdot \vec{V}) - (\nabla \cdot \nabla) \vec{V} \tag{1-11}$$

where $\vec{V}$ can be either $\vec{E}$ or $\vec{B}$. Because of eq. (1-7) and (1-8), we can simplify eq. (1-11).

$$\nabla \times \nabla \times \vec{V} = - \nabla^2 \vec{V} \tag{1-12}$$

Now, taking the rotation of eq. (1-9) gives

$$\nabla \times \nabla \times \vec{E} = \nabla \times \left(- \frac{\partial \vec{B}}{\partial t} \right) \tag{1-13}$$

Using eq. (1-12), eq. (1-13) becomes

$$- \nabla^2 \vec{E} = - \frac{\partial}{\partial t} \nabla \times \vec{B}$$

$$= - \frac{\partial}{\partial t} \left(\mu_o \epsilon_o \frac{\partial^2 \vec{E}}{\partial t^2} \right) \qquad \text{from eq. (1-10)}$$

or

$$\nabla^2 \vec{E} = \mu_o \epsilon_o \frac{\partial^2 \vec{E}}{\partial t^2} \tag{1-14}$$

Similarly, taking the rotation of eq. (1-10) yields

$$\nabla^2 \vec{B} = \mu_o \epsilon_o \frac{\partial^2 \vec{B}}{\partial t^2} \tag{1-15}$$

Eq. (1-14) and (1-15) are the vector wave equations. Just as a reminder to us, a one dimensional scalar wave equation is given (in the z - direction, say) mathematically by

$$\frac{\partial^2 f}{\partial z^2} = \frac{1}{v^2} \frac{\partial^2 f}{\partial t^2} \tag{1-16}$$

where f is the scalar wave function and v is the wave propagation velocity. Thus, comparing (1-14), (1-15) with (1-16) $\vec{E}$ and $\vec{B}$ each satisfies in the Cartesian coordinates a 3 - dimensional vector wave equation with the same wave velocity

$$c \equiv \frac{1}{\sqrt{\mu_o \epsilon_o}} \tag{1-17}$$

which is the velocity of light in free space.

Let us now use the one-dimensional scalar wave equation (1-16) to illustrate the nature of waves. Any function $f(z, t)$ of the type $f(t \pm z/v)$ or their linear combination is a solution to the wave equation (1-16). This can be shown by substitution into eq. (1-16). The functions $f_1(t - z/v)$ and $f_2(t + z/v)$ represent waves propagating in the positive and negative z directions, respectively. This is just a mathematical consequence. We give a simple explanation. Consider the function

$$f(z, t) = f_1 \ (t - z/v) \tag{1-18}$$

We keep in mind that z is position and t is time. Let the phase of $f(t - z/v)$ be some definite value,

i.e.

$$t - \frac{z}{v} = t' \tag{1-19}$$

where t' is any definite value. Eq. (1-19) represents a family of parallel lines in the z - t plane for different values of t' (Fig. 1-3a). Let us consider one of the lines given by eq. (1-19). It means that for a fixed value of t', the function $f(z, t)$ of eq. (1-18) will have the same value $f_1(t')$ whatever the changes in z and t are, so long as they are constrained by eq. (1-19); i.e. so long as the combined changes of z and t follow eq. (1-19), $f(z, t)$ will always have the same value $f_1(t')$.

Now, eq. (1-19) can be re-written as

$$z = v(t - t') \tag{1-20}$$

When z and t changes according to eq. (1-19) or (1-20), the rate of change of z with respect to t is, from (1-20),

$$\frac{dz}{dt} = v \tag{1-21}$$

Thus, the rate of change (velocity) is v and the direction of change is in the positive z direction because t (time) always changes in the positive direction. Thus, the value $f_1(t')$ is constrained to move along the line given by eq. (1-19) in the positive z direction at the velocity v, Fig. (1-3b). The reason why $f_1(t')$ should move at all is because t (time) changes continuously so that z has to follow according to eq. (1-19). (Nothing will move if we could FREEZE t and z at some definite value.)

Similarly, another value of $f(z, t)$, say $f_1(t'')$ moves along the line

$$t - z/v = t'' \tag{1-22}$$

with the same velocity v in the same positive z direction, and so on; i.e. every point of the function $f(z, t)$ given by eq. (1-18) moves with the velocity v in the same positive z - direction without changing the form of the function. Fig. (1-3c) shows the propagation of the projection $f_1(t, z = z_o)$ moving from $z = z_o$ to $z = z_o'$ while Fig. (1-3d) shows the propagation of the projection $f_1(t = t_o, z)$ moving from $t = t_o$ to $t = t_o'$. Similarly, the reader can analyse the motion of $f(z, t) = f(t + z/v)$ in the negative z - direction. This is left as an exercise.

We now use only the spatial coordinate to describe the motion (Fig. 1-4a). Referring to eq. (1-20), for a constant t', the wave function $f_1(t')$ moves along the z - axis. When the time changes from t_1 to t_2, $f(t')$, always a constant, moves from P_1 to P_2 with velocity v.

More generally, the distance $0P_1$ represents a plane perpendicular to the z - axis at P_1 (Fig. 1-4b) because $0P_1 = \vec{r} \cdot \hat{z}$ is the projection of the position vector $\vec{r}$ of every point of the plane on the z - axis. The equation of this plane is $\vec{r} \cdot \hat{z} = v(t_1 - t')$ ($\equiv 0P_1$). When P_1 moves to P_2 at the velocity v, this plane $\vec{r} \cdot \hat{z} = v(t_1 - t')$ moves into the plane

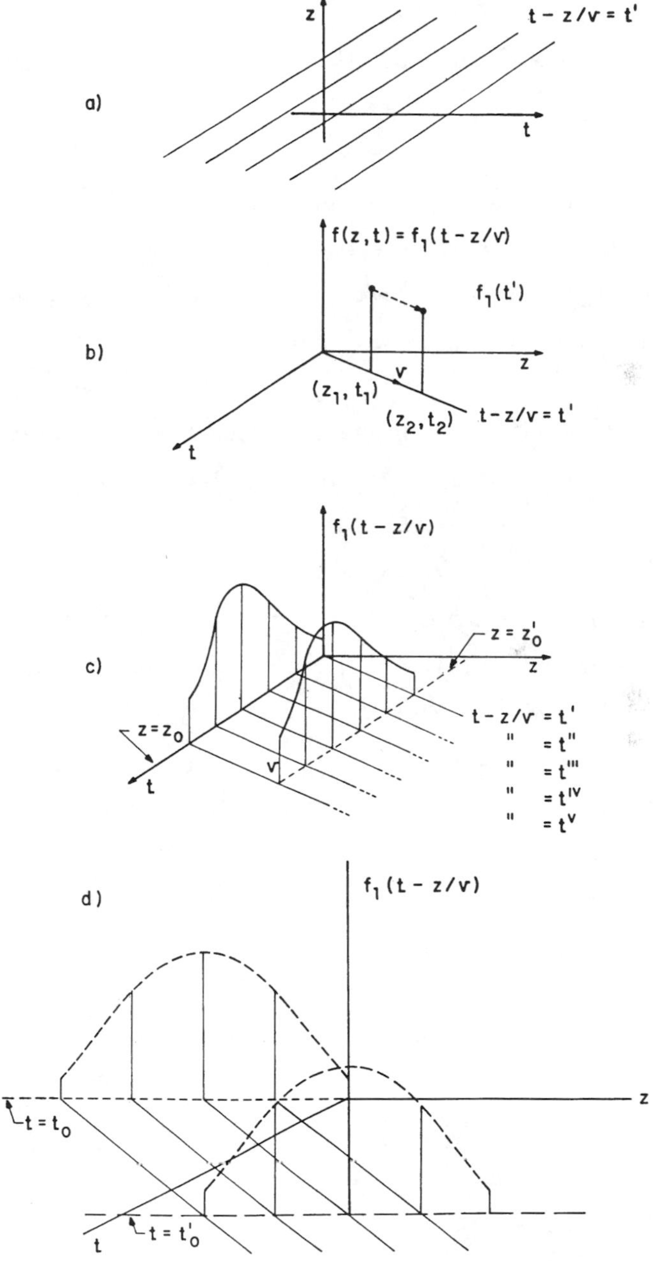

Fig. 1-3. Pictorial visualization of a one dimensional wave propagation.

$\vec{r} \cdot \hat{z} = v(t_2 - t') \equiv (OP_2)$ at the same velocity. $f_1(t')$ is constant everywhere on these two planes because t' is constant. Thus, we conclude that in general

$$\vec{r} \cdot \hat{z} = v(t - t') \tag{1-23}$$

is a plane perpendicular to the z -axis. This plane moves in the positive z -direction at a velocity v. At every point on this moving plane, the wave function $f_1(t')$ is a constant. We call the moving plane a plane wavefront and $f_1(t - z/v)$ a plane wave function.

The plane wavefront is a plane of constant phase (i.e. t' = constant). For another value of the phase, say t", the plane is shifted with respect to eq. (1-23); i.e. the new plane is

$$\vec{r} \cdot \hat{z} = v(t - t'')$$

and all over this plane, the value of the wave function is $f_1(t'')$, a constant. $f_1(t'')$ still propagates in the z - direction at the velocity v, Fig. 1-4(c). Thus, we conclude that for a plane wave function $f_1(t - z/v)$, all the planes of constant phase

(i.e. $t - z/v = t' =$ constant or $t - \vec{r} \cdot \hat{z}/v = t' =$ constant)

propagate in the positive z - direction with velocity v. At every point on each moving plane, corresponding to one fixed phase t', the value of the wave function is the constant $f_1(t')$.

Similarly, $f_2(t + z/v)$ is a plane wave function propagating in the negative z - direction.

In optics, we often use wave functions of the form $f(\omega t \pm z/v)$ or $f(\omega t \pm \vec{k} \cdot \vec{r})$. Both are of the same form as eq. (1-18). For instance, we consider $f(\omega t - kz)$.

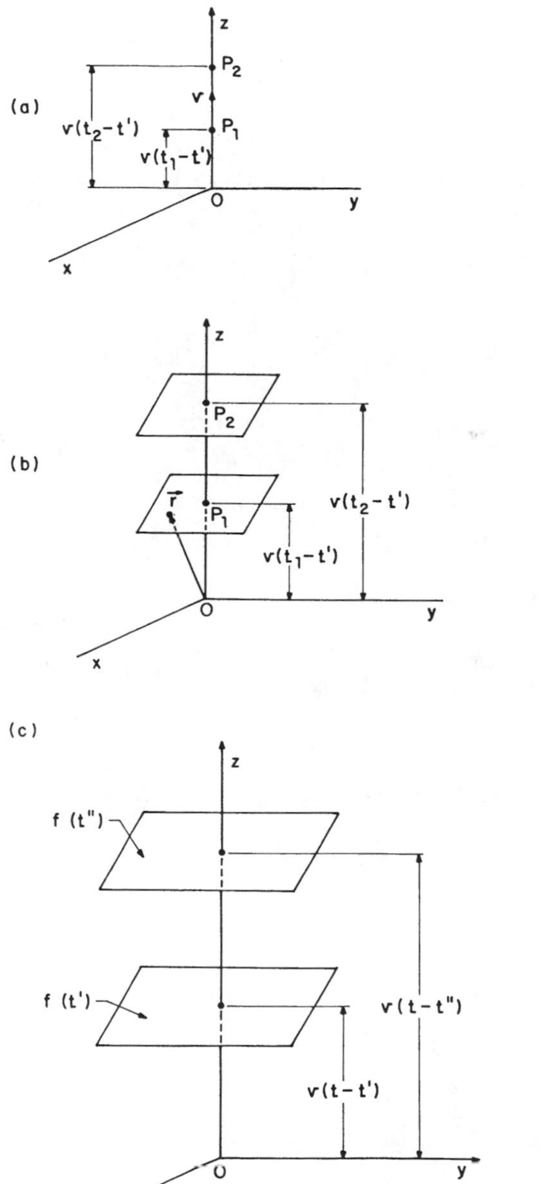

Fig. 1-4. Pictorial visualization of one dimensional plane wave propagation in space.

$$\omega t - kz = \omega\left(t - \frac{z}{(\omega/k)}\right)$$

$$\equiv \omega(t - z/v) \tag{1-24}$$

if
$$v \equiv \omega/k \tag{1-25}$$

For a monochromatic wave (ω = constant) we thus have

$$f(\omega t - kz) = f(\omega(t - z/v))$$

which is of the same form as eq. (1-18). Similarly, consider $f(\omega t - \vec{k}\cdot\vec{r})$, for example.

$$\omega t - \vec{k}\cdot\vec{r} = \omega\left(t - \frac{\vec{k}\cdot\vec{r}}{\omega}\right) \tag{1-26}$$

If we rotate the coordinate axes such that the z - axis coincides with the $\vec{k}$ axis, eq. (1-26) becomes

$$\omega t - \vec{k}\cdot\vec{r} = \omega\left(t - \frac{k\hat{z}\cdot\vec{r}}{\omega}\right) \tag{1-27}$$

Again, if $v \equiv \omega/k$, eq. (1-27)

becomes

$$\omega t - \vec{k}\cdot\vec{r} = \omega\left(t - \frac{\vec{r}\cdot\hat{z}}{v}\right) \tag{1-28}$$

$$= \omega t' \tag{1-28'}$$

where we have set eq. (1-28) equal to a constant $\omega t'$.

$$\therefore \quad \omega\left(t - \frac{\vec{r}\cdot\hat{z}}{v}\right) = \omega t'$$

or
$$\vec{r}\cdot\hat{z} = (t - t') \tag{1-29}$$

which is identical to eq. (1-23), i.e. $\omega t - \vec{k}\cdot\vec{r} = \omega t'$ is a moving plane in the direction of $\vec{k}$ ($\hat{z}$ in eq. (1-29)) at the velocity $v = \omega/k$. On this plane, the wave function has a constant value $f_1(\omega t')$. A similar analysis applies to $f(\omega t + \vec{k}\cdot\vec{r})$.

In conclusion, whenever we see a function of the form (in the <u>Cartesian coordinates</u>)

$$f(\omega t - \vec{k}\cdot\vec{r})$$

and $f(\omega t - k z)$ etc.

where ω = constant (monochromatic) they represent plane wave functions which are solutions to the wave equation. The plane wave front of the wave function propagates in the direction of $\vec{k}$ (or $\hat{z}$) (i.e. $\vec{k}$ (or $\hat{z}$) is the normal to the plane) at the velocity ω/k. This velocity is called the wave velocity. On the plane wave front, the wave function has a constant value. A similar argument applies to wave fronts propagating in the (- k) direction.

In the case of the 3-D vector wave equations (1-14) and (1-15), the plane wave solutions are of the form $\vec{E}(\omega t \pm \vec{k} \cdot \vec{r})$ and $\vec{B}(\omega t \pm \vec{k} \cdot \vec{r})$ in the Cartesian coordinates with ω = constant. On the plane wave front

$$\omega t \pm \vec{k} \cdot \vec{r} = \omega t' \qquad \text{(see eq. (1-28))}$$

the vector wave functions $\vec{E}$ and $\vec{B}$ are constant vectors. For simplicity in the rest of this chapter, we simply use scalar wave functions to describe different concepts of waves.

Customarily, in optics, the wave function is expressed in one of the following harmonic forms: $\cos(\omega t \pm \vec{k} \cdot \vec{r})$, $\sin(\omega t \pm \vec{k} \cdot \vec{r})$, $e^{\pm i(\omega t \pm \vec{k} \cdot \vec{r})}$ are the same functions with $\vec{k} \cdot \vec{r}$ replaced by kz (or ky or kx).

§1.3 Spherical waves

If the wave function $f(\vec{r}, t)$ is a function of r and t only, where r is the radial position in space, it is convenient to express the 3-D wave equation in spherical coordinates (r, Θ, φ). From vector calculus

$$\nabla^2 f = \frac{1}{r^2}\frac{\partial}{\partial r}\left(r^2 \frac{\partial f}{\partial r}\right) + \frac{1}{r^2 \sin\Theta}\frac{\partial}{\partial \Theta}\left(\sin\Theta \frac{\partial f}{\partial \Theta}\right) + \frac{1}{r^2 \sin^2\Theta}\frac{\partial^2 f}{\partial \phi^2} \qquad (1\text{-}30)$$

Since $f(\vec{r}, t) = f(r, t)$, independent of Θ and φ , (1-31)

eq. (1-30) becomes

$$\nabla^2 f = \frac{1}{r^2}\frac{\partial}{\partial r}\left[r^2 \frac{\partial f}{\partial r}\right] \equiv \frac{1}{r}\frac{\partial^2}{\partial r^2}\left[rf \right] \tag{1-32}$$

Substituting eq. (1-32) into the 3-D wave equation $\nabla^2 f = \frac{1}{v^2}\frac{\partial^2 f}{\partial t^2}$, we have

$$\frac{1}{r}\frac{\partial^2}{\partial r^2}(rf) = \frac{1}{v^2}\frac{\partial^2 f}{\partial t^2}$$

or

$$\frac{\partial^2}{\partial r^2}(rf) = \frac{1}{v^2}\frac{\partial^2 (rf)}{\partial t^2} \tag{1-33}$$

Eq. (1-33) is in the form of a one-dimensional wave equation whose wave function is (rf). Thus, the solution is

$$rf(r, t) = g\left[t \pm \frac{r}{v}\right] \tag{1-34}$$

Or in the form of $(\omega t - \vec{k} \cdot \vec{r})$, one has

$$rf(r, t) = g(\omega t \pm kr) \qquad \text{(independent of } \Theta \text{ and } \phi)$$

Hence

$$f(r, t) = \frac{g(\omega t \pm kr)}{r} \tag{1-35}$$

In eq. (1-35), $g(\omega t - kr)$ is a wave function whose wave front is a spherical surface, (note that we are now in the spherical coordinate; thus $g(\omega t - kr)$ does not represent a plane wave) because it propagates with a velocity $v = \omega/k$ in all directions of r. The direction of propagation is that of increasing r; i.e. the spherical wave front is diverging from the origin. Similarly, the wave function $g(\omega t + kr)$ has a spherical wave front converging to the origin.

The total solution to the 3-D wave equation is the wave function $f(r, t)$ given by eq. (1-35). It means that $f(r, t)$ are spherical wave fronts whose amplitude decreases linearly with r.

§1.4 Wave vector, phase velocity, group velocity

(A) We work now in the Cartesian coordinates. The wave function is given by $f(\omega t - \vec{k} \cdot \vec{r})$.

$$\text{Def. wave vector} \equiv \vec{k}$$

$\vec{k}$ is always perpendicular to the plane wavefront (§1-2, eq. 1-28')

$$\omega t - \vec{k} \cdot \vec{r} = \text{constant} \qquad (1\text{-}36)$$

(The same orthogonal relationship applies to a spherical wavefront or any wave front.) Since the propagation velocity of the wave front is the wave velocity

$$v = \frac{\omega}{k} \qquad (k \equiv |\vec{k}|$$

we have

$$k = \frac{\omega}{v} \qquad (1\text{-}37)$$

In optics, when the wave $f(\omega t - \vec{k} \cdot \vec{r})$ propagates in a dielectric medium,

$$v = \frac{c}{n} \qquad (1\text{-}38)$$

where n is the refractive index. Substituting eq. (1-38) into (1-37) yields

$$k = \frac{\omega}{c} n \qquad \text{(in a dielectric)} \qquad (1\text{-}39)$$

If the wave is in vacuum, n = 1,

$$k_o = \frac{\omega}{c} \qquad \text{(in vacuum)} \qquad (1\text{-}40)$$

(B) When two monochromatic waves of slightly different frequencies come together, we can combine them using the superposition principle. This principle requires that the wave equation be linear. (Linear means the wave function f and its derivatives all appear to the first order; there is no term in the wave equation containing f^2, $\left(\frac{\partial f}{\partial z}\right)^2$, ...etc.) Then any linear combination of all different f's is also a solution.

Assume that the two waves E_1, E_2 with frequencies ω_1 and ω_2, have the same amplitude. Assume that they also propagate in the same direction z.

$$E_1 = E_o\, e^{-i(\omega_1 t - k_1 z)} \tag{1-41}$$

$$E_2 = E_o\, e^{-i(\omega_2 t - k_2 z)} \tag{1-42}$$

Their superposition yields the total field which is the summation of the two

$$E = E_1 + E_2 \tag{1-43}$$

We define

$$\bar{\omega} \equiv \frac{1}{2}(\omega_1 + \omega_2) \tag{1-44}$$

$$\Delta\omega \equiv \omega_1 - \omega_2 \tag{1-45}$$

$$\bar{k} \equiv \frac{1}{2}(k_1 + k_2) \tag{1-46}$$

$$\Delta k \equiv k_1 - k_2 \tag{1-47}$$

Substituting eq. (1-44 to 47 and 41, 42) into eq. (1-43), (the reader can do this <u>exercise</u> in calculation) yields

$$E = 2E_o\, e^{-i(\bar{\omega}t - \bar{k}z)} \cos\left[\frac{\Delta\omega}{2}t - \frac{\Delta kz}{2}\right] \tag{1-48}$$

There are now two propagating wave functions in eq. (1-48), $e^{-i(\bar{\omega}t - \bar{k}z)}$ and $\cos\left[\frac{\Delta\omega}{2}t - \frac{\Delta k}{2}z\right]$. The former propagates with a mean frequency $\bar{\omega}$ and mean wave vector $\bar{k}$ so that its velocity is

$$\bar{v}_p = \bar{\omega}/\bar{k} \tag{1-49}$$

while the second with a frequency $\frac{\Delta\omega}{2}$, and wave vector $\frac{\Delta k}{2}$ at a velocity

$$v_g = \frac{\Delta\omega}{\Delta k} \tag{1-50}$$

Special condition:

$$\left.\begin{array}{l}\Delta\omega/\bar{\omega} \ll 1\\ \Delta k/\bar{k} \ll 1\end{array}\right\} \tag{1-51}$$

Under the condition (1-51), the frequency of the cosine term in eq. (1-48) is much lower than the mean frequency $\bar{\omega}$ of the exponential term. We thus interpret eq. (1-48) as having a carrier plane wave of mean frequency $\bar{\omega}$ (from $e^{-i(\bar{\omega}t - \bar{k}z)}$) whose amplitude is also a wave $\left[\cos\left(\frac{\Delta\omega}{2}t - \frac{\Delta k}{2}z\right)\right]$ propagating with a much lower <u>modulating</u> frequency $\Delta\omega/2$. The propagation velocity of the carrier wave is the <u>phase velocity</u> v_p (eq. 1-49) and that of the modulating amplitude is the <u>group velocity</u> v_g (eq. 1-50).

(C) When a group of monochromatic plane waves travelling in the same z-direction are superimposed together, the total field is the summation of all of them. If their frequencies differ continuously across a width $\Delta\omega$, the summation becomes an integral:

$$E(z, t) = \int_{\Delta\omega} E_o(\omega) e^{-i(\omega t - kz)} d\omega$$

$$= \int_{\Delta\omega} E_o(\omega) e^{-i(\bar{\omega} t - \bar{k} z)} e^{-i(\omega - \bar{\omega})t} \cdot e^{-i(k - \bar{k})z} d\omega \tag{1-52}$$

where $\bar{\omega}$ and $\bar{k}$ are the mean frequency and wave vector, respectively, and $\Delta\omega$ denotes the frequency interval around $\bar{\omega}$. Eq. (1-52) can be regrouped:

$$E(z, t) = A(z, t) e^{-i(\bar{\omega} t - \bar{k} z)} \tag{1-53}$$

where

$$A(z, t) \equiv \int_{\Delta\omega} E_o(\omega) e^{-i[(\omega - \bar{\omega})t - (k - \bar{k})z]} d\omega$$

$$= \int_{\Delta\omega} E_o(\omega) e^{-i\left[(\omega - \bar{\omega}) \left(t - \frac{k - \bar{k}}{\omega - \bar{\omega}} z\right)\right]} d\omega \tag{1-54}$$

Again, as in the case of eq. (1-48), we interpret eq. (1-53) as having a carrier plane wave ($e^{-i(\bar{\omega} t - \bar{k} z)}$) whose plane of constant phase propagates in the z-direction at the wave velocity of $\bar{\omega}/\bar{k}$ and a variable amplitude wave A(z, t) given by eq. (1-54). It is easier to interpret the latter by rewriting it as

$$A(z, t) = \int_{\Delta\omega} E_o(\omega) e^{-i\left[\omega_g\left(t - \frac{z}{v_g}\right)\right]} d\omega \tag{1-55}$$

where
$$\omega_g \equiv \omega - \bar{\omega} \tag{1-56}$$

$$v_g \equiv \frac{\omega - \bar{\omega}}{k - \bar{k}} \tag{1-57}$$

Eq. (1-55) reveals itself as a superposition of a group of wave functions of frequencies ω_g. On the surface

$$t - \frac{z}{v_g} = 0 \tag{1-58}$$

(cf. discussion around eq. (1-23) A(z, t) is maximum everywhere.) This plane surface of maximum A(z, t) advances at the group <u>velocity</u> v_g given by eq. (1-57).

If $\Delta\omega/\bar{\omega} \ll 1$, $\Delta k/\bar{k} \ll 1$,

$$v_g = \frac{\omega - \bar{\omega}}{k - \bar{k}} \approx \frac{d\omega}{dk} \tag{1-59}$$

This is true for most spectral lines and laser pulses in which the width $\Delta\omega$ is narrow.

But in the case of ultrashort pulses in which $\Delta\omega$ is no longer much smaller than $\bar{\omega}$, we need to reconsider v_g. In fact, v_g is not a constant because of the dispersion of the different frequency components. (Dispersion means a different propagating velocity (or index) for a different frequency in a medium.) Thus all the waves propagate at different velocities giving rise to what one calls "<u>group velocity dispersion</u>".

(D) In view of the group velocity dispersion, we need to consider a more general expression for v_g. "Inspired" by eq. (1-57) and (1-59), we can in general make the following Taylor's expansion:

$$\omega = \bar{\omega} + \left(\frac{d\omega}{dk}\right)_{k=\bar{k}} (k - \bar{k}) + \frac{1}{2}\left(\frac{d^2\omega}{dk^2}\right)_{k=\bar{k}} (k - \bar{k})^2 + \ldots \tag{1-60}$$

We see that eq. (1-60) gives the normal group velocity v_g (eq. 1-59) if

$$|k - \bar{k}| \ll \bar{k} \tag{1-61}$$

or $$\Delta k \ll \bar{k}$$

which is the condition for the validity of eq. (1-59). Under this condition, eq.(1-60) becomes (by keeping the first two terms)

$$\omega \simeq \bar{\omega} + \left(\frac{d\omega}{dk}\right)_{k=\bar{k}}(k - \bar{k})$$

$$\Delta\, v_g \equiv \frac{\omega - \bar{\omega}}{k - \bar{k}} = \left(\frac{d\omega}{dk}\right)_{k=\bar{k}} \tag{1-62}$$

If eq. (1-61) is not valid, i.e. if $|k - \bar{k}|$ becomes larger, we need to consider the higher order terms in eq. (1-60). If $|k - \bar{k}|$ and $|\omega - \bar{\omega}|$ are still not very large, adding the 3rd term is sufficient.

$$\omega \simeq \bar{\omega} + \left(\frac{d\omega}{dk}\right)_{k=\bar{k}}(k - \bar{k}) + \frac{1}{2}\left(\frac{d^2\omega}{dk^2}\right)_{k=\bar{k}}(k - \bar{k})^2$$

$$v_g \equiv \frac{\omega - \bar{\omega}}{k - \bar{k}} = \left(\frac{d\omega}{dk}\right)_{k=\bar{k}} + \frac{1}{2}\left(\frac{d^2\omega}{dk^2}\right)_{k=\bar{k}}(k - \bar{k}) \tag{1-63}$$

Let $$v_{go} \equiv \left(\frac{d\omega}{dk}\right)_{k=\bar{k}} \tag{1-64}$$

$$\equiv \text{constant}$$

Eq. (1-63) becomes

$$v_g = v_{go} + \frac{1}{2}\left(\frac{d^2\omega}{dk^2}\right)_{k=\bar{k}}(k - \bar{k}) \tag{1-65}$$

Thus, v_g is a function of k. Physically, this is already explained as being due to the dispersion. From eq. (1-39)

$$k = \frac{\omega}{c}\, n(\omega) \tag{1-66}$$

because dispersion means n is a function of ω. Because of eq. (1-66), v_g is also a function of frequency.

Definition: <u>Group velocity spread (Δv_g)</u>

From eq. (1-65),

$$\Delta v_g \equiv v_g - v_{go} = \frac{1}{2}\left(\frac{d^2\omega}{dk^2}\right)_{k=\bar{k}} (k - \bar{k}) \tag{1-67}$$

In the case of the propagation of an ultrashort laser pulse (ps to fs) in a medium of index $n(\omega)$, eq. (1-67) tells us that there is a spread in the pulse's spatial extent of the order of $(\Delta v_g)t$ where t is the time of propagation.

Closing remarks

We have tried to give an elementary look at the wave nature of the electromagnetic fields $\vec{E}$ and $\vec{B}$ starting from the Maxwell's equations. The pictorial descriptions of the propagation of $\vec{E}$ and $\vec{B}$ fields and part of the wave propagation represent the author's "unorthodox" way of interpretation. Hopefully, they will give the reader a better feeling. More rigorous mathematical descriptions can be found in advanced texts (e.g. Born and Wolf).

CHAPTER II

THE LASER

§2.1 Definition of laser oscillator

The laser (light amplification by stimulated emission of radiation) is essentially a resonator that contains a light amplifier whose amplification process is stimulated emission (Fig. 2-1). Such combination becomes an oscillator when it is sustained to operate in equilibrium, giving out a beam of light from one or more ends of the resonator. The condition under which the oscillator operates in equilibrium is

$$\text{gain} = \text{loss}$$

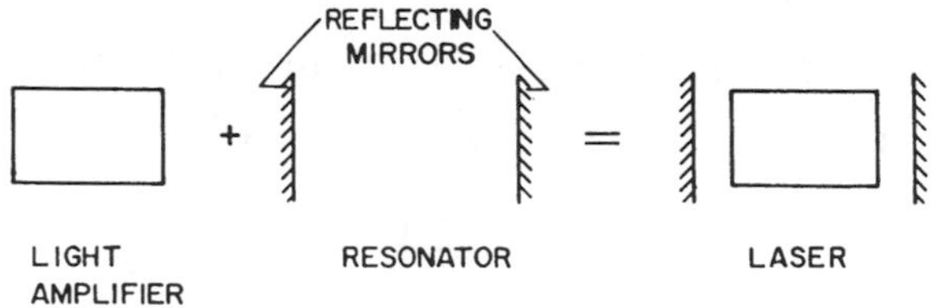

Fig. 2-1. Schematic definition of a laser oscillator.

More precisely let us consider Fig. 2-2. Assume that a pulse of light beam of energy E_o starts at $t = 0$, $z = 0$ and propagates through the amplifying medium. When it reaches the mirror at the right end, it is partially reflected back into the amplifier, reaching the mirror at the left end and again partially reflected back towards the starting point $z = 0$, thus making

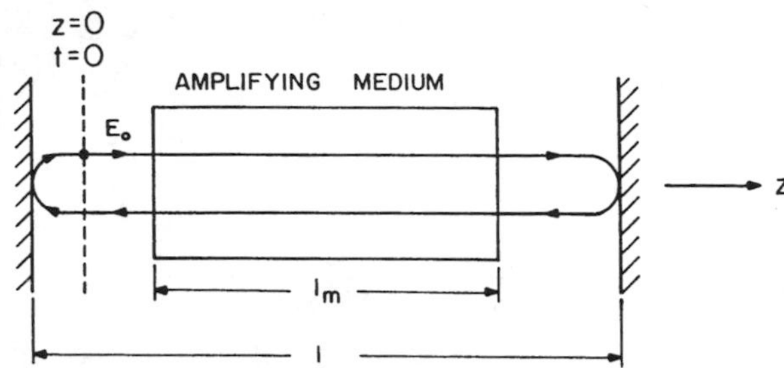

Fig. 2-2. Schematic illustration of a light beam making a round trip in a cavity.

a round trip. If the laser operation is to be sustained in equilibrium, the net gain and net loss during any round trip cycle should be equal. In other words, the energy in the light pulse remains unchanged after any round trip.

i.e.
$$E\ (\text{round trip}) = E_o \tag{2-1}$$

But
$$E\ (\text{round trip}) = E_o e^{g \cdot 2l_m}\, e^{-\alpha_c \cdot 2l} \tag{2-2}$$

where g is the gain cofficient (cm^{-1}) of the amplifying medium of length l_m and α_c is the total loss coefficient (cm^{-1}) of the oscillator. No saturation is assumed. The factor 2 in the exponentials means "round trip". Here, we have assumed exponential gain and exponential loss, the latter being a description of most decay processes governed by Beer's law (i.e. exponential decay). They also imply that the energy gain (loss) per unit length is proportional to the energy,

i.e.
$$\pm \frac{dE}{dz} \propto E$$

or
$$\begin{pmatrix} + \\ - \end{pmatrix} \frac{dE}{dz} = \begin{pmatrix} g \\ \alpha_c \end{pmatrix} E$$

$$E(z) = \begin{cases} E_o e^{gz}, & \text{gain.} \\ E_o e^{-\alpha_c z}, & \text{loss.} \end{cases}$$

Going back to eq. (2-1) and (2-2), we have

$$E_o e^{g 2l_m}\, e^{-\alpha_c 2l} = E_o$$

or $$2(gl_m - \alpha_c l) = 0$$

$$gl_m = \alpha_c l \quad \text{(threshold condition)} \tag{2-3}$$

Discussion:

(1) Loss α_c

The coefficient α_c is essentially a lumped factor representing loss through

(a) transmission of the two mirrors

(b) scattering, absorption and diffraction at the two mirrors

(c) absorption, scattering and diffraction in the amplifying medium

(d) reflection, scattering, absorption and diffraction at the two ends of the amplifying medium.

In principle, these losses can be measured experimentally either separately or together. We take this opportunity to link the loss α_c with the quality factor Q which is commonly used in electrical engineering. In a passive resonator (no gain), Q is defined as

$$Q \equiv \omega \frac{E}{\frac{-dE}{dt}}$$

where E is the radiation energy stored in the resonator at time t; $-\frac{dE}{dt}$ is the average energy dissipation per second, and ω is the angular frequency of the radiation, $\omega = 2\pi\nu$. (ν is the frequency in Hertz).

Thus, $$\frac{dE}{dt} = -\frac{\omega}{Q} E$$

and $$E = E_o e^{-\frac{\omega}{Q} t} \tag{2-4}$$

where E_o is the energy in the cavity at t = 0. Eq. (2-4) illustrates the Beer's law and the role of Q in the law. Assume now that we inject a radiation pulse of energy E_o into the empty cavity at t = 0 and let it bounce back and forth between the mirrors N times. The energy in the pulse after N round trips (in t seconds) is

$$E = E_o e^{-N \cdot 2l \cdot \alpha_c}$$

where $$N \equiv \frac{t}{2l/c},$$

$$\frac{2l}{c} = \text{one round trip time.}$$

Hence $$E = E_o e^{-\frac{t}{2l/c} \cdot 2l \cdot \alpha_c}$$

$$= E_o e^{-c\alpha_c t} \quad (2\text{-}4')$$

Comparing with eq. (2-4), we have $\frac{\omega}{Q} = c\alpha_c$ (2-4")

From either equation (2-4) or (2-4"), we see that the higher the value of Q is, the lower is the loss, and vice versa. Thus, changing the Q means changing the loss; hence the name Q-switching for some lasers (Chapter VIII).

Thus, eq. (2-3) becomes

$$Q = \frac{\omega l}{g l_m c} = \frac{2\pi l}{g l_m \lambda} \quad \text{(threshold condition)} \quad (2\text{-}4\text{-}1)$$

where $$\lambda \equiv \text{wavelength}$$

$$= \frac{c}{\nu}$$

(2) Gain g

The gain coefficient g is due to stimulated emission between two energy levels of the active medium. Because emission is always accompanied by the possibility of absorption, one has to consider the two together and the following section discusses such phenomena.

§2.2 Stimulated emission

Stimulated emission is a general phenomenon in nature. It is the result of the interaction of an electromagnetic wave with matter. Whenever there is an electromagnetic wave of an appropriate frequency interacting with a material with some appropriate energy levels, there will be stimulated emission. Thus, one can say that any material might become a laser material (i.e. amplifying medium) under some appropriate conditions. However, the reality is much more complicated. Because there are many more loss mechanisms than gain (stimulated emission), one has to prepare the medium carefully in order that the gain can overcome the loss. We shall consider first of all the simplest system in thermal equilibrium so as to illustrate the relationship between emission and absorption. This analysis was first published by Einstein in 1917 (Physickalische Zeitschrift, 18, 121 (1917)).

Fig. 2-3 shows such a simple system, a two level system, labelled 1 and 2. The energy of level i (i = 1,2) with respect to an arbitrary reference is E_i. The two levels could be any pair of energy levels (be they electronic, vibrational of rotational) of a material that can be coupled by an electromagnetic field of an appropriate frequency ν such that

$$h\nu = E_2 - E_1 \tag{2-5}$$

When such a purely 2-level system (which we now call atom) has an energy E_1 with respect to an arbitrary reference, we say that the atom is in level 1. If it has an energy E_2, it is in level 2. When it is in level 1, it can absorb radiation of frequency ν that satisfies eq. (2-5). When it is in level 2, it can emit radiation through either spontaneous emission or

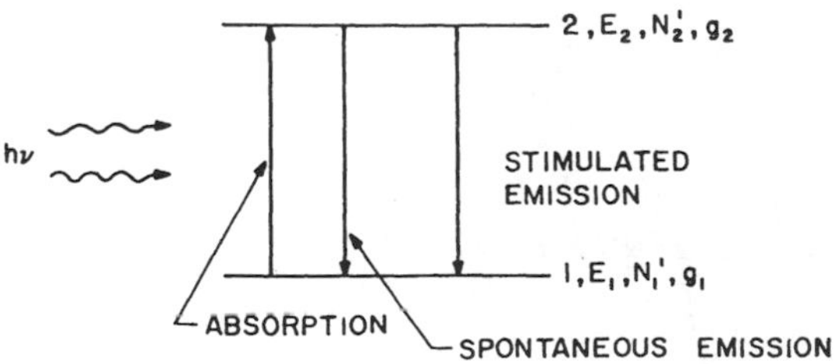

Fig. 2-3. Two-level system showing absorption and emissions of radiation.

stimulated emission. The stimulated emission is a function of the radiation density that is present in the region where the atom is situated. The emitted radiation has a frequency ν. Even when there is no radiation present in the region, the atom will emit a photon $h\nu$ by itself. Analysis by quantum mechanics shows that this is a purely quantum effect that can be viewed essentially as the interaction of the random vacuum fluctuation with the atom "stimulating" the latter to emit a photon "randomly" if it is already excited into level 2. The statistical nature of this "random" spontaneous emission will be seen immediately in what follows. We note here that all the above statements can be rigorously calculated by quantum mechanics. Interested readers can consult more advanced books on quantum electrodynamics or quantum optics.

We make the following assumptions:

(a) Each of the levels is degenerate with a <u>multiplicity</u> g_i $(i = 1,2)$. Definition: <u>multiplicity</u> g_i $(i = 1,2) \equiv$ The number of states with the same energy E_i $(i = 1,2)$. Each state in level i has an equal number of atoms N_i'.

(b) There are a total of N_{tot} two-level atoms each having a discrete energy E_i , $(i = 1,2)$.

Thus,
$$g_1 N_1' + g_2 N_2' = N_{tot} \tag{2-6}$$

(c) These atoms bathe in an electromagnetic radiation field within an isolated region.

(d) The radiation field is monochromatic with frequency ν satisfying eq. (2-5).

(e) The whole isolated (atoms + field) system is in thermal equilibrium.

(f) Energy exchange is through absorption and emission only. Hence, there is no atomic collision. Fig. 2-3 summarizes these assumptions.

Now, thermal equilibrium (assumption (e)) means that the net rate of dynamic change of the isolated system is zero. (Note that Nature is always in a dynamic state; i.e. everything in the microscopic scale is in motion unless the temperature becomes T = 0 K which itself is defined as a state of matter whose microscopic constituents are at rest.) That is, if we consider only the number of atoms N_1 in level 1, the net rate of change of $N_1 = 0$;

or
$$\left(\frac{\partial N_1}{\partial t}\right)_{net} = 0 \tag{2-7}$$

where
$$N_i \equiv g_i N_i' \qquad i = 1,2 \tag{2-8}$$

The variation in N_1 is caused by the absorption of photons from level 1 into level 2 (decrease of N_1) and the emission of radiation from level 2 back to level 1 (decrease of N_2 leading to the increase of N_1) by assumption (f).

i.e.
$$0 = \left(\frac{\partial N_1}{\partial t}\right)_{net} = -\left(\frac{\partial N_1}{\partial t}\right)_{abs.} + \left(\frac{\partial N_2}{\partial t}\right)_{spon.em.} + \left(\frac{\partial N_2}{\partial t}\right)_{st.em.} \tag{2-9}$$

In eq. (2-9), the negative sign means "decrease" while the positive sign means "increase". The abreviations "abs.", "spon. em." and "st. em." mean absorption, spontaneaous emission and stimulated emission, respectively. Now, common sense suggests us the following linear relationship.

$$\left(\frac{\partial N_1}{\partial t}\right)_{abs.} \propto \rho(\nu) g_1 N_1'$$

$$\left(\frac{\partial N_2}{\partial t}\right)_{st.\ em.} \propto \rho(\nu) g_2 N_2'$$

$$\left(\frac{\partial N_2}{\partial t}\right)_{spon.\ em.} \propto g_2 N_2' \quad \text{(independent or } \rho(\nu)\text{)}$$

where $\rho(\nu)$ is the radiation density at frequency ν in the interaction region.

i.e.
$$\left(\frac{\partial N_1}{\partial t}\right)_{abs} = B_{12}\rho(\nu) g_1 N_1' \tag{2-10}$$

$$\left(\frac{\partial N_2}{\partial t}\right)_{\text{st. em.}} = B_{21}\rho(\nu)g_2N_2' \tag{2-11}$$

$$\left(\frac{\partial N_2}{\partial t}\right)_{\text{spon. em.}} = A_{21}g_2N_2' \tag{2-12}$$

where A_{21}, B_{12}, B_{21}, are the constants of proportionality. These linear relationships are indeed valid in a more rigorous quantum mechanical analysis so long as the radiation density is not high. Substituting (2-10, 11, 12) into (2-9),

$$-B_{12}\rho(\nu)g_1N_1' + B_{21}\rho(\nu)g_2N_2' + A_{21}g_2N_2' = 0$$

or

$$\frac{g_2N_2'}{g_1N_1'} \equiv \frac{N_2}{N_1} = \frac{B_{12}\rho(\nu)}{B_{21}\rho(\nu) + A_{21}} \tag{2-13}$$

But from thermodynamics, in thermal equilibrium,

$$\frac{N_2}{N_1} = \frac{g_2}{g_1}\exp\left(-\frac{E_2-E_1}{kT}\right) \tag{2-14}$$

where T is the temperature, and k, the Planck's constant. Equating (2-13) and (2-14) and solving for $\rho(\nu)$ gives

$$\rho(\nu) = \frac{(A_{21}/B_{21})}{\left(\frac{g_1}{g_2}\right)\left(\frac{B_{12}}{B_{21}}\right)\exp\left(\frac{h\nu}{kT}\right) -1} \tag{2-15}$$

where $h\nu = E_2 - E_1$ from eq. (2-5).

A radiation field in thermal equilibrium with an atomic system means that it is a radiation field from a black body. Such a field has a Boltzmann distribution given by Boltzmann's law:

$$\rho(\nu) = \frac{8\pi\nu^2}{c^3}\,\frac{h\nu}{(e^{h\nu/kT})_{-1}} \tag{2-16}$$

Comparing eq. (2-15) and (2-16), we obtain

$$\frac{A_{21}}{B_{21}} = \frac{8\pi\nu^2}{c^3}h\nu \tag{2-17}$$

$$B_{12} = \frac{g_2}{g_1} B_{21} \tag{2-18}$$

Discussion:

(1) A_{21}, B_{12}, B_{21} are called the Einstein A, B coefficients.

(2) If $g_2 = g_1 = 1$ (i.e. no degeneracy) eq. (2-18) gives

$$B_{12} = B_{21}. \tag{2-18'}$$

From eq. (2-10) and (2-11), we see that $B_{12}\rho(\nu)$ and $B_{21}\rho(\nu)$ are the probability per second of absorption and stimulated emission, respectively. Therefore, if there is no degeneracy, i.e. if we have only two discrete levels,

$$B_{12}\rho(\nu) = B_{21}\rho(\nu) \qquad \text{from eq. (2-18')}$$

Or, the probabilities per second of absorption and stimulated emission are equal. This is a very important result. It means that, physically, because there is also the contribution of spontaneous emission, the total down transition probability per second (2 → 1) is always greater than the up transition (1 → 2) probability per second. Since thermal equilibrium requires the total rate of change of N_1 and N_2 to be equal, N_1 will always be greater than N_2 unless $T = \infty$ (see below).

(3) Pure spontaneous emission can be described, using eq. (2-12), as

$$-\frac{\partial N_2}{\partial t} = A_{21} N_2$$

assuming $\rho(\nu) = 0$ and there are N_{20} atoms excited to level 2 at $t = 0$. The solution of this equation is

$$N_2 = N_{20} \exp(-A_{21}t)$$

or

$$N_2 = N_{20} \exp\left(-\frac{t}{\tau_{21}}\right) \tag{2-19}$$

where $\tau_{21} \equiv (A_{21})^{-1}$ can be considered as the decay time or lifetime of level 2, as shown in Fig. 2-4. This decay time characterizes the

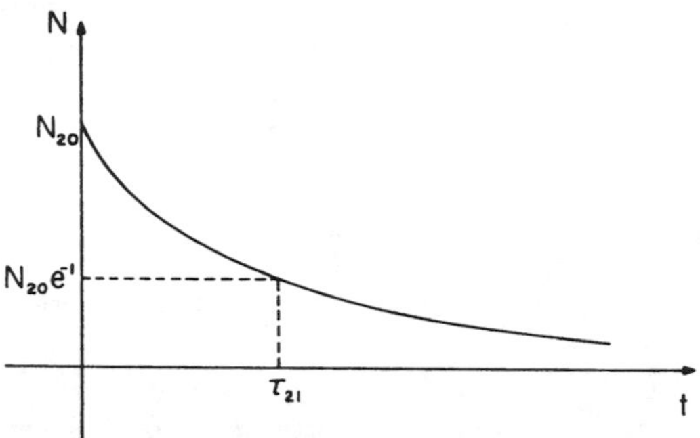

Fig. 2-4. Spontaneous decay of level 2.

statistical nature of spontaneous emission. It is also called the fluorescence lifetime.

(4) Stimulated emission is a coherent process; i.e. the emitted radiation is in phase with the radiation that stimulates the emission.

(5) If we just ask what the total rate of change of N_1 is without imposing the condition of thermal equilibrium, we have, from eq. (2-10), (2-11), (2-12),

$$\frac{\partial N_1}{\partial t} = - B_{12}\rho(\nu)N_1 + B_{21}\rho(\nu)N_2 + A_{21} N_2 \tag{2-20}$$

Using eq. (2-17) and (2-18), this becomes

$$\frac{\partial N_1}{\partial t} = - \frac{g_2}{g_1} B_{21}\rho(\nu)N_1 + B_{21}\rho(\nu)N_2 + A_{21}N_2 \tag{2-21}$$

Normally, spontaneous emission is rather weak and can be neglected as compared to stimulated emission. Eq. (2-20) thus becomes, by setting $A_{21}N_2 \approx 0$,

$$\frac{\partial N_1}{\partial t} = B_{21}\rho(\nu) \left(N_2 - \frac{g_2}{g_1} N_1 \right) \tag{2-22}$$

This equation says that the net rate of change of the number of atoms in level 1 is equal to the probability per second of down transition ($B_{21}\rho(\nu)$) multiplied by the net number of atoms leaving level 2 which is $N_2 - \frac{g_2}{g_1} N_1$.

§2.3 Level broadening

In reality, the physical world is more complicated. The atoms will interact with their environment. Even if they are isolated as in the previous assumptions, they still interact with the photons. Any of these interactions will lead to a width in a transition, i.e. broadening. For example, instead of having N_2 atoms in level 2 all having the same energy E_2, we have now, because of the broadening, the situation shown in Fig. 2-5, where $g(\nu,\nu_o)$ is the distribution function or lineshape function. The net modification in the transition is shown in Fig. 2-6. The width of level 2 represents the relative change with respect to level 1 during the transition, be it absorption or emission. That's why level 1 is not broadened. Now, $N(\nu) = N_2 g(\nu,\nu_o)$ by definition. Integrating with respect to frequency, we have

$$N_2 \equiv \int_o^\infty N(\nu)d\nu = \int_o^\infty N_2 g(\nu, \nu_o)d\nu$$

$$= N_2 \int_o^\infty g(\nu, \nu_o)d\nu$$

Hence

$$\int_o^\infty g(\nu, \nu_o)d\nu = 1 \qquad (2\text{-}23)$$

i.e. the area under $g(\nu, \nu_o)$ equals 1 as shown in Fig. 2-7. Physically, this can be understood as follows. The number of atoms in level 1 capable of absorbing photons in the range $h\nu$ to $h(\nu + d\nu)$ is (see Fig. 2-5)

$$N_1(\nu)d\nu = N_1 g(\nu, \nu_o)d\nu \qquad (2\text{-}24)$$

Similarly, the number of atoms in level 2 capable of emitting photons in the range $h\nu$ to $h(\nu + d\nu)$ is

$$N_2(\nu)d\nu = N_2 g(\nu , \nu_o)d\nu \qquad (2\text{-}25)$$

From eq. (2-24) and (2-25), we see that (omitting ν_o for compactness)

$g(\nu)$ = probability of absorption or emission per unit frequency

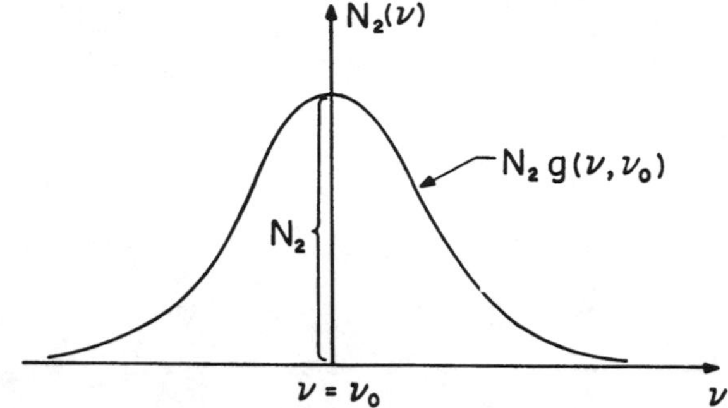

Fig. 2-5. Broadening of level 2.

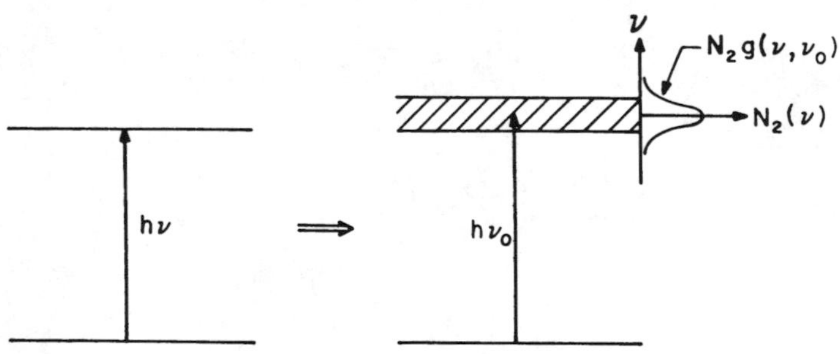

Fig. 2-6. Level broadening.

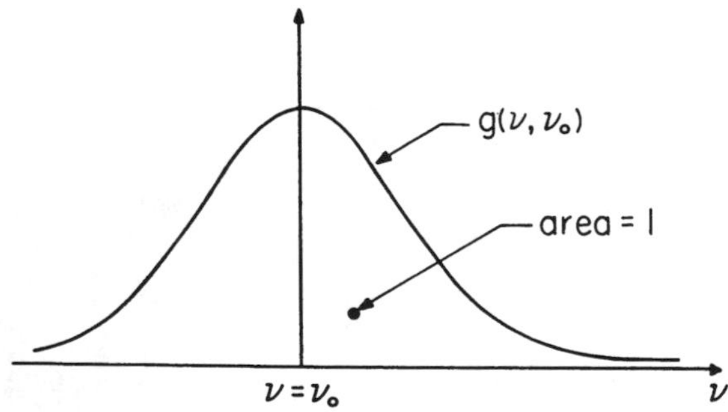

Fig. 2-7. The distribution function $g(\nu, \nu_0)$.

and

$$g(\nu)d\nu = \text{probability of absorption or emission of photons in the range } h\nu \text{ and } h(\nu + d\nu)$$

$$\int_0^\infty g(\nu)d\nu = \text{total probability of transitions (absorption and emission) between } \nu = 0 \text{ and } \nu = \infty.$$

And of course such a total probability in our isolated system must be equal to 1, which is equation (2-23).

Physically, there are two classes of broadening, homogeneous and inhomogeneous.

Definition: homogeneous broadening

All atoms, during a transition, are affected in an identical way and therefore the transitions of all the atoms are broadened identically.

Definition: inhomogeneous broadening

Each atom's transition frequency is shifted to a different and distinct extent so that the total broadening of the transitions is a "combination" of all these individual shifts.

Examples of homogeneous broadening:

a) Natural linewidth

From quantum mechanics, we know that a photon possesses a momentum. When an atom radiates a photon through spontaneous emission there is a back reaction exerted on the atom by the photon. This leads to an atomic recoil thus creating an uncertainty in the position of the electron in the atom. This is equivalent to saying that there is an uncertainty in the energy of level 2. Using the Heisenberg uncertainty principle, we have

$$(\Delta E)(\Delta t) \sim h$$

Now, in order to make a measurement of the transition by spontaneous emission, we need to measure the emitted photon. Since the lifetime of level 2 is τ_{21}, it takes about τ_{21} seconds to make a measurement.

i.e.
$$\Delta t = \tau_{21}$$
$$\Delta E \sim \frac{h}{\tau_{21}}$$

i.e. the energy of the emitted photon can only be measured to an accuracy of $\Delta E \sim h/\tau_{21}$,

or to within a width of

$$\Delta\omega = \frac{\Delta E}{h} \sim \frac{1}{\tau_{21}}$$

(b) Collision broadening

If there is no collision, an ensemble of excited atoms will emit a long train of electromagnetic wave. Whenever there is a collision, the emission process is momentarily terminated so that the emitted wave train is shortened, i.e. the long wave train is truncated. From wave theory, any truncated wave will have spectral sidebands. This means that there is a spectral width. Such collision broadening takes place implicitly at constant temperature.

(c) Thermal broadening

Because temperature is really a measure of the average kinetic energy of all the particles in a system, changing temperature means changing the particle's kinetic energy. Its effect on transition broadening is essentially due to collisions; the higher the temperature is, the more collisions there are and the broader the transition is.

Examples of inhomogeneous broadening

(a) Doppler broadening

This is due to the atomic motion at a velocity v. Thus, each atomic transition is Doppler shifted from

$$\nu_o \to \nu_o \left(1 + \frac{v}{c}\right) \quad (\text{if } v \ll c)$$

Because the ensemble of atoms normally has a distribution in the velocity space, v is different for different atoms so that the individual shifts of the transitions of all the atoms are different. The broadening is thus a "combination" of all these shifts.

(b) Broadening due to crystal inhomogeneity

In crystalline lasing materials (the newly invented verb "lase" is now used popularly to mean the action of emitting laser radiation) the active atoms are essentially ions doped uniformly into the lattice of a host crystal. For example, Cr^{+++} ions in a ruby crystal which is aluminum oxide or Nd^{+++} ions in glass or in YAG (Yittrium aluminum garnet.) These active ions are not distributed perfectly uniformly in the host. Microscopically, an ion might "see" a different local environment due to crystal defects, random variations of dislocations, lattice strain etc. Thus, each ion will experience a different local static electric field which will shift the transition by a different extent. (We might call this a local Stark shift). The combined result is the combination of all the different shifts resulting in an inhomogeneous broadening.

Mathematical form of $g(\nu, \nu_o)$

It is known that homogeneous broadening has a Lorentian distribution while inhomogeneous broadening has a Gaussian distribution.

i.e.

$$g(\nu, \nu_o) = \begin{cases} \dfrac{\Delta\nu}{2\pi}\left[(\nu - \nu_o)^2 + \left(\dfrac{\Delta\nu}{2}\right)^2\right]^{-1} & \text{(Lorentian)} \\ \dfrac{2}{\Delta\nu}\left(\dfrac{\ln 2}{\pi}\right)^{1/2}\exp\left[-\left(\dfrac{\nu - \nu_o}{\Delta\nu/2}\right)^2 \ln 2\right] & \text{(Gaussian)} \end{cases} \quad (2\text{-}26)$$

This is shown graphically in Fig. 2-9 where the peak values of the two distributions are

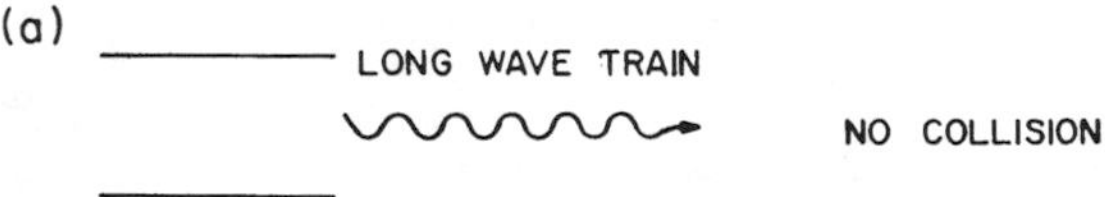

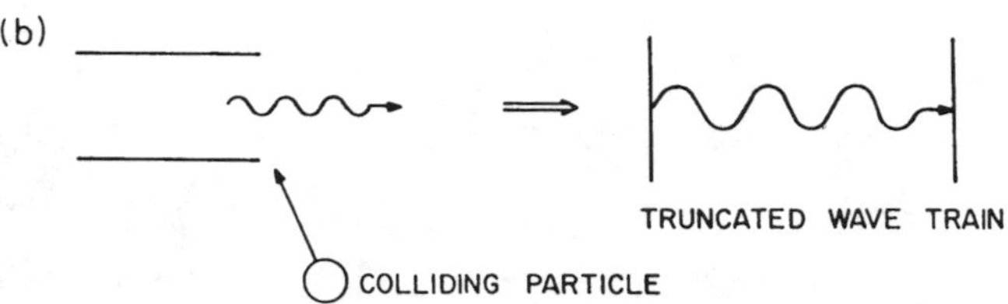

Fig. 2-8. Collision broadening.

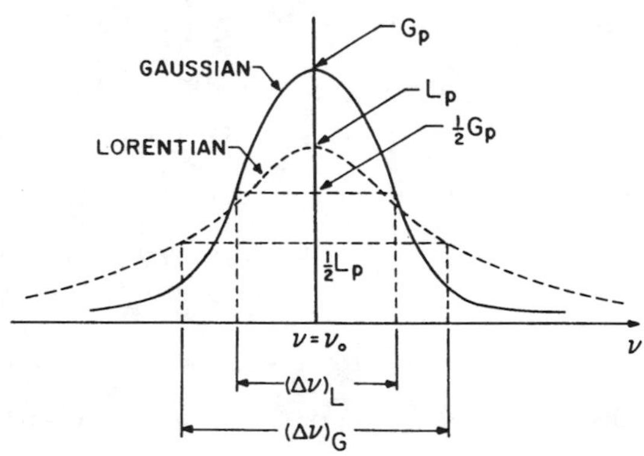

Fig. 2-9. Schematic drawing comparing Lorentian and Gaussian distributions.

$$L_p \equiv g(\nu_o)_L = \frac{2}{\pi\Delta\nu} \qquad \text{(Lorentian)} \qquad (2\text{-}27)$$

$$G_p \equiv g(\nu_o)_G = \frac{2}{(\Delta\nu)\pi}(\pi \ln 2)^{1/2} \qquad \text{(Gaussian)} \qquad (2\text{-}28)$$

§2.4 <u>Consequence of broadening</u>

The width $\Delta\nu$ of a transition is much broader than the width $d\nu$ of the laser line (See §2.8 for more discussion.) For example, in the case of ruby laser, $\Delta\nu \sim 0.5$ nm and $d\nu \sim 0.01$ to 0.001 nm. The aforementioned isolated two level system in thermal equilibrium with the radiation field will then operate in a slightly different way. This is shown in Fig. 2-10. The laser is forced to operate at the frequency ν_ℓ of width $d\nu$, while the lineshape (distribution) function of the transition is given by $g(\nu, \nu_o)$ of width $\Delta\nu$. ($d\nu \ll \Delta\nu$). Thus, the net number of atoms interacting with the laser radiation at ν_ℓ of width $d\nu$

= (net number of atoms ready to leave level 2) x (probability of stimulated emission from ν_ℓ to $\nu_\ell + d\nu$)

$= (N_2 - \frac{g_2}{g_1} N_1) \cdot g(\nu_\ell, \nu_o)d\nu$ (see explanation of eq. 2-22 and meaning of $(g(\nu, \nu_o))$.

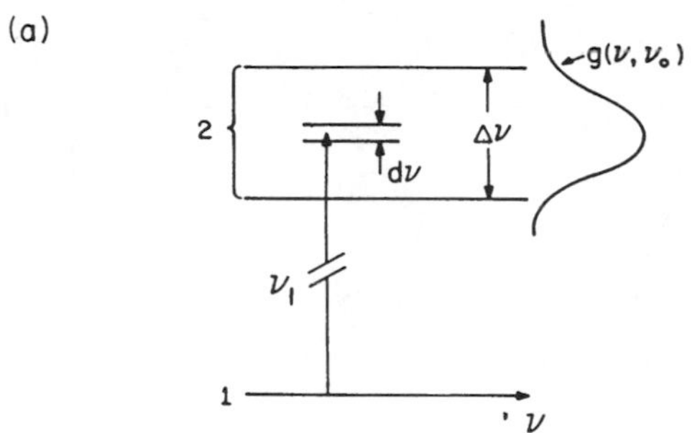

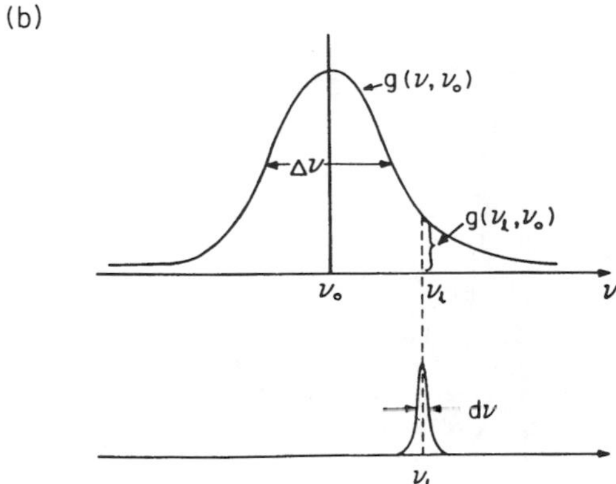

Fig. 2-10. Lasing transition at frequency ν_ℓ of width $d\nu$ in a broad lineshape $g(\nu, \nu_o)$ of width $\Delta\nu$.

Replacing the term $(N_2 - \frac{g_2}{g_1} N_1)$ in eq. (2-22) by the above expression, we obtain the equation describing the rate of change of N_1 when interacting with the laser field:

$$\frac{\partial N_1}{\partial t} = B_{21}\rho(\nu_\ell)\ (N_2 - \frac{g_2}{g_1} N_1) g(\nu_\ell, \nu_o) d\nu$$

Definition: <u>number density</u> $n_i \equiv \frac{N_i}{V}$, $i = 1,2$; and $V \equiv$ volume

With this definition, dividing the above expression by $(-V)$, we obtain

$$\frac{\partial n_1}{\partial t} = B_{21}\rho(\nu_\ell) d\nu\ g(\nu_\ell , \nu_o) \left[\frac{g_2}{g_1} n_1 - n_2 \right] \tag{2-29}$$

But $\quad -\frac{\partial n_1}{\partial t} \equiv$ net rate of decrease of the density of atoms in level 1

$=$ net absorption

$=$ net rate of decrase of the laser radiation's photon density

$$= -\frac{\partial}{\partial t} \left[\frac{\rho(\nu_\ell) d\nu}{h\nu_\ell} \right]$$

where it is assumed that all the laser photons have the same frequency ν_ℓ (because $d\nu \ll \Delta\nu$). Substituting into eq. (2-29), we get

$$-\frac{\partial \rho(\nu_\ell)}{\partial t} = \rho(\nu_\ell) h\nu_\ell B_{21} g(\nu_\ell , \nu_o) \left[\frac{g_2}{g_1} n_1 - n_2 \right] \tag{2-30}$$

Consider that the laser photons pass through a slab of active material of width dx, as shown in Fig. 2-11. If c denotes the speed of the radiation in the material the passage time is $dt = \frac{dx}{c}$

$$\therefore \quad -\frac{\partial \rho(\nu_\ell)}{\partial t} \rightarrow -\frac{\partial \rho(\nu_\ell)}{\partial x/c}$$

$$\therefore \quad \text{eq. (2-30)} \rightarrow$$

$$-\frac{\partial \rho(\nu_\ell)}{\partial x} = \rho(\nu_\ell) h\nu_\ell B_{21} g(\nu_\ell , \nu_o) \left[\frac{g_2}{g_1} n_1 - n_2 \right] \frac{1}{c}$$

Integrating →

$$\rho(\nu_\ell) = \rho_o(\nu_\ell) \exp\left\{ -h\nu_\ell g(\nu_\ell , \nu_o) B_{21} \left[\frac{g_2}{g_1} n_1 - n_2 \right] \frac{x}{c} \right\} \tag{2-31}$$

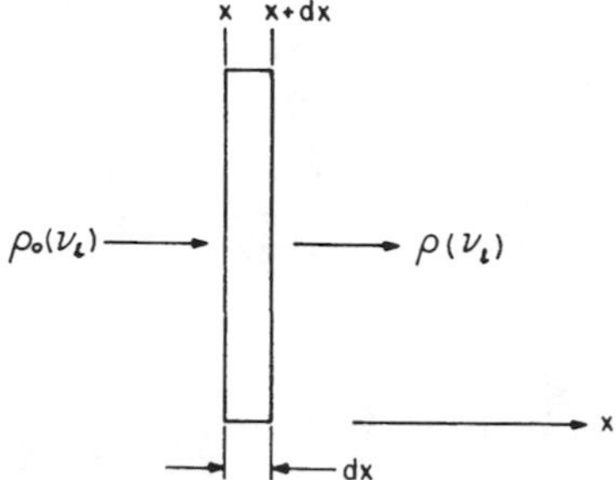

Fig. 2-11. Single pass amplification or attenuation through a general thin slab of optical medium.

Definition: <u>absorption coefficient</u>

$$\alpha(\nu_\ell) \equiv \left[\frac{g_2}{g_1} n_1 - n_2 \right] \sigma_{21}(\nu_\ell) \qquad (2\text{-}32)$$

where $\sigma_{21}(\nu_\ell) \equiv$ <u>stimulated emission cross section</u>

$$\equiv \frac{h\nu_\ell g(\nu_\ell , \nu_o)}{c} B_{21} \qquad (2\text{-}33)$$

Substituting into eq. (2-31) and using eq. (2-32) we have

$$\rho(\nu_\ell) = \rho_o(\nu_\ell)\exp[- \alpha(\nu_\ell)x] \qquad (2\text{-}34)$$

This is an expected result by inspection of Fig. 2-11. It is the Beer's law if $\alpha(\nu_\ell) > 0$. However, if $\alpha(\nu_\ell) < 0$, the result will be

$$\rho(\nu_\ell) = \rho_o(\nu_\ell)\exp[+ |\alpha(\nu_\ell)|x]$$

which means amplification of the laser radiation at ν_ℓ.

(Note: similar to the definition of eq. (2-33), we can also define the <u>absorption cross section</u> as

$$\sigma_{12}(\nu_\ell) \equiv \frac{h\nu_\ell g(\nu_\ell , \nu_o)}{c} B_{12} \qquad (2\text{-}35)$$

Using the relationship between B_{12} and B_{21}, i.e. $B_{21} = \frac{g_1}{g_2} B_{12}$, comparing eq. (2-33) and (2-35),

$$\frac{\sigma_{12}}{\sigma_{21}} = \frac{g_2}{g_1}$$

If $$g_2 = g_1 = 1, \quad \Rightarrow \quad \sigma_{12} = \sigma_{21}$$

This means that in a strictly two level system without degeneracy, the absorption and stimulated emission cross sections are equal. (This relationship can also be obtained more rigorously using quantum mechanics.)

§2.5 Impossibility of having gain (i.e. $\alpha(\nu_\ell) < 0$) in a two level system in thermal equilibrium

This can be seen by noting that in thermal equilibrium, from eq. (2-14),

$$\frac{N_2}{N_1} = \frac{g_2}{g_1} \exp\left\{ - \frac{(E_2 - E_1)}{kT} \right\} < \frac{g_2}{g_1} \qquad (2\text{-}36)$$

Hence $$N_2 < N_1 \left(\frac{g_2}{g_1} \right) \qquad (2\text{-}36')$$

or $$N_1 \frac{g_2}{g_1} - N_2 > 0$$

or $$n_1 \frac{g_2}{g_1} - n_2 > 0$$

Thus $$\alpha(\nu_\ell) > 0 \qquad \text{Using eq. (2-32)}$$

and equation (2-34) is always an exponentially decreasing function. That is, there can be no net amplification of radiation in a two level system in thermal equilibrium.

Even if one increases the temperature of the system so as to increase N_2/N_1 (see eq. 2-36), the best one can have is $N_2 = N_1$ at $T = \infty$. This leads to (by eq. (2-36) and (2-32)) $\alpha(\nu_\ell) = 0$ and the result is saturation; i.e. no gain and no loss [$\rho(\nu_\ell) = \rho_o(\nu_\ell)$ from eq. 2-34].

It is customary to use the temperature as a parameter to characterize the active medium. Thus, if by some means, one can make $\alpha(\nu_\ell) < 0$, in eq. (2-36') one will have $\left(N_1 \frac{g_2}{g_1} - N_2 \right) < 0$.

This is equivalent to making $T < 0$ in eq. (2-36). One thus says that the active medium has achieved a <u>negative temperature</u>. (We should keep in mind that, physically, a negative temperature is impossible to achieve.)

Under such a situation of negative temperature, $N_2 > \frac{g_2}{g_1} N1$, and we say that the <u>population is inverted</u>, or we have an <u>inverted system</u>. In the case of $g_2 = g_1 = 1$, $N_2 > N_1$; i.e. <u>population inversion</u> means that there are more atoms in level 2 than level 1 in a 2 level system without degeneracy. <u>In order to achieve population inversion, one has to pump the system by some means other than heating</u>.

§2.6 Pumping

One can imagine using some magic means to raise atoms from level 1 to level 2 so as to reach population inversion. However, if the system is to be in equilibrium, isolated and dilute (no collision), the only means to raise the atoms from level 1 to level 2 is by absorption of radiation. We thus fall back to our initial conditions set forth in the beginning of this chapter, i.e. thermal equilibrium with the radiation at the transition frequency. And we already know that inversion is impossible.

In practice, one pumps atoms into level 2 via some other levels. One can either raise atoms from level 1 into level 2 via a third level (<u>3 - level system</u>) or raise atoms from a third, irrelevant level into level 2 via a fourth also irrelevant level, (<u>4 - level system</u>). Most known laser systems operate with a 4 - level system while 3 - level systems are used to a much lesser extent because of the inefficiency to attain inversion (see below).

§2.7 Rate Equations Approach

We now go into some detail of 3 - and 4 - level systems using the intuitive rate equation approach. The rate equation approach essentially balances the rate of gain and loss of particles (atoms) in different levels and also of the laser photons. Monochromaticity of the laser is assumed while longitudinal and spatial distributions of the laser radiation inside the laser cavity are ignored. This means that we are assuming that the photon density is uniform inside the cavity so that we can analyse only the change along the axis of the laser. Such an approximation is reasonable because the spatial (longitudinal and transverse) distribution of the laser radiation (or modes) inside the cavity depends mostly on the geometry of the cavity and in particular, the size, shape and aperture of the end mirrors. The active medium is just an amplifier that enhances the radiation densities inside those electromagnetic modes satisfying the boundary conditions inside the cavity. Such a decoupling of the active medium and the cavity makes life easier in the analysis of laser oscillation. One can study first of all the energy and particle number balances (rate equation analysis), and the mode structures independently. After that, the result can be matched giving the realistic results.

We now concentrate on the rate equations. The result should give us broad features of the average power, peak power, laser temporal pulse envelope and threshold conditions etc.

(a) Idealized three level system

The system is shown in Fig. 2-12. One pumps atoms in level 1 into level 3 which is rather broad. This pumping could be the absoprtion of some appropriate radiation from a flashlamp, another laser, etc. It could also be achieved by collision, chemical reaction etc. Whatever it is, let's ignore now the detail and assume simply that atoms have been raised from level 1 into level 3 at a rate W_p. On reaching level 3, they decay rapidly onto level 2 and stay there for a long enough time to allow laser

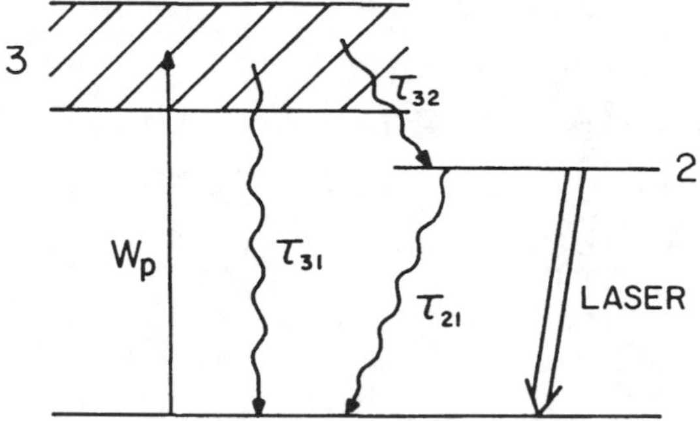

Fig. 2-12. 3-level system.

action to occur between level 2 and 1. We still keep the previous assumptions that the system is isolated and in equilibrium, and that no collision among the atoms is allowed. Thus, the decay mechanism of levels 3 and 2 are through spontaneous emission whose lifetimes are τ_{31}, τ_{32} and τ_{21}. The above statement of the rapid decay from 3 → 2 means the following additional assumptions:

$$\tau_{31} \gg \tau_{32} \tag{2-37}$$

$$\tau_{21} \gg \tau_{32} \tag{2-38}$$

Hence
$$n_3 \approx 0 \tag{2-39}$$

If $n_{tot} \equiv$ total number of atoms in the system

we have
$$n_1 + n_2 + n_3 = n_{tot}$$

i.e.
$$n_1 + n_2 \approx n_{tot} \tag{2-40}$$

Under these assumptions, we derive the following <u>rate equations</u>.

$$\frac{\partial n}{\partial t} = -\gamma\sigma_{21}c\phi n - \frac{(\gamma - 1)n_{tot} + n}{\tau f} + W_p(n_{tot} - n) \tag{2-41}$$

and
$$\frac{\partial \phi}{\partial t} = \sigma_{21}c\phi n - \frac{\phi}{\tau_c} + S \tag{2-42}$$

where
$n \equiv$ <u>inversion density</u>

$$\equiv n_2 - \frac{g_2}{g_1} n_1 \tag{2-43}$$

$$\gamma \equiv 1 + \frac{g_2}{g_1} \tag{2-44}$$

$\tau_f \equiv \tau_{21}$, and $\tau_c \equiv$ photon lifetime in the cavity (see below) and $\phi \equiv$ photon density (cm^{-3}). S is the spontaneous emission rate.

Derivation:

From the definition of the stimulated emission cross section σ_{21} (eq. 2-33), we have

$$B_{21} = \frac{c}{h\nu g(\nu)} \sigma_{21}(\nu) \tag{2-45}$$

Now, $\rho(\nu) \equiv$ laser radiation density/frequency

$= h\nu\phi \cdot$ (net probability of stimulated emission/frequency)

$= h\nu\phi \cdot g(\nu)$

Hence $B_{21}\rho(\nu) = c\sigma_{21}(\nu)\phi$ (2-46)

Now, $\frac{\partial n_1}{\partial t}$ = net stimulated emission rate

+ spontaneous emission rate

- pumping rate of level 1 into level 3

$$= \left(n_2 - \frac{g_2}{g_1} n_1 \right) B_{21}\, \rho(\nu) + \frac{n_2}{\tau_{21}} - W_p n_1 \tag{2-47}$$

The first term comes from eq. (2-22).

$$\frac{\partial n}{\partial t} \equiv \frac{\partial}{\partial t}\left(n_2 - \frac{g_2}{g_1} n_1 \right)$$

$$= - \frac{\partial n_1}{\partial t}\left(1 + \frac{g_2}{g_1} \right) \tag{2-48}$$

$$\left\{ \text{since } n_1 + n_2 \approx n_{tot} = \text{const.}; \ \frac{\partial n_1}{\partial t} = - \frac{\partial n_2}{\partial t} \right\}$$

Substituting eq. (2-46) and (2-47) into eq. (2-48), and using eq (2-43) and (2-44), we obtain eq. (2-41).

To obtain the photon rate equation,

$$\frac{\partial\phi}{\partial t} = \text{rate of creation of photon density - rate of loss}$$

$$= \text{net rate of stimulated emission + rate of spontaneous emission - rate of total loss in the cavity}$$

$$= \left(n_2 - \frac{g_2}{g_1} n_1 \right) B_{21}\rho(\nu) + S - \frac{\phi}{\tau_c} \tag{2-49}$$

where S is the spontaneous emission rate and τ_c is a decay constant of the photon density. τ_c can be related to Q and α_c of the cavity by the following consideration. Assuming there is no emission and absorption and a photon density ϕ_o is injected by some means into the laser cavity. ϕ will decay through all the possible loss mechanisms mentioned in the beginning of the chapter

Hence
$$\frac{\partial\phi}{\partial t} = -\frac{\phi}{\tau_c}$$

$$\phi = \phi_o e^{-t/\tau_c} \tag{2-50}$$

But
$$E = E_o e^{-\frac{\omega}{Q}t} \quad \text{(eq. 2-4)}$$

the two equations are identical because the photon distribution is assumed uniform in the cavity. Hence, using also eq. (2-4')

$$\tau_c = \frac{Q}{\omega} = \frac{1}{\alpha_c c} \tag{2-51}$$

τ_c is called <u>photon lifetime</u> in the cavity. Using eq. (2-22), we can rewrite eq. (2-49):

$$\frac{\partial\phi}{\partial t} = \sigma_{21} c\phi n - \frac{\phi}{\tau_c} + S$$

which is eq. (2-42).

(b) <u>Idealized 4 - level system</u>

This is shown in Fig. (2-13). Again, we ignore for the moment the detail of pumping. In addition to the assumption of equilibrium, no collision and

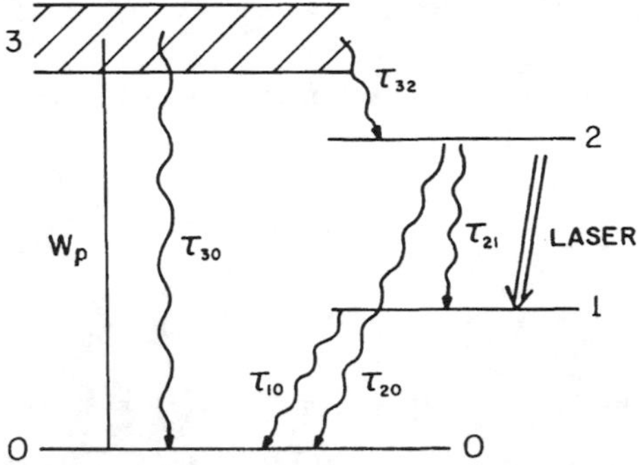

Fig. 2-13. 4-level system.

isolation, we further assume

$$\tau_{30} \gg \tau_{32}$$

$$\tau_{21} \gg \tau_{32}$$

$$\tau_{20} \gg \tau_{32}$$

$$\tau_{10} \approx 0$$

i.e. after a fast pumping of atoms from level 0 to level 3, the atoms decay very rapidly into level 2 and wait there till laser action takes place between $2 \rightarrow 1$. Once arriving at level 1, the atoms "immediately" decay back to level zero. Hence,

$$n_3 \approx 0$$

$$n_1 \approx 0 \tag{2-52'}$$

and $\quad n_0 + n_1 + n_2 + n_3 = n_{tot}$ becomes

$$n_0 + n_2 = n_{tot} = \text{constant} \tag{2-52}$$

where $\quad n_{tot} \equiv$ total number density of atoms in the system.

Definition: $\dfrac{1}{\tau_f} \equiv \dfrac{1}{\tau_{21}} + \dfrac{1}{\tau_{20}}$

Again, we derive the following rate equations.

$$\frac{\partial n}{\partial t} = W_p(n_{tot} - n) - n\sigma_{21}\phi c - \frac{n}{\tau_f} \tag{2-53}$$

and

$$\frac{\partial \phi}{\partial t} = \sigma_{21}c\phi n - \frac{\phi}{\tau_c} + S \tag{2-54}$$

Derivation:

$$\frac{\partial n_2}{\partial t} = \text{pumping - net stimulated emission - spontaneous emission}$$

$$= W_p n_0 - \left(n_2 - \frac{g_2}{g_1} n_1 \right) \sigma_{21} \phi c - \frac{n_2}{\tau_{21}} - \frac{n_2}{\tau_{20}}$$

Since $$n \equiv n_2 - \frac{g_2}{g_1} n_1 \approx n_2 \qquad (\text{since } n_1 \approx 0) \qquad (2\text{-}55)$$

$$\therefore \quad \frac{\partial n}{\partial t} = W_p n_0 - n\sigma_{21}\phi c - \frac{n_2}{\tau_f}$$

which is eq. (2-53) after using eq. (2-52) and (2-55).

Eq. (2-54) can be derived in an identical way as that leading to eq. (2-42) and is left as an exercise to the reader.

§2.8 Threshold Oscillation

Now eq. (2-34) and (2-32) show that the gain of the active medium should be given by

$$g = -\alpha(\nu_\ell) = \left(n_2 - \frac{g_2}{g_1} n_1 \right) \sigma_{21}(\nu_\ell) \qquad (2\text{-}55')$$

But the threshold oscillation condition is (eq. 2-3)

$$g l_m = \alpha_c l$$

hence $$-\alpha(\nu_\ell) l_m = \alpha_c l$$

$$\left(n_2 - \frac{g_2}{g_1} n_1 \right) \sigma_{21}(\nu_\ell) l_m = \alpha_c l$$

After some re-arrangement, we obtain

$$n \equiv n_2 - \frac{g_2}{g_1} n_1 = \frac{8\pi\nu^2 \tau_{21} l}{\tau_c g(\nu_\ell\, , \nu_o) c^3 l_m} \qquad (2\text{-}56)$$

where eq. (2-51), (2-33) and (2-17) have been used. Eq. (2-56) shows that at threshold oscillation, the inversion n is inversely proportional to

$g(\nu, \nu_o)$. If the laser is allowed to oscillate at any frequency ν_ℓ within the transition line width of $g(\nu, \nu_o)$, then once the active medium is pumped, all these frequencies will compete with one another to extract energy from the gain medium. The frequency at $\nu = \nu_o$ will first reach the threshold condition because the probability of net stimulated emission which is $g(\nu, \nu_o)$ is largest at $\nu = \nu_o$; i.e. at the peak of $g(\nu, \nu_o)$. If the system is maintained at threshold, once it starts oscillating at $\nu = \nu_o$, the other frequencies will be suppressed to oscillate because the energy in the gain medium is extracted by $\nu = \nu_o$ and there is not enough left for the other frequencies to overcome the loss. The laser thus oscillates at $\nu = \nu_o$ with a very narrow width. <u>That's why laser lines are very narrow</u>. Fig. (2-14) shows schematically that at the peak of the gain curve, gain just overcomes loss and laser action takes place only within the very narrow width $d\nu$.

One can of course force the laser to oscillate at any frequency $\nu \neq \nu_o$ within the width of $g(\nu, \nu_o)$ by introducing more loss to all frequencies except the one of interest. Then, net gain will build up only for the frequency of interest and it will oscillate first and will be maintained at threshold so that the other frequencies cannot oscillate.

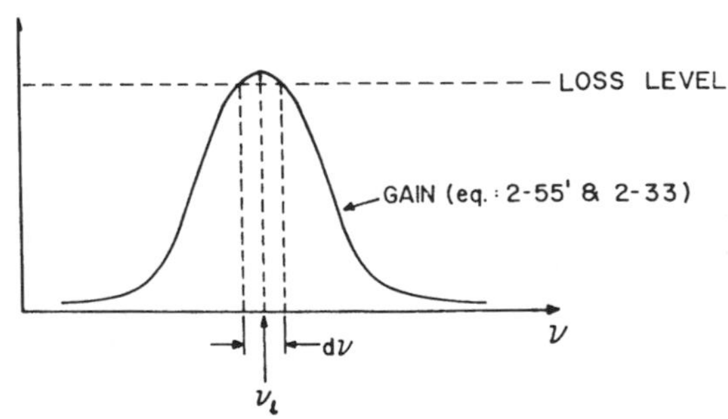

Fig. 2-14. Threshold oscillation leading to a very sharp lasing line.

It is of course desirable to have an efficient laser; this means as low an inversion n as possible. From eq. (2-56), n is inversely proportional to τ_c, the photon density lifetime. That is, the longer τ_c is, the lower n will be. Long τ_c means small α_c according to eq. (2-51); i.e. low loss.

We now ask the question, "what should the pumping rate W_p be in order to maintain the laser at threshold oscillation?" For a three level system, the answer is

$$W_p(\text{min.}) = \frac{g_2}{\tau_f g_1} \tag{2-57}$$

and for a 4 - level system, there is no minimum W_p ; i.e. any W_p is good.

Proof: (A) 3 - level system

Assumptions: (1) low threshold value:

$$\phi \approx 0$$

(2) steady state condition:

$$\frac{\partial n}{\partial t} = 0$$

Thus, eq. (2-41) becomes

$$0 = -0 - \frac{(\gamma - 1)n_{tot} + n}{\tau_f} + W_p(n_{tot} - n)$$

Re-arranging terms, using eq. (2-44). we obtain

$$\frac{n}{n_{tot}} = \frac{W_p\tau_f - g_2/g_1}{W_p\tau_f + 1} \tag{2-58}$$

In order to have an inversion, we should have

$$\frac{n}{n_{tot}} \geq 0$$

From eq. (2-58), this means

$$W_p\tau_f - g_2/g_1 \geq 0$$

or

$$W_p \geq \frac{g_2}{\tau_f g_1}$$

and the minimum pumping rate is

$$W_p(\text{min.}) = \frac{g_2}{\tau_f g_1} \qquad (2\text{-}59)$$

which is eq. (2-57). We can now take note of some of the disadvantages of a three-level system.

Disadvantage 1:

From eq. (2-59), we see that to obtain low W_p, one has to have a long τ_f; i.e. a long fluorescence lifetime for level 2.

Disadvantage 2:

At minimum inversion,

$$n = 0$$

i.e.

$$n_2 - \frac{g_2}{g_1} n_1 = 0$$

or

$$n_2 = \frac{g_2}{g_1} n_1 \qquad (2\text{-}60)$$

If

$$g_2 = g_1 = 1, \qquad n_2 = n_1 = \frac{1}{2} n_{tot} \qquad (2\text{-}61)$$

This means that for non-degenerate levels 1 and 2, one needs to pump half of the total atoms into level 2 in order to just reach minimum threshold and this means a lot of pumping, i.e. not efficient.

(B) 4 - level systems

The same assumptions of

$$\phi \approx 0$$

$$\frac{\partial n}{\partial t} = 0$$

are used. Thus, eq. (2-53) becomes

$$0 = W_p(n_{tot} - n) - 0 - \frac{n}{\tau_f}$$

$$\frac{n}{n_{tot}} = \frac{W_p}{W_p + \frac{1}{\tau_f}} \qquad (2\text{-}62)$$

the right hand side of eq. (2-62) is <u>always</u> positive. Hence,

$$\frac{n}{n_{tot}} > 0 \quad \text{(always)}$$

i.e. there is no minimum pump rate. Any W_p will induce an inversion in an idealized four level system. Physically, this is evident because n_1 was assumed zero (eq. (2-52')) so that whenever there is a slight population n_2 in level 2, it is inverted with respect to level 1.

§2.9 Threshold pump power

Essentially, at threshold oscillation, the pumping is to maintain $\phi \approx 0$ for the laser output; i.e. the pumping is mainly used to compensate for the loss in the cavity.

Assumption: the only loss in the cavity is through fluorescence or spontaneous emission.

If $P_{eff} \equiv$ effective pump power

$P_f \equiv$ fluorescence power

Our assumption means

$$P_{eff} = P_f = \frac{n_2(th) \cdot h\nu}{\tau_f}$$

where $n_2(th)$ is the threshould population of level 2. For idealized 3 - and 4 - level systems we have

$$P_{eff} = \begin{cases} \dfrac{h\nu}{\tau_f} \cdot \dfrac{n_{tot}}{2} & \text{(3 - level, using eq. (2-61))} \quad (2\text{-}64) \\[2ex] \dfrac{h\nu}{\tau_f}\, n & \text{(4 - level, using eq. (2-55))} \quad (2\text{-}65) \end{cases}$$

§2.10 Above threshold oscillation and gain saturation

Very often, a laser oscillates with a photon flux above that of threshold; i.e. $\phi \neq 0$ but $\frac{\partial n}{\partial t} = 0$ (still). This means that the steady state condition is still valid but in the presence of a strong ϕ.

<u>3 - level system</u>:

The rate eq. (2-41) now becomes

$$0 = -\gamma\,\sigma_{21}c\phi n - \frac{(\gamma - 1)n_{tot} + n}{\tau_f} + W_p\,(n_{tot} - n)$$

solving for n, we obtain

$$n = n_{tot}\left[W_p - \frac{\gamma - 1}{\tau_f}\right]\left(\gamma\,\sigma_{21}\,c\phi + W_p + \frac{1}{\tau_f}\right)^{-1} \quad \text{(3-level)} \qquad (2\text{-}66)$$

<u>4 - Level system</u>:

The rate equation (2-53) becomes

$$0 = W_p\,(n_{tot} - n) - n\,\sigma_{21}\phi c - \frac{n}{\tau_f}$$

solving for n, we obtain

$$n = n_{tot} \cdot \frac{W_p}{W_p + \sigma_{21}\phi c + \frac{1}{\tau_f}} \quad \text{(4 - level)} \qquad (2\text{-}67)$$

We now define the <u>small gain coefficient</u> g_o. This follows from the expression of the gain coefficient g given by (using eq. 2-32):

$$g = -\alpha(\nu_\ell) = \left(n_2 - \frac{g_2}{g_1}n_1\right)\sigma_{21}\,(\nu_\ell)$$

or $g = n\sigma_{21}$ (using eq. 2-43) (2-68)

<u>Definition</u>: $g_o \equiv g$ (at $\phi = 0$)

$$= n\,(\phi = 0)\,\sigma_{21} \qquad \text{(from eq. (2-68))}$$

$$= \begin{cases} \sigma_{21}n_{tot}\left(W_p - \frac{\gamma - 1}{\tau_f}\right)\left(W_p + \frac{1}{\tau_f}\right)^{-1} & \text{(3-level)} \quad (2\text{-}69) \\ \sigma_{21}n_{tot}W_p\left(W_p + \frac{1}{\tau_f}\right)^{-1} & \text{(4-level)} \quad (2\text{-}70) \end{cases}$$

From eq. (2-69) and (2-70), we see that g_o depends only on the material parameters (σ_{21} and τ_f) and the pumping rate W_p. This small signal gain is what an active medium has when it is pumped above threshold and when the laser action is inhibited (e.g. by blocking the light inside the cavity, increasing the loss significantly during a period of time i.e. Q-switching, etc.). If feedback is restored at some moment, ϕ in the resonator will increase exponentially in the beginning; i.e. the increase follows

$$e^{g_o x}$$

As soon as the ϕ becomes appreciable, g_o becomes g where

$$g = n\,\sigma_{21} \qquad \text{(from eq. 2-68)}$$

$$= \begin{cases} \sigma_{21} n_{tot}\left(W_p - \dfrac{\gamma - 1}{\tau_f}\right)\left[\left(W_p + \dfrac{1}{\tau_f}\right)\left(\dfrac{\gamma\sigma_{21}c\phi}{W_p + \dfrac{1}{\tau_f}} + 1\right)\right]^{-1} & \text{(3-level)} \quad (2\text{-}71) \\ \sigma_{21} n_{tot} W_p \left[\left(W_p + \dfrac{1}{\tau_f}\right)\left(1 + \dfrac{\sigma_{21}\phi c}{W_p + \dfrac{1}{\tau_f}}\right)\right]^{-1} & \text{(4-level)} \quad (2\text{-}72) \end{cases}$$

Using eq. (2-69) and (2-70), we simplify eq. (2-71) and (2-72) into:

$$g = \begin{cases} g_o\left(1 + \dfrac{\gamma\sigma_{21}\phi c}{W_p + \dfrac{1}{\tau_f}}\right)^{-1} & \text{(3-level)} \quad (2\text{-}73) \\ g_o\left(1 + \dfrac{\sigma_{21}\phi c}{W_p + \dfrac{1}{\tau_f}}\right)^{-1} & \text{(4-level)} \quad (2\text{-}74) \end{cases}$$

Definition: **intensity** $I \equiv c\phi h\nu \left(\dfrac{\text{Joules}}{\text{cm}^2\text{sec}}\right)$ (2-75)

Definition: Saturation intensity $\equiv I_s$

$\equiv$ intensity at which $g = \frac{1}{2} g_o$

Since g is given by eq. (2-73) and (2-74), we have, at $I = I_s$,

$$\frac{1}{2} g_o = \begin{cases} g_o\left(1 + \dfrac{\gamma\sigma_{12}\phi_s c}{W_p + \dfrac{1}{\tau_f}}\right)^{-1} & \text{(3-level)} \quad (2\text{-}76) \\ g_o\left(1 + \dfrac{\sigma_{21}\phi_s c}{W_p + \dfrac{1}{\tau_f}}\right)^{-1} & \text{(4-level)} \quad (2\text{-}77) \end{cases}$$

where ϕ_s = saturation photon density at $I = I_s$

From eq. (2-75), using (2-76) and (2-77), one obtains

$$I \equiv c\phi_s h\nu$$

$$= \begin{cases} \left(W_p + \dfrac{1}{\tau_f} \right) \dfrac{h\nu}{\gamma\sigma_{21}} & \text{(3-level)} \quad (2\text{-}78) \\[2ex] \left(W_p + \dfrac{1}{\tau_f} \right) \dfrac{h\nu}{\sigma_{21}} & \text{(4-level)} \quad (2\text{-}79) \end{cases}$$

Substituting eq. (2-78), (2-79) into eq. (2-73), (2-74), we have the "universal" relation:

$$g = \frac{g_o}{1 + I/I_s} \quad \text{(both 3- and 4-level systems)} \qquad (2\text{-}80)$$

<u>**Definition**</u>: $\tau_{st} \equiv$ <u>**stimulated emission lifetime**</u>

The notion of stimulated emission lifetime is sometimes used. We explain it as follows. Assume that a population n_2 is generated in level 2 and left alone in a gain medium. Its evolution decay in time will include fluorescence and stimulated emission:

$$\frac{\partial n_2}{\partial t} = -\frac{n_2}{\tau_f} - \frac{n_2}{\tau_{st}} \qquad (2\text{-}81)$$

But

$$\frac{\partial n_2}{\partial t} = -\frac{n_2}{\tau_f} - \text{Stimulated emission}$$

$$= \frac{n_2}{\tau_f} - B_{21}\rho(\nu)n_2 \qquad (2\text{-}82)$$

Eq. (2-81) and (2-82) gives

$$\frac{1}{\tau_{st}} = B_{21}\rho(\nu) = c\sigma_{21}\phi \quad \text{(from eq. 2-46)}$$

$$\tau_{st} = \frac{1}{\sigma_{21}c\phi} = \frac{h\nu}{\sigma_{21}I} \quad \text{(from eq. 2-75)} \qquad (2\text{-}83)$$

Physically, eq. (2-83) shows that the stimulated emission lifetime of level 2 is inversely proportional to both the stimulated emission cross section and intensity at a fixed frequency. The larger σ_{21} and I are, the shorter

τ_{st} will be. Thus in the presence of strong stimulated emission (large σ_{21} and I),

$$\tau_f \gg \tau_{st}$$

and we can neglect the first term (spontaneous emission) on the right hand side of eq. (2-81). In other words, in most practical laser calculations, spontaneous emission is neglected. (Cf. eq. (2-22) and the discussion therein.)

§2.11 <u>Output power calculation</u>

We now calculate the output power from a laser oscillator taking into account possible saturation in the gain medium. We start by considering amplification of intensity in a single pass through the gain medium. Oscillation condition is then imposed to obtain the equation for the output power.

(A) <u>Single pass amplification</u>

We first ask what the amplification of a beam of light is after one single pass through the amplifying medium of the laser. Referring to eq. (2-34) and (2-55') and Fig. (2-11), we see that the gain in photon density $\rho(\nu_\ell)$ across an amplifying "slab" of thickness dx (Fig. 2-15(a) is

$$\rho(x + dx) - \rho(x) \equiv d\rho(x)$$

$$= (gdx)\ \rho(x) \qquad (2\text{-}84)$$

where g, the gain, is given by eq. (2-55'). It is understood that the laser operates at the frequency ν_ℓ so that we omit writing it in the argument of $\rho(x)$. From eq. (2-75), the intensity

$$I \equiv c\phi h\nu = c\rho \qquad (2\text{-}85)$$

Thus, multiplying eq. (2-84) by c and using (2-85), we have

$$dI(x) = (gdx)I(x) \qquad (2\text{-}86)$$

If we include the loss in (2-86), we have

$$dI(x) = (g - \alpha_c)(dx)I(x) \qquad (2\text{-}87)$$

where α_c is the loss coefficient (see §2-1). Eq. (2-87) becomes

$$\frac{dI(x)}{dx} = (g - \alpha_c)\ dx \tag{2-88}$$

Note that g is now given by eq. (2-80) which we rewrite as follows:

$$g = \frac{g_o}{1 + I/I_s} \tag{2-89}$$

Substituting eq. (2-89) into (2-88), we have

$$\frac{dI(x)}{dx} = \frac{g_o I(x)}{1 + I(x)/I_s} - \alpha_c I(x) \tag{2-90}$$

which is the <u>amplifier equation</u>. The analytical solution is not available. Numerical integration is normally required.

The condition for <u>small signal gain</u> can be defined as:

$$I(x) \ll I_s \tag{2-91}$$

$$g \simeq g_o \tag{2-92}$$

Eq. (2-90) becomes

$$\frac{dI(x)}{dx} \simeq g_o I(x) - \alpha_c I(x)$$

or

$$I(x) = I_o e^{(g_o - \alpha_c)x} \tag{2-93}$$

which is essentially eq. (2-2). Here, I_o is the intensity at x = 0, and x is the position at a point along the amplifier's axis (Fig. 2-15(b)). This shows also that the introduction in §2-1 pertains only to the case of small signal gain.

(B) <u>Oscillation</u>

We now allow oscillation to take place and simplify the laser cavity by assuming that the ends of the amplifying medium also act as reflectors, plane parallel in the present analysis. This is shown in Fig. 2-15(c). The mirror at x = 0 is assumed to be one that takes care of the total loss in the cavity (cf. §2-1), while the mirror at $x = \ell$ is the output mirror.

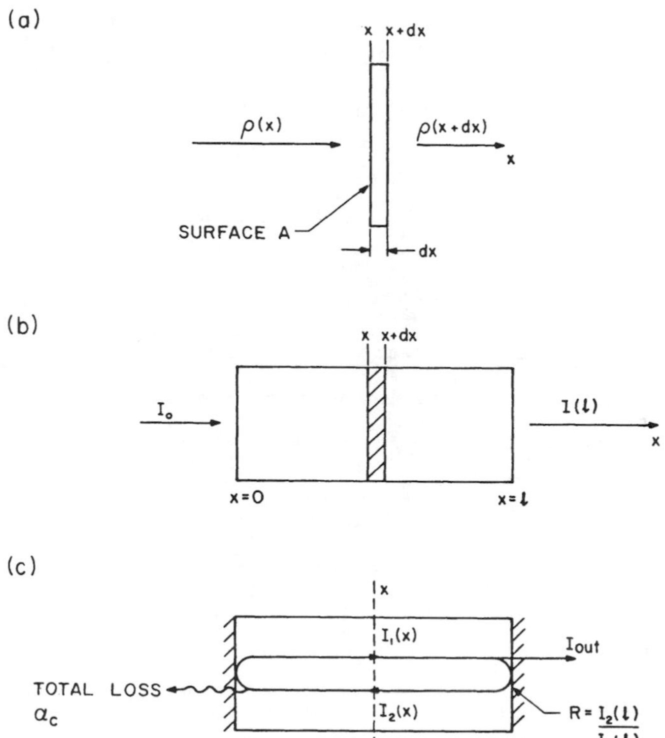

Fig. 2-15. (a) Single pass amplification through a thin slab of gain medium.
(b) Single pass amplification through a general gain medium.
(c) Above-threshold oscillation analysis.

A beam of light "circulates" in the cavity as shown shematically in Fig. 2-15(c). $I_1(x)$ is the beam propagating towards the right side. When it reaches the mirror, part of it will be transmitted and part reflected. All other losses are incorporated in the other mirror. Hence the reflectivity of the mirror at $x = \ell$ is

$$R = \frac{I_2(\ell)}{I_1(\ell)} \tag{2-94}$$

where I_2 is the reflected beam intensity. When I_2 reaches the mirror at $x = 0$, part of it is lost into the mirror. This loss represents the total loss α_c. The reflected beam should be equal to I_1 under the oscillation condition. We note that at any position x, the gain medium "sees" a total intensity I_T which is the sum of I_1 and I_2.

i.e. $$I_T(x) = I_1(x) + I_2(x) \tag{2-95}$$

The loss "seen" by each beam is α_c. The gain "seen" by each beam is

$$g_1 = g_2 = \frac{g_o}{1 + I_T/I_s} \tag{2-96}$$

Note that it is the total intensity that saturates the gain medium at any point in the medium and thus we use I_T in the expression of the two gains. The coupled equations governing each beam for one pass are then (using eq. (2-90))

$$\frac{dI_1(x)}{dx} = g_1 I_1(x) - \alpha_c I_1(x) \tag{2-97}$$

$$-\frac{dI_2(x)}{dx} = g_2 I_2(x) - \alpha_c I_2(x) \tag{2-98}$$

The reason why we put a negative sign on the left hand side of eq. (2-98) is that $I_2(x)$ propagates in the negative "x" direction. It can be seen that the mean intensity I in the cavity is constant, where

mean intensity $$I(x) \equiv \sqrt{I_1(x) I_2(x)} \tag{2-99}$$

This can be proved by multiplying eq. (2-97) by $I_2(x)$ and (2-98) by I_1 and take the difference (exercise). The result is

$$\frac{dI(x)}{dx} = 0 \tag{2-100}$$

so that $$I(x) = \text{constant} \tag{2-101}$$

One can express the power output of the laser in terms of I, R and A. This is left as an exercise. The result is

Power P (Joule/sec)

$$= AI\left(\frac{1 - R}{R^{1/2}}\right) \tag{2-102}$$

The precise values of I and P depend on the solution of the coupled eq.

(2-97) and (2-98). Numerical solution is possible once g_o and I_s are known. We won't go into any detail here.

Closing remark

This chapter intends to give an elementary physical account of the operation of a laser oscillator. It is not meant to be complete. Interested readers should consult specialized books devoted entirely to laser operation.

CHAPTER III

SNELL'S LAW, FRESNEL EQUATIONS, BREWSTER ANGLE AND CRITICAL ANGLE

We now consider how the Brewster angle window works and what makes the end prism totally reflecting. We need only to consider the idealized case of sending a plane monochromatic electromagnetic wave, say a well collimated laser beam, across an interface separating two different isotropic materials, as shown in Fig. 3-1. The medium in which the incident beam propagates is assumed transparent to the particular frequency of the EM wave. The second medium could either be a transparent one or a reflecting one. But let us consider now only the case in which the second medium is also transparent.

§3.1 Reflection and refraction at boundaries

Considering only the boundary conditions under which an electromagnetic wave crosses an interface separating two isotropic media,

i.e. (a) the tangential components of the electric field $\vec{E}$ is continuous

(b) the tangential components of $B/\vec{\mu}$ (or $\vec{H}$) is continuous,

one can derive the following relationships. (cf. any optics text)

(1) The relations that show the directions of the reflected and transmitted beams,

i.e. (a) the incident, reflected and transmitted beams are in the same plane, called the plane of incidence.

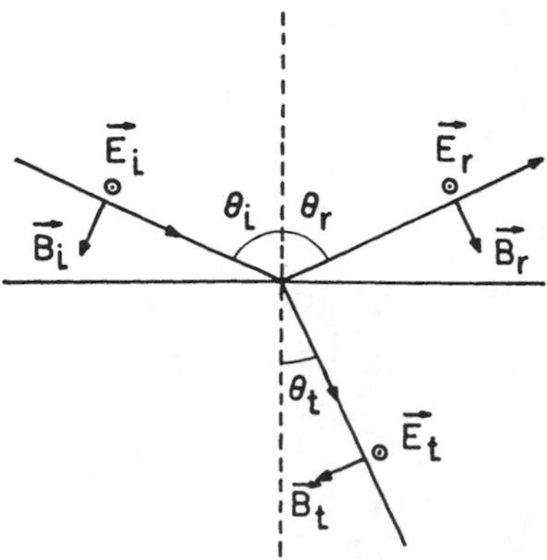

Fig. 3-1. Fresnel reflection and refraction across a boundary. The electric vector of the incident wave is perpendicular to the plane of incidence.

(b) $\Theta_i = \Theta_r$ (3-1)

and (c) $n_i \sin\Theta_i = n_t \sin\Theta_t$ (3-2)

the second relation (eq. (3-2)) is called <u>Snell's law</u> of refraction.

(2) The relations that give the amplitudes of the reflected and transmitted electric fields relative to the incident electric field, i.e. the <u>Fresnel's equations</u>. They are:

$$r_\perp \equiv \left(\frac{E_{or}}{E_{oi}}\right)_\perp = \frac{n_i\cos\Theta_i - n_t\cos\Theta_t}{n_i\cos\Theta_i + n_t\cos\Theta_t} = -\frac{\sin(\Theta_i - \Theta_t)}{\sin(\Theta_i + \Theta_t)} \tag{3-3}$$

$$r_\parallel \equiv \left(\frac{E_{or}}{E_{oi}}\right)_\parallel = \frac{n_t\cos\Theta_i - n_i\cos\Theta_t}{n_i\cos\Theta_t + n_t\cos\Theta_i} = \frac{\tan(\Theta_i - \Theta_t)}{\tan(\Theta_i + \Theta_t)} \tag{3-4}$$

$$t_\perp \equiv \left(\frac{E_{ot}}{E_{oi}}\right)_\perp = \frac{2n_i\cos\Theta_i}{n_i\cos\Theta_i + n_t\cos\Theta_t} = \frac{2\sin\Theta_t\cos\Theta_i}{\sin(\Theta_i + \Theta_t)} \tag{3-5}$$

$$t_\parallel \equiv \left(\frac{E_{ot}}{E_{oi}}\right)_\parallel = \frac{2n_i\cos\Theta_i}{n_i\cos\Theta_t + n_t\cos\Theta_i} = \frac{2\sin\Theta_t\cos\Theta_i}{\sin(\Theta_i + \Theta_t)\cos(\Theta_i - \Theta_t)} \tag{3-6}$$

where E_{oj} ($j = i,r,t$) is the amplitude of the electric field, and, $\perp(\parallel)$ means the incident $\vec{E}$ field is perpendicular (parallel) to the plane of incidence as shown in Fig. 3-1 (Fig. 3-2). We assume that the reflected and transmitted electric fields are in the directions shown in Fig. 3-1 and 3-2. The Fresnel's equations will correct the situation if the assumed direction is not right. For instance, eq. (3-3) shows that $r_\perp$ is negative. It means that the amplitude E_{or} changes sign with respect to the incident field. The monochromatic fields of the incident (i), reflected (r) and transmitted (t) fields are:

$$\vec{E}_i = \vec{E}_{oi} \cos(\vec{k}_i \bullet \vec{r} - \omega t) \tag{3-7}$$

$$\vec{E}_r = \vec{E}_{or} \cos(\vec{k}_r \bullet \vec{r} - \omega t + \epsilon_r) \tag{3-8}$$

$$\vec{E}_t = \vec{E}_{ot} \cos(\vec{k}_t \bullet \vec{r} - \omega t + \epsilon_t) \tag{3-9}$$

where ϵ_r and ϵ_t are constant phases with respect to the incident field. The proof of the Fresnel equations is straightforward and can be found in any optics text and we shall not derive them here. We only want to point out some useful features of these equations. Fig. 3-3 shows four curves representing the variation of $r_\perp$, $r_\parallel$, $t_\perp$ and $t_\parallel$ as one changes θ_i, under the condition that $n_i < n_t$ where the n's are the indices of refraction. Note

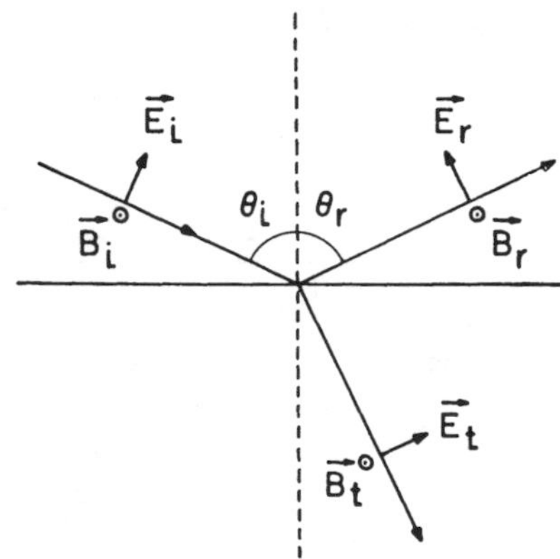

Fig. 3-2. Fresnel reflection and refraction across a boundary. The electric vector of the incident wave is in the plane of incidence.

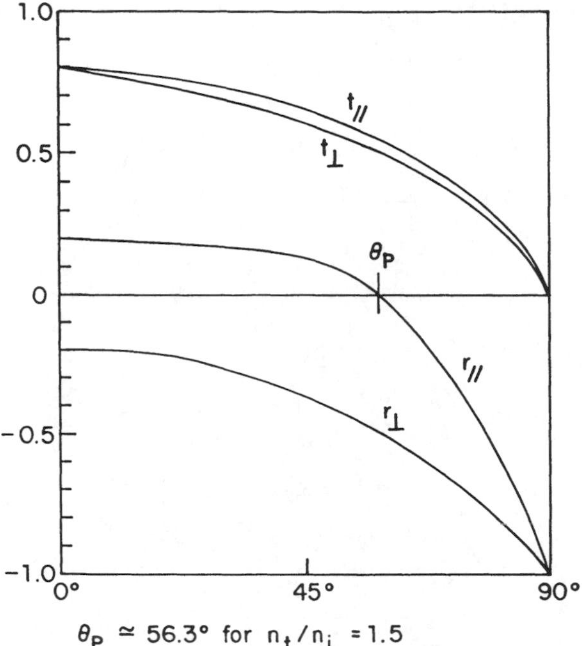

Fig. 3-3. Schematic relations between $r_\perp$, $r_\parallel$, $t_\perp$, $t_\parallel$ and the angle of incidence (horizontal scale) for a material boundary in which $n_t/n_i = 1.5$.

that the t's are positive, while $r_\perp$ is negative and $r_\parallel$ starts from a positive value, passes through zero and then becomes negative.

What is interesting here is the following. (Note first of all the definition of r's and t's in eqs. 3-3 to 3-6.)

(a) $r_\perp < 0$ (eq. 3-3) for all θ_i and θ_t; i.e. the reflected electric field amplitude is 180° out of phase with respect to the incident electric field amplitude.

$(\because -1 = e^{\pm i\pi})$

(b) $r_\parallel = 0$ at $\theta_i = \theta_B$

where θ_B is defined as the polarizing angle or <u>Brewster angle</u>.

From eq. (3-4), $r_{\parallel} = 0$ means

either $\tan(\theta_i - \theta_t) = 0$ (3-10)

or $\tan(\theta_i + \theta_t) = \infty$ (3-11)

Eq. (3-10) is impossible because $\theta_i \neq \theta_t \neq 0$ in general. Thus eq. (3-11) has to be valid.

i.e. $\theta_i + \theta_t = \pi/2$

or $\theta_B + \theta_t = \pi/2$ $(\because \theta_i = \theta_B)$ (3-12)

Using Snell's law (eq. 3-2)

$$n_i \sin\theta_B = n_t \sin\theta_t = n_t \sin\left(\frac{\pi}{2} - \theta_B\right) = n_t \cos\theta_B$$

$$\therefore \quad \boxed{\tan\theta_B = \frac{n_t}{n_i}} \tag{3-13}$$

This is an important result that is used very often by optics or laser users. (Note: the relation is valid also in the case of refraction from a denser medium, i.e. $n_i > n_t$. The angle θ_B' in this case is such that $\theta_B + \theta_B' = \frac{\pi}{2}$).

However, experimentally, one cannot measure the field E. Rather, it is the power (Joules/sec) of the radiation that one measures. The power is related to the intensity or irradiance (Joules/m^2 sec) by a simple factor, the cross sectional area of the beam of radiation. The intensity is the averaged value of the Poynting vector, defined in electromagnetic theory as

$$\vec{S} = \vec{E} \times \vec{B}/\mu \tag{3-14}$$

Thus, referring to Fig. 3-4, where a beam of light of finite cross section is reflected and transmitted at an interface, one can prove the following using the Fresnel equations.

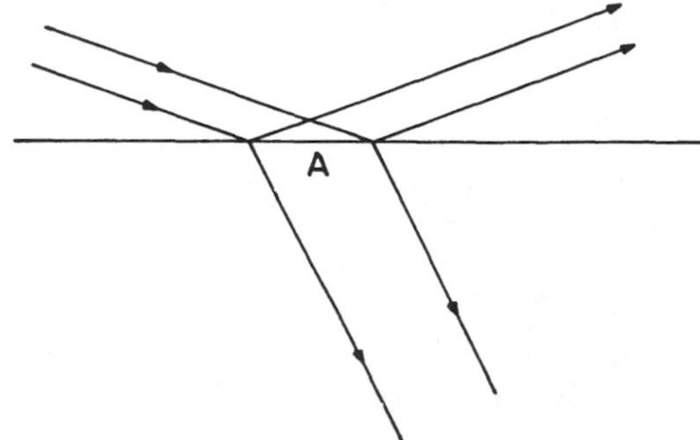

Fig. 3-4. Reflectance and transmittance across a boundary.

(a) The reflectance R

$$\equiv \frac{\text{reflected power}}{\text{incident power}} = \frac{E_{or}^2}{E_{oi}^2} = r^2 \tag{3-15}$$

(b) The transmittance T

$$= \frac{\text{transmitted power}}{\text{incident power}}$$

$$= \left(\frac{n_t \cos\Theta_t}{n_i \cos\Theta_i}\right) t^2 \tag{3-16}$$

(c) $R + T = 1$ (3-17)

which is the law of conservation of energy, assuming no other loss exists at the boundary.

Fig. 3-5 shows the variation of $R_\perp$, $R_\parallel$, $T_\perp$ and $T_\parallel$ as a function of incident angle Θ_i for $n_t > n_i$. Note again $R_\parallel = 0$ at $\Theta_i = \Theta_B$, at the Brewster angle. No electromagnetic power can be reflected at the Brewster angle.

§3.2 Taking advantage of the Brewster angle and the features of the reflectance and transmittance

(a) Since at $\Theta_i = \Theta_B$, the reflectance $R = 0$, one often sets the transmitting window of a linearly polarized laser beam at the

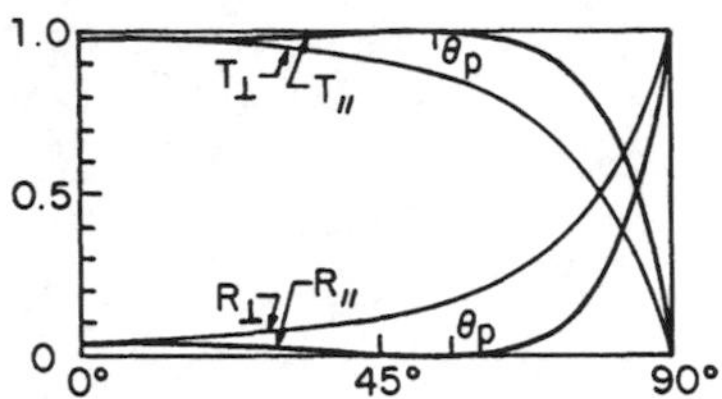

Fig. 3-5. Variation of $R_\perp$, $R_\parallel$, $T_\perp$, $T_\parallel$ as a function of the angle of incidence Θ_i (horizontal scale).

Brewster angle, so that the polarization of the laser is in the plane of incidence as shown in Fig. 3-6. This eliminates the loss by reflection, so that almost all the energy in the laser beam is transmitted. (There are still scattering, absorption and diffraction losses.)

(b) By stacking a good number of plane parallel plates at the Brewster angle, an unpolarized laser beam can be made polarized by passing through this stack, as shown in Fig. 3-7. All the radiation polarized (or whose electric field is) perpendicular to the plane of incidence will be successively partially reflected while the other component parallel to the plane of incidence will be transmitted without any reflection. Thus, after many such passages through the stack, the tramsmitted beam is polarized linearly in

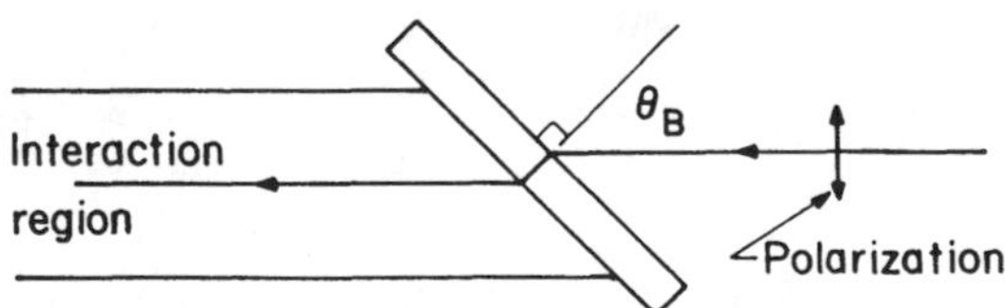

Fig. 3-6. Transmission through a window set at the Brewster angle Θ_B. The polarization (linear) of the incident wave is in the plane of incidence.

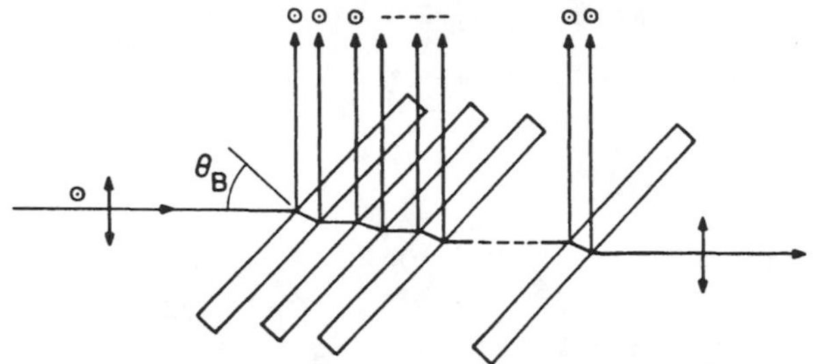

Fig. 3-7. A stack of windows set at the Brewster angle acts as a polarizer.

the plane of incidence to a good degree of accuracy. We define the degree D of linear polarization

as
$$D = \frac{P_{\parallel}}{P_{\perp}} \tag{3-18}$$

where $P_{\parallel}$ ($P_{\perp}$) is the power of the radiation in the same laser beam whose polarization is parallel (perpendicular) to the plane of incidence.

Let's try to make a simple calculation. Assuming that in the incident laser beam, $P_{\parallel}^{(in)} = P_{\perp}^{(in)} \equiv P_o$. After transmitting through one plate set at the Brester angle, $P_{\parallel}$ is unchanged while there have been two reflections at both the interfaces of the plate. Note that the two $R_{\perp}$'s at the two interfaces are both equal to $\sin^2(\theta_B - \theta_t)$ at the Brewster angle as can be seen from eq. (3-3) and the condition $\theta_i + \theta_t = \pi/2$ at $\theta_i = \theta_B$. Hence, the power of the $\perp$ polarization in the transmitted beam, $P_{\perp}^{(1\ plate)}$, is

$$P_{\perp}^{(1\ plate)} = P_o(1 - R_{\perp})^2 \tag{3-19}$$

the square means two reflections at both the interfaces. Thus, for the transmission through N plates, we have

$$P_{\perp}^{(N\ plates)} = P_o\ (1 - R_{\perp})^{2N} \quad (3\text{-}20)$$

Now, $R_{\perp} = \sin^2(\theta_B - \theta_t)$

and $\theta_B + \theta_t = \pi/2 \Rightarrow \theta_t = \frac{\pi}{2} - \theta_B$

$\because \quad R_{\perp} = \sin^2(2\theta_B - \pi/2)$

$= \cos^2(2\theta_B) = (\cos^2\theta_B - \sin^2\theta_B)^2$

$\because \quad \tan\theta_B = \dfrac{n_t}{n_i}$

$$R_{\perp} = \left(\frac{n_i^2 - n_t^2}{n_i^2 + n_t^2}\right)^2 \quad (3\text{-}21)$$

Note that for internal reflection, $n_i > n_t$ · while for external reflection, $n_t > n_i$ · Both leads to the same expression for R (eq. 3-21). Substituting this expression into eq.(3-20), we have

$$P_{\perp}^{(N\ plate)} = P_o\left[\frac{4n_i^2\ n_t^2}{\left(n_i^2 + n_t^2\right)^2}\right]^{2N} \quad (3\text{-}22)$$

Let $$D^{(N)} = \frac{P_{\parallel}}{P_{\perp}}\ (N\ plates)$$

$$\therefore D^{(N)} = \left[\frac{\left(n_i^2 + n_t^2\right)^2}{4n_i^2\ n_t^2}\right]^{2N}$$

$$D^{(N)} = \left[\frac{1}{4}\left(\frac{n_i}{n_t} + \frac{n_t}{n_i}\right)^2\right]^{2N} \quad (3\text{-}23)$$

e.g.1. glass/air, $n_i \doteqdot 1$, $\quad n_t = 1.5$

$N = 1, \quad D^{(1)} = 1.38$

$N = 10, \ D^{(10)} = 24.6$

$N = 20, \ D^{(20)} = 627.5$

e.g.2. Ge/air, $n_i \doteq 1$, $n_t \doteq 4$ for 10μ radiation (CO_2 laser)

$N = 1, \quad D^{(1)} \doteq 20.4$

$N = 3, \quad D^{(3)} \doteq 8{,}500.$

It is evident that polarizer of this type is efficient when the material has a high index of refraction, such as Ge at 10μ. Thus, in manipulating CO_2 laser beams $(10{,}6\mu)$, this type of Ge polarizer is often used because it is efficient and other types of inexpensive polarizers are not easily available.

(c) By setting the end windows (or ends, in the case of solids) of the active material inside a laser cavity at the Brewster angle, competition will favor the $\|$ - component to lase because it has no reflection loss in passing through the ends of the active medium. If the gain is not too high, the $\perp$ - component will be suppressed completely so that the laser output is linearly polarized in the $\|$ - direction.

§3.3 Critical angle and total internal reflection

We now turn to the condition of $n_i > n_t$; i.e. the radiation passes from a denser medium to a less dense medium. Fig. 3-8. All the consequences of the boundary conditions mentioned at the beginning of this chapter, namely,

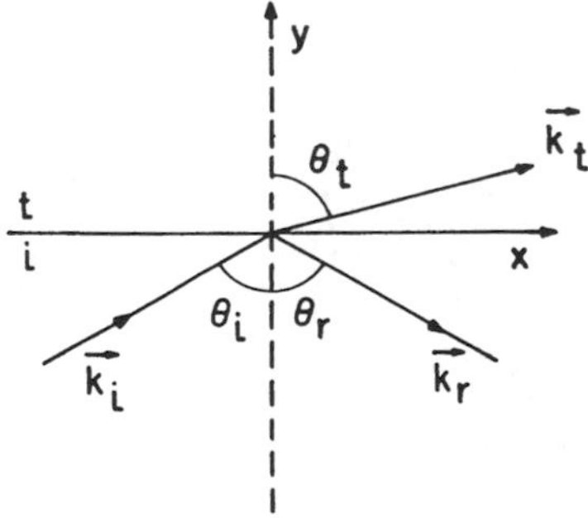

Fig. 3-8. Reflection and refraction at a boundary when the incident medium is denser than the transmission region.

$\theta_i = \theta_r$; the Snell's law; $\vec{k}_i$, $\vec{k}_r$, $\vec{k}_t$ lie in the same plane of incidence; and the Fresnel equations are still valid. In particular, we show in Fig. 3-9 the curves for $r_\perp$ and $r_\parallel$ as a function of θ_i. We can note the following two features

(a) $r_\parallel$ passes through zero at $\theta_i = \theta_B'$.

(b) both $r_\perp$ and $r_\parallel$ reach unity at $\theta_i = \theta_c$ and stay at unity for further increase of θ_i. θ_c is called the <u>critical angle</u>. At $\theta_i \geq \theta_c$ <u>total internal reflection</u> takes place.

θ_B' is again the Brewster angle. Using the same procedure as in the case of $n_i < n_t$, one can obtain (exercise for the reader)

$$\tan\theta_B' = \frac{n_t}{n_i} \qquad (n_t < n_i) \tag{3-24}$$

and

$$\theta_B + \theta_B' = \pi/2 \tag{3-25}$$

The critical angle can be calculated through the condition that

$$r_\perp = t_\parallel = 1 \qquad \text{at } \theta_i = \theta_c.$$

Using either eq. (3-3) or (3-4), one finds

$$\theta_t = \pi/2 \qquad \text{at } \theta_i = \theta_c \tag{3-26}$$

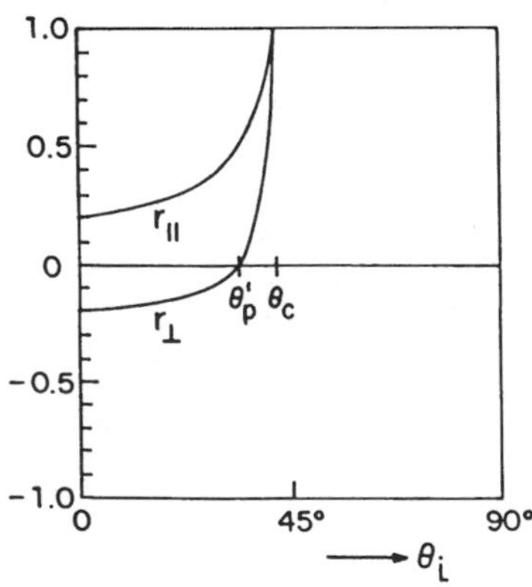

Fig. 3-9. $r_\parallel$ and $r_\perp$ as a function of the angle of incidence θ_i when the incident medium is denser.

and Snell's law leads to

$$\sin\theta_c = \frac{n_t}{n_i} \tag{3-27}$$

It is interesting to note that when $\theta_i \geq \theta_c$ at which total internal reflection takes place, it doesn't mean that the electromagnetic field is now completely confined in the incident medium. In fact, the field penetrates into the transmission region exponentially. This can be seen as follows. Under the plane wave approximation, the transmitted field is

$$E_t = E_{ot} \exp\{i(\vec{k}_t \bullet \vec{r} - \omega t + \epsilon_t)\}$$

where ϵ_t is a phase factor relative to the incident field. This phase factor depends only on n_i, n_t and θ_i, and can be obtained by calculating $t_\parallel$ using eq. (3-6). Referring back to Fig. 3-8, where every thing happens in the plane of incidence, which is defined as the x-y plane, we can decompose the transmitted wave vector $\vec{k}_t$ into $\left(k_x\hat{i} + k_y\hat{j}\right)$. Using Snell's law, we obtain

$$E_t = E_{ot} \exp(-\alpha y) \exp\{i(\beta x - \omega t + \epsilon_t)\} \tag{3-28}$$

where

$$\alpha = k_t \sqrt{\left(\frac{n_i}{n_t}\right)^2 \sin^2\theta_i - 1} \tag{3-29}$$

$$\beta = k_t \frac{n_i}{n_t} \sin\theta_i \tag{3-30}$$

Equation (3-28) shows that the transmitted field E_t does penetrate into the y - direction (transmission region) through the factor $\exp(-\alpha y)$. Moreover, it has a propagating component $\exp\{i(\beta x - \omega t + \epsilon_t)\}$ that propagates only in the x - direction, i.e. along the surface of the interface, but on the transmission side. E_t as given by eq. (3-29) is called the <u>evanescent wave</u>. It doen't have any propagating component in the y - direction, hence no transmitted ray going <u>into</u> the transmitting medium.

Fig. 3-10 shows several examples of total internal reflection.

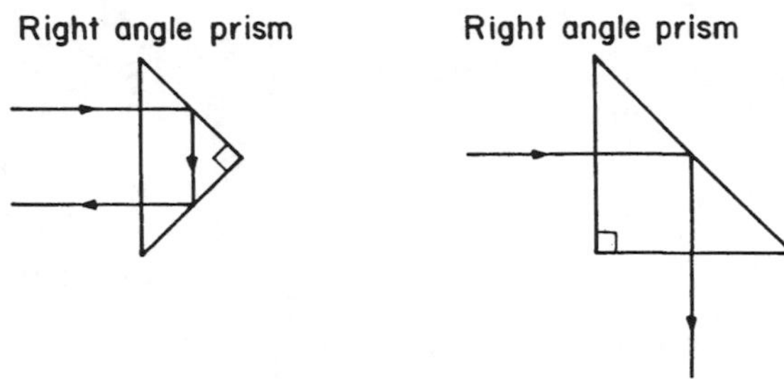

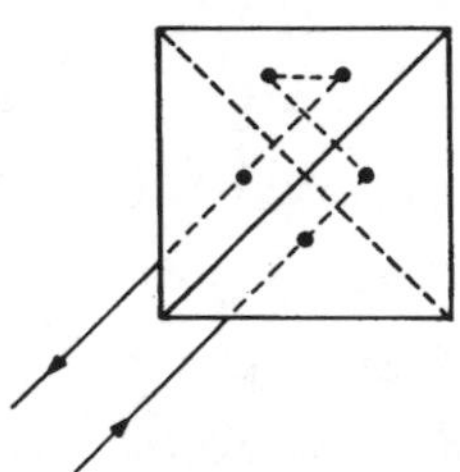

Exagerated evanescent field

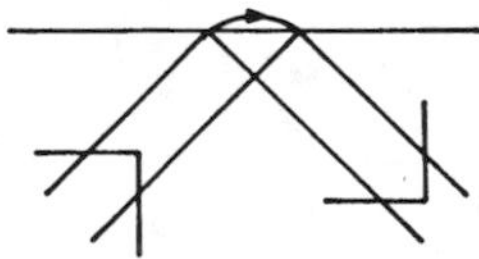

Fig. 3-10. Examples of total internal reflection at work.

§3.4 Demonstration of some important results of the Fresnel equations

Some of the important features of the result of the Fresnel equations are:

(a) If $n_i < n_t$, $R \to 1$ as $\Theta_i \to \pi/2$

(i.e. the reflectance becomes unity at grazing incidence)

(b) If $n_i < n_t$, $R_\parallel = 0$ at $\Theta_i = \Theta_B$

i.e. the reflectance of the $\parallel$ - component is zero at the Brewster angle

(c) If $n_i > n_t$, $R = 1$ at $\theta_i \geq \theta_c$. That is, when the incident angle is equal to or greater than the critical angle, one has total internal reflection. However, there is an evanescent wave at the interface. It is emphasized because of its important applications in fiber optics, integrated optics etc.

The following simple experiment gives a rather interesting demonstration of the above three points in front of a class of students.

Necessary equipment:

- one He-Ne laser, about 2mW, preferably linearly polarized.
- a support for the laser
- a small plastic bottle of alcohol (ethanol) with a squirting tip. (Note: Acetone, methanol etc. would also do, but the vapor in a closed class room will not be good for health.) A variation is to use a pipette or a glass tube.
- a hollow glass tube, outer diameter less than $\frac{1}{2}$ cm.
- a right angle prism, at least 2.5 cm x 2.5 cm x 2.5 cm.
- a lab-jack or any appropriate support for the prism so that the height can be adjusted.
- a small shallow dish to collect the alcohol.
- any piece of rectangular glass plate; e.g. a microscopic substrate. However, the longer the plate is the easier it is to perform the demonstration
- a piece of white paper, about the size of a page of this book or larger.

<u>Preparation</u>

Draw a line on the output surface of the laser head and just below the output hole. This line should be perpendicular to the laser's polarization direction. Tape several layers of masking tapes on the output surface of the laser head such that the tape's edge lines up with the line just drawn. Fig. (3-11). Rotate the laser around its axis so that the polarization is almost vertical (or the tapes almost horizontal). Fix the laser on the support. Put every thing on a cart and wheel them to the class room. The demonstration can be done on the cart or on a class room table.

Demonstration procedure

Warning: Never point the laser beam towards anybody's eyes nor should one ever look into a laser beam. The following order is not important but I happen to like it personally.

1) With the demonstrator facing the class, the cart in front of him, the laser is pointed towards one of the side walls (left or right, depending on the convenience). Turn on the laser and ask the class to look at the red laser spot on the wall. It is important that the height of the laser beam be set above the head level of the class to avoid any accidental pointing of the laser beam into anyone's eye.

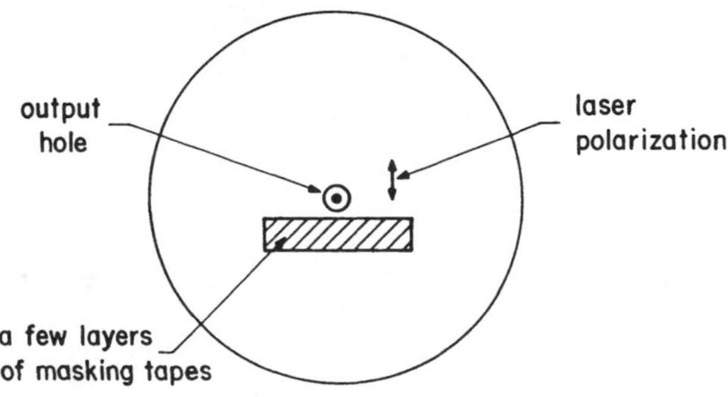

Fig. 3-11. Schematic set up for a demonstration.

2) Put the glass plate into the laser beam so that its surface is nearly perpendicular to the beam. Make a remark that the intensity of the spot on the wall is almost not altered because the reflectance from the glass surfaces is about 8%.

3) Now increase quickly the angle of incidence such that the reflected light beam also appears on the side wall. To have a steady hold of the plate, touch the shorter side of the plate on the output surface of the laser head. Note that the transmitted beam's intensity is still very strong and appears unaffected.

4) Now, carefully and slowly, increase the angle of incidence until the reflected spot is very near the transmitted spot. Note that the transmitted beam weakens rather quickly while the reflected beam increases in intensity. If the demonstrator is careful, he should be able to make the transmitted beam almost disappeared. This is a remarkable phenomenon because the transmitted beam, initially almost unaffected, can be reduced so significantly.

5) Now hold the glass plate with its shorter side supported by the horizontal edge of the layers of masking tape on the output surface of the laser head (Fig. 3-12). Tell the class that the glass plate will be turned with the pivot at the edge of the masking tape, and then explain why with such an orientation the reflected light is $R_{\parallel}$.

6) Start with the plate at almost grazing incidence and ask the class to watch the movement of the reflected spot. Gradually decrease the angle of incidence and the reflected spot will rise and scan across the ceiling. When the angle of incidence approaches the Brewster angle, the intensity of the laser spot on the ceiling decreases and at $\theta_i = \theta_B$, there is no reflection. Further

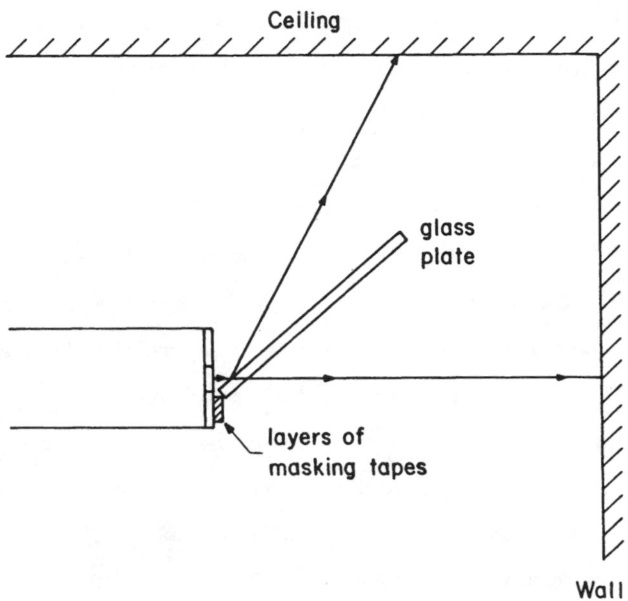

Fig. 3-12. Another view of the schematic set up for the same demonstration.

reduction of the angle of incidence makes the reflected beam reappear on the ceiling. Again, it is remarkable that the reflected beam, while very strong at grazing incidence, is zero at the Brewster angle.

7) Put away the glass plate, set the prism inside the shallow disk and put them on the lab-jack.

8) Adjust the lab-jack so that the prism intercepts the laser beam. Let the laser beam enter from one side of the right-angled surface and is totally internally reflected at the hypotenuse surface. Turn the cart slightly such that the hypotenuse surface of the prism faces the class.

9) Turn off the light in the room. Note a small spot of scattering light on the hypotenuse surface where the laser beam is totally

reflected internally (Fig. 3-13). However, holding the white paper in front of this spot doesn't show any transmitted beam of the laser in any direction. The scattered light shows that the light penetrates across the interface and is scattered by the microscopic irregularities of the surface or dust.

10) Notice that the evanescent field has a penetration term $e^{-\alpha y}$ in the direction normal to and outward from the prism surface (Fig. 3-13), and a propagation term $\exp[i(\omega t - kx)]$ propagating along the surface in the x - direction. Hold the glass tube vertically and touch the prism surface at where the light is totally reflected. Immediately, there is a horizontal "fan" of light coming out of the glass tube (see Fig. 3-14). This is because the evanescent field penetrates into the tube and the

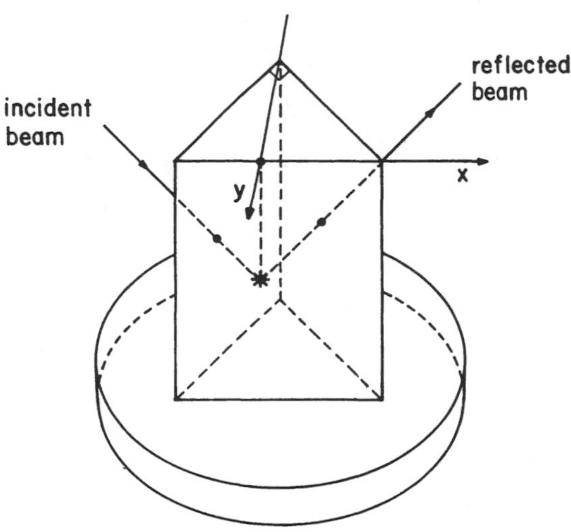

Evanescent field ~ $e^{-\alpha y}\ e^{-i(\omega t - kx)}$

Fig. 3-13. A demonstration of total internal reflection and evanescent wave.

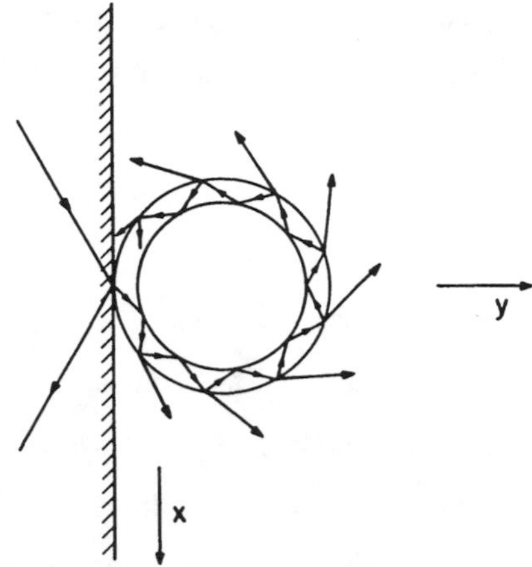

Fig. 3-14. Frustrated total internal reflection.

propagation component "guides" the beam around the tube through multiple reflections and transmissions, resulting in a horizontal fan of light[1,2]. This shows that the evanescent field really has penetrated across the interface. (see section 3.5 for more discussion.)

11) Put away the glass tube. Hold the white page of paper at the side of the prism (on the downstream side of the x - direction, Fig. 3-13). Squirt some alcohol on the surface where the light is totally reflected. A bright splash of light coming from the evanescent field propagating along the surface hits the paper. This is another demonstration of the penetration and propagation nature of the evanescent field.

---

1 S.L. Chin, "Optical Fan Levelling System", Canadian Patent No. 9555393, Oct. 1974.

2 S.L. Chin and K.A. Mace, "Optical Fan Levelling system", U.S. Patent No. 3984154, Oct. 1975.

12) A bit of light show: Touch the wet tip of the alcohol squirter at the spot where there is total internal reflection and observe the reflected beam (Fig. 3-13) on a screen (or the white page of paper). Beautiful interference fringes of different forms appear and disappear dynamically. Interference of the evanescent wave in the thin alcohol film on the prism surface and the evaporation process cause all these.

§3.5 Making good use of the evanescent field

The evanescent field (eq. 3-28, 29) has been put into good use in optoelectronics and photonics. One popular application is to couple laser (optical) beams in and out of thin films for integrated optics etc. This is shown in Fig. (3-15). By pressing the prisms on to the thin film, the air space between the prism and the thin film is reduced allowing the evanescent field to penetrate into the film and propagate inside the film

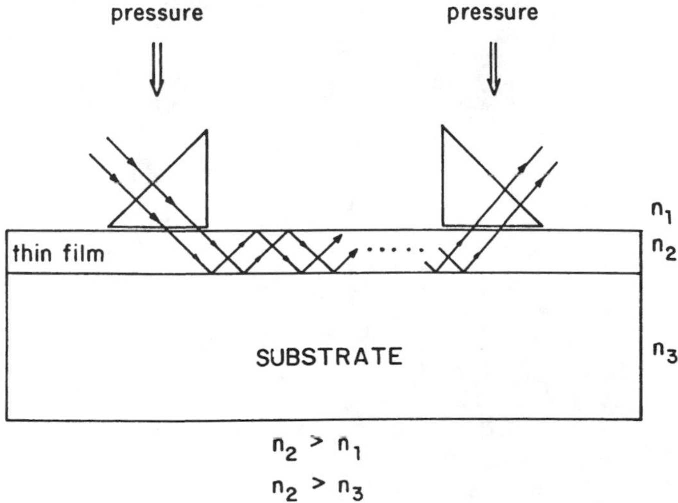

Fig. 3-15. Application of total internal reflection and evanescent wave.

again by total internal reflection ($\because n_2 > n_1$, $n_2 > n_3$). This possibility is already shown in §3.4, demonstration #10. Many other uses are applied to different situations but the principle of frustrating the total internal reflection allowing the evanescent field to go through is the same. For instance, one wants to obtain some light out of an optical fibre through which a certain light wave propagates. Since light propagation through fibers is essentially by total internal reflection (more rigorous description by solving the Maxwell equations can be found in any fiber optics book), touching the bare fibre surface with another transparent material of an appropriate index will induce some leakage through the frustation of the total internal reflection in the fiber.

CHAPTER IV

RESONATOR, A GEOMETRICAL VIEW

§4.1 Introduction

So far, our discussion of the laser operation is based on balancing the gain and loss in the laser cavity, implicitly assuming that the laser beam inside the cavity will automatically bounce back and forth between the two mirrors. In reality, this is not a rule, but a special situation. This can be understood by referring to Fig. 4-1 showing a beam bouncing back and forth between two parallel plane mirrors. This is only a very special case in which the beam is perpendicular to both the surfaces of the two mirrors. Any other inclination of the light beam with respect to the mirror surfaces will lead to a walk-off and become a total loss, i.e. no laser oscillation is possible, as shown in Fig. 4-2. Of course, one immediately thinks of using a better system, namely, a pair of concave mirrors. This is indeed true but it doesn't mean that any pair of concave mirrors will allow a stable laser oscillation. Without a proper choice of the radii of curvature of the mirrors and the distance between them, the light beam in the cavity will still walk-off. We shall study in this chapter the condition under which a pair of mirror will allow a stable laser oscillation. The approach is based on geometrical optics because it is much easier to understand. More rigorous diffraction approach can be found in any more advanced laser textbook.

Fig. 4-1. A cavity bounded by two plane parallel mirrors.

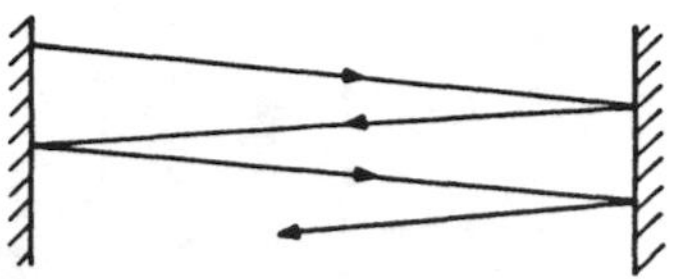

Fig. 4-2. A light beam walks off the cavity bounded by a pair of plane parallel mirrors.

§4.2 General considerations

We look at the most general laser cavity bounded by two concave mirrors whose radii of curvature can be any positive or negative values as shown in Fig. 4-3. It is well-known in geometrical optics that the focal length of a concave mirror of radius of curvature R is $f = R/2$. Hence, the effect of such a mirror on a beam of light is equivalent to that of a lens with focal length $f = R/2$. As such, we can make an equivalence between Fig. 4-3 and Fig. 4-4. In Fig. 4-3, a beam of light is assumed to start at point P just in front of R_2 towards R_1 and be reflected to the point R just in front of R_1. This is equivalent to passing the beam of light through a lens of focal length $f_1 = R_1/2$, as shown in Fig. 4-4. From point R in Fig. 4-3, the beam keeps on propagating towards R_2, and is reflected to the point T just in front of R_2. This is equivalent to the continuation of the ray PQR towards S and T in Fig. 4-4, passing the equivalent lens of focal length $f_2 = R_2/2$. This operation completes a round trip of the beam between the two mirrors. For an oscillation to take place, the beam in the cavity will bounce back and forth for an infinite number of times in principle. This is equivalent

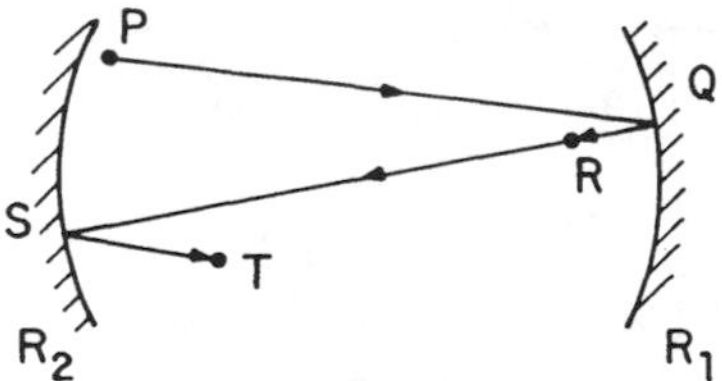

Fig. 4-3. A cavity bounded by two spherical mirrors.

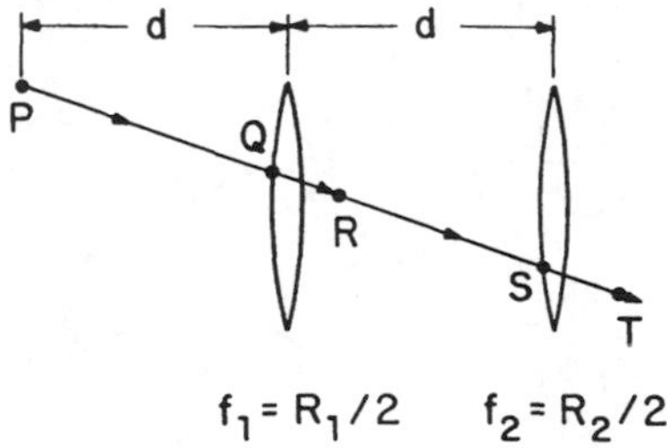

Fig. 4-4. Propagation through a pair of lenses which is equivalent to the round trip reflection between two spherical mirrors.

to having a beam of light passing through an infinite pairs of equivalent lenses in series, as shown in Fig. 4-5. The condition under which there exists such an oscillation is that the beam does not walk-off the mirrors; i.e. the beam is forever bouncing within the bounds of the mirrors. This is equivalent to saying that the beam in Fig. 4-5 always comes back and passes through the lens series to infinity.

§4.3 Case of one lens

We shall now make use of geometrical optics in the matrix form to analyse Fig. 4-5. We start by analysing the case of a beam passing through a single lens. The analysis is based on two assumptions: paraxial ray and very thin lens. Paraxial ray means that the beam is confined to a narrow region

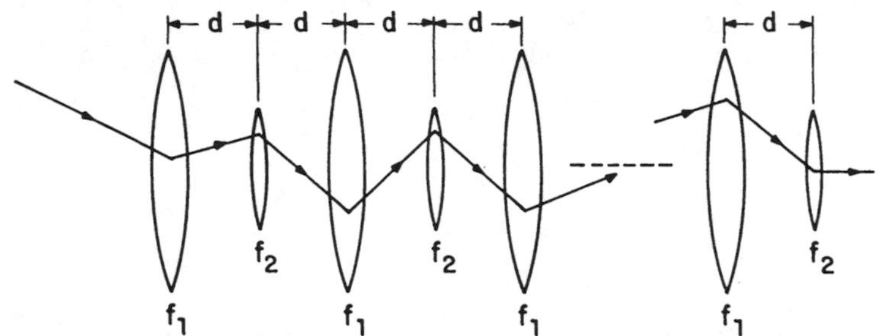

Fig. 4-5. Multiple reflections between two spherical mirrors are equivalent to the transmission through many pairs of lenses.

around the axis of the lens. It means the angle between the beam and the axis of the lens is small.

Fig. 4-6 shows an exaggerated situation of a parallel beam of light incident on a thin lens and exit toward the focal plane where the light focuses. The incident beam at the input position i of the lens is characterized by r_i, r_i' where r_i is the vertical distance from point i to the axis of the lens (z-axis) while

$$r_i' \equiv \frac{dr_i}{dz}$$

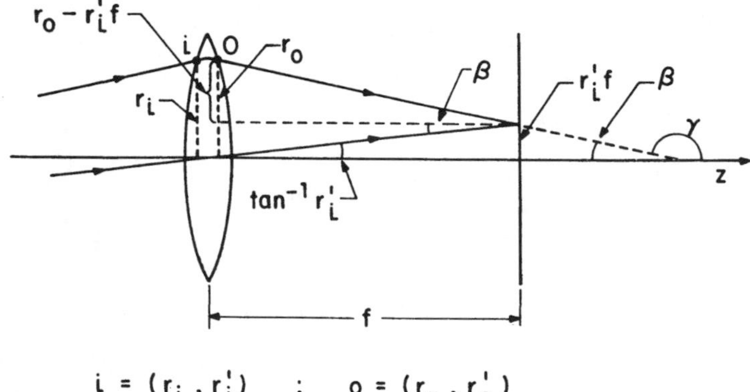

Fig. 4-6. Transmission through a lens.

is the slope of the ray at r_i with respect to the z - axis. Similarly, at the output side of the lens, the output beam at point O is characterized by r_o and r_o' which are similarly defined. We shall find the relationship between (r_i, r_i') and (r_o, r_o'). It is obvious that

$$r_o = r_i \tag{4-1}$$

because of the thin lens approximation. Inspecting Fig. 4-6, we see that

$$r_o' \equiv \frac{dr_o}{dz} = \tan\gamma = \tan(\pi - \beta)$$

$$= -\tan\beta = -\frac{r_o - r_i' f}{f} \tag{4-2}$$

Rewriting eq. (4-1) and (4-2):

$$r_o = r_i + 0 \bullet r_i'$$

$$r_o' = \left(-\frac{1}{f}\right) r_i + r_i'$$

i.e.

$$\begin{pmatrix} r_o \\ r_o' \end{pmatrix} = \begin{pmatrix} 1 & 0 \\ -\frac{1}{f} & 1 \end{pmatrix} \begin{pmatrix} r_i \\ r_i' \end{pmatrix} \tag{4-3}$$

Eq. (4-3) is the matrix form representation of a paraxial ray passing through a thin lens, showing the relationship of the ray at the two sides (i and O) of the lens. We need also the matrix representation of the propagation in free space.

Fig. 4-7 shows a ray passing through two positions z_1 and z_2 in space. The "coordinates" of the ray at z_1 and z_2 are (r_i, r_i') and (r_o, r_o'). We see immediately that

$$r_o = r_i + r_i' d$$

$$r_o' = r_i'$$

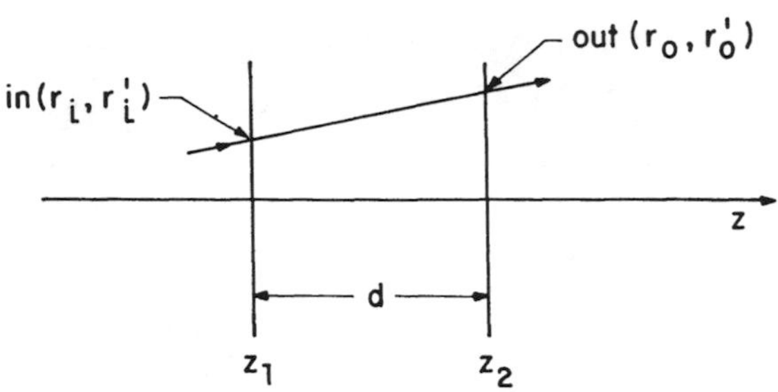

Fig. 4-7. Transmission through space.

In matrix form, they become

$$\begin{pmatrix} r_o \\ r_o' \end{pmatrix} = \begin{pmatrix} 1 & d \\ 0 & 1 \end{pmatrix} \begin{pmatrix} r_i \\ r_i' \end{pmatrix} \tag{4-4}$$

Thus, combining the above two cases, we have the situation shown in Fig. 4-8, in which we have reversed the propagation direction for the sake of understanding the solution. The "coordinates" at 0 are thus

$$\begin{pmatrix} r_o \\ r_o' \end{pmatrix} = \begin{pmatrix} \text{lens} \\ \text{matrix} \end{pmatrix} \begin{pmatrix} \text{free} \\ \text{space} \\ \text{matrix} \end{pmatrix} \begin{pmatrix} r_i \\ r_i' \end{pmatrix}$$

Using eq. (4-3) and (4-4),

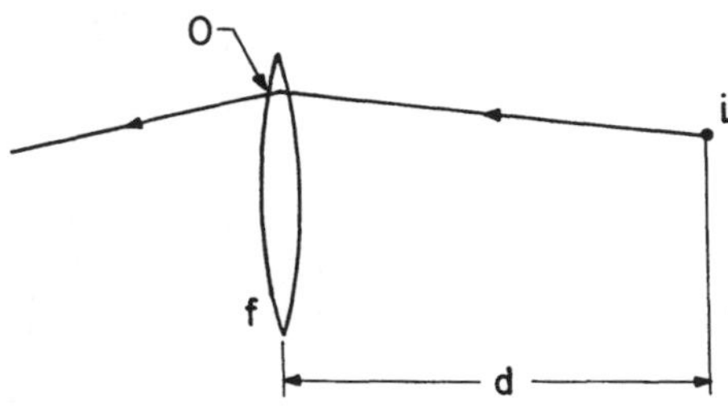

Fig. 4-8. Transmission through space and a lens.

$$\begin{pmatrix} r_o \\ r_o' \end{pmatrix} = \begin{pmatrix} 1 & 0 \\ -\frac{1}{f} & 1 \end{pmatrix} \begin{pmatrix} 1 & d \\ 0 & 1 \end{pmatrix} \begin{pmatrix} r_i \\ r_i' \end{pmatrix}$$

$$= \begin{pmatrix} 1 & d \\ -\frac{1}{f} & -\frac{d}{f} + 1 \end{pmatrix} \begin{pmatrix} r_i \\ r_i' \end{pmatrix} \tag{4-5}$$

§4.4 Case of two lenses and equivalence to one round trip in the cavity

As mentioned in §4.2, having a beam of light making one round trip between the mirrors of the laser cavity is equivalent to passing the beam through two equivalent lenses, as shown in Fig. 4-9. The light beam starts again from the right at point i and exits the second lens at point 0. The "coordinates" of the beam at 0 are thus

$$\begin{pmatrix} r_o \\ r_o' \end{pmatrix} = \begin{pmatrix} \text{lens } f_2 \end{pmatrix} \begin{pmatrix} \text{free} \\ \text{space} \\ d \end{pmatrix} \begin{pmatrix} \text{lens} \\ f_1 \end{pmatrix} \begin{pmatrix} \text{free} \\ \text{space} \\ d \end{pmatrix} \begin{pmatrix} r_i \\ r_i' \end{pmatrix}$$

It is now straightforward to calculate the resultant matrix using eq. (4-3), (4-4) and (4.5). The detailed calculation is left as an exercise for the readers, and the result is

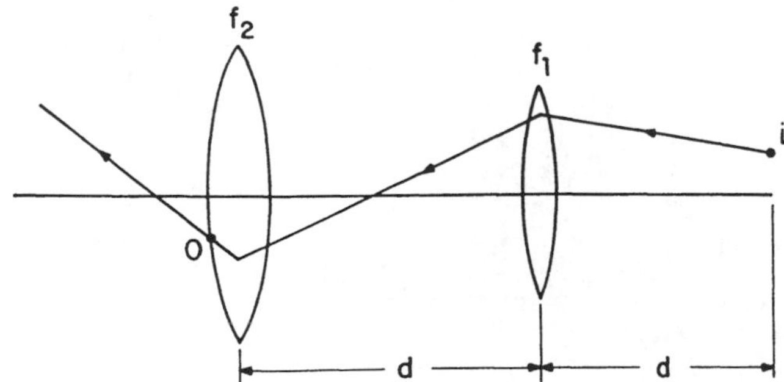

Fig. 4-9. Round trip reflection between two spherical mirrors is equivalent to the transmission through space and two lenses.

$$\begin{pmatrix} r_o \\ r_o' \end{pmatrix} = \begin{pmatrix} A & B \\ C & D \end{pmatrix} \begin{pmatrix} r_i \\ r_i' \end{pmatrix} \tag{4-6}$$

where
$$A = 1 - \frac{d}{f_2} \tag{4-7}$$

$$B = d\left(2 - \frac{d}{f_2}\right) \tag{4-8}$$

$$C = -\frac{1}{f_1} - \frac{1}{f_2}\left(1 - \frac{d}{f_1}\right) \tag{4-9}$$

$$D = -\frac{d}{f_1} + \left(1 - \frac{d}{f_2}\right)\left(1 - \frac{d}{f_1}\right) \tag{4-10}$$

§4.5 General case of a biperiodic lens series and the condition for a stable resonator

A biperiodic lens series shown in Fig. 4-5 which is equivalent to laser oscillation in a cavity between two mirrors (§4.2) can now be considered. Each unit section of the series is a pair of lenses shown in §4.4. We consider a general section, say the sth section. Fromg eq. (4-6), we have

$$\begin{pmatrix} r_{s+1} \\ r_{s+1}' \end{pmatrix} = \begin{pmatrix} A & B \\ C & D \end{pmatrix} \begin{pmatrix} r_s \\ r_s' \end{pmatrix} \tag{4-11}$$

where i → s and 0 → s + 1, and A, B, C, D are given by eq. (4-7) to (4.10).

i.e.
$$r_{s+1} = A\, r_s + B\, r_s' \tag{4-12}$$

$$r_{s+1}' = C\, r_s + D\, r_s' \tag{4-13}$$

Eq. (4-12) →
$$r_s' = \frac{1}{B}\left(r_{s+1} - A\, r_s\right) \tag{4-14}$$

Increasing s by unity, we have

$$r_{s+1}' = \frac{1}{B}\left(r_{s+2} - A\, r_{s+1}\right) \tag{4-15}$$

Substituting eq. (4-13) into it,

$$C\, r_s + D\, r_s' = \frac{1}{B}\left(r_{s+2} - A\, r_{s+1}\right) \tag{4-16}$$

Using eq. (4-14) and simplify, eq. (4-16) becomes

$$r_{s+2} - (D + A)\, r_{s+1} + (AD - BC) r_s = 0 \tag{4-17}$$

But
$$\begin{vmatrix} A & B \\ C & D \end{vmatrix} = 1$$

as can be verified using eq. (4-7) to (4-10). Thus eq. (4-17) simplifies to

$$r_{s+2} - 2\, b r_{s+1} + r_s = 0 \tag{4-18}$$

where
$$b \equiv \frac{1}{2}\,(A + D) \tag{4-19}$$

We try the following solution for eq. (4-18):

$$r_s = r_o\, e^{is\Theta} \tag{4-20}$$

Substitute into eq. (4-18), we have, after some slight simplification,

$$e^{i2\Theta} - 2\, b e^{i\Theta} + 1 = 0 \tag{4-21}$$

i.e.
$$e^{i\Theta} = b \pm i\sqrt{1 - b^2} \tag{4-22}$$

Similarly, $e^{-is\Theta}$ is also a solution of eq. (4-18). The general solution of eq. (4-18) is thus a linear combination of $e^{-is\Theta}$ and $e^{-is\Theta}$, which can be represented by a sine function.

Thus,

$$r_s = r_{max} \sin(s\Theta + \delta) \tag{4-23}$$

where r_{max} and δ are some appropriate real quantities. The condition of stable laser oscillation means that the light beam will always pass through the biperiodic lens series (§4.2). That is to say, eq. (4-23) should always oscillate between two appropriate values of (r_{max}) and ($-\, r_{max}$). This demands that Θ be real. From eq. (4-22), this means

$$|b| \leq 1 \tag{4-23}$$

i.e. $$\left|\frac{1}{2}(A + D)\right| \leq 1 \qquad \text{(from eq. (4-19))}$$

or $$\left|1 - \frac{d}{f_1} - \frac{d}{f_2} + \frac{d^2}{2f_1f_2}\right| \leq 1 \qquad (4\text{-}24)$$

After some simplification, which we leave as an exercise for the readers, we obtain,

$$0 \leq \left(1 - \frac{d}{2f_1}\right)\left(1 - \frac{d}{2f_2}\right) \leq 1 \qquad (4\text{-}25)$$

Definition: g parameter

$$g \equiv 1 - \frac{d}{2f} \qquad \text{for a lens} \qquad (4\text{-}26)$$

Since $f = R/2$,

$$g \equiv 1 - \frac{d}{R} \qquad \text{for a mirror} \qquad (4\text{-}27)$$

Eq. (4-25) becomes

$$0 \leq g_1 \, g_2 \leq 1 \qquad (4\text{-}28)$$

which is thus the condition for a stable resonator where

$$g_1 = 1 - \frac{d}{R_1} \qquad (4\text{-}29)$$

$$g_2 = 1 - \frac{d}{R_2} \qquad (4\text{-}30)$$

The condition (4-28) is shown graphically in Fig. 4-10. The shaded regions are regions of stable resonators, including the origin and the two axes where $g_1 \, g_2 = 0$, and also any point on the hyperbolas where $g_1 \, g_2 = 1$. Note: the signs of R_1 and R_2 are such that they are positive if the inner surfaces are concave as shown in Fig. (4-3). Otherwise, it is negative.

Examples:

(a) Plane parallel mirrors.

$$R_1 = R_2 = \infty$$

$$g_1 = g_2 = 1$$

$$g_1 \, g_2 = 1 \qquad \Rightarrow \qquad \text{stable}$$

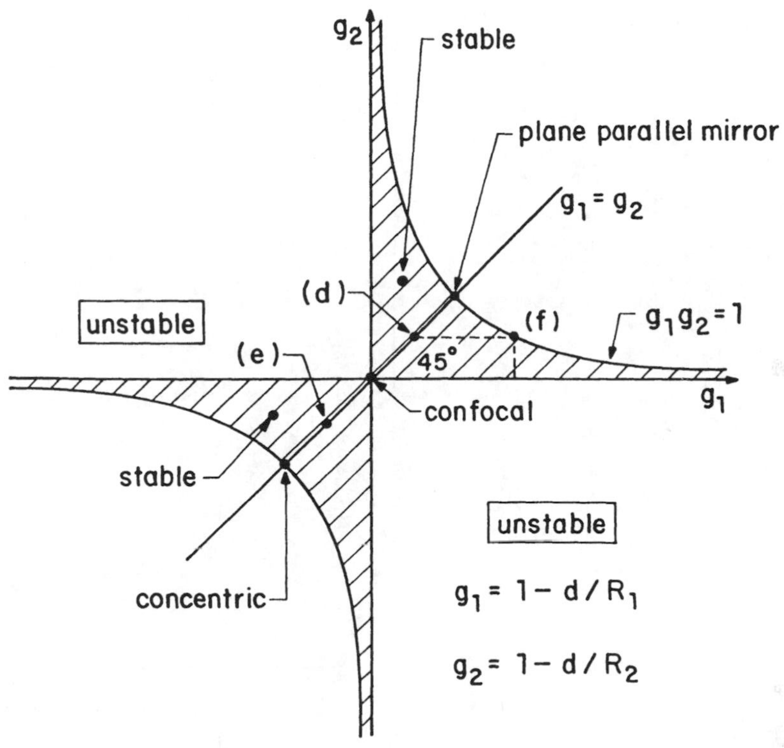

Fig. 4-10. Stability curves for a resonator.

(b) Confocal resonator

$R_1 = R_2 = d$

$g_1 = g_2 = 0 \quad \Rightarrow \quad$ stable

(c) Concentric resonator

$R_1 = R_2 \; d/2$

$g_1 = g_2 = -1$

$g_1 \, g_2 = 1 \quad \Rightarrow \quad$ stable

(d) $R_1 = R_2 = 2d$

$g_1 = g_2 = 0.5$

$g_1 \, g_2 = 0.25 \quad \Rightarrow \quad$ stable

(e) $R_1 = R_2 = \frac{2}{3}d$

$g_1 = g_2 = -0.5$

$g_1 g_2 = 0.25 \quad \Rightarrow$ stable

(f) $R_1 = -d \quad , \quad R_2 \geq 2d$

$g_1 = 2 \quad , \quad g_2 \leq \frac{1}{2}$

$g_1 g_2 \leq 1 \quad \Rightarrow$ stable

This last example shows that one can have a stable resonator even if one of the mirrors is convex, so long as the other mirror compensates for the negative effect (diverging) of the convex surface.

(g) $R_1 > +d \quad , \quad R_2 < +d$

Hence $\frac{d}{R_1} < 1$ and $\frac{d}{R_2} > 1$

$g_1 = 1 - \frac{d}{R_1} > 0$ and $g_2 = 1 - \frac{d}{R_2} < 0$

Result: $g_1 g_2 < 0$, i.e. unstable.

i.e. Even if R_1 and R_2 are concave, it is not a guarantee that the resonator will be stable.

CHAPTER V

PARAXIAL GAUSSIAN WAVE PROPAGATION AND MODES

The real laser beam coming out of a resonator does not have a uniform intensity across it. Different definite intensity distributions across the beam can be created by controlling the geometrical parameters of the laser cavity. Each distribution corresponds to the solution of the wave equation satisfying the appropriate boundary conditions defined by the geometrical parameters. The most popular laser beam is a cylindrical beam emitted by a laser resonator bounded by two circular mirrors whose g parameters satisfy the stability conditions (Chapter IV). Each solution of the wave equation is called a mode of the laser resonator. For pedagogical reasons, we start by considering the simplest transverse mode which is a Gaussian spherical wave and its propagation and then come back to the discussion of cavity modes. The discussion on the propagation of a Gaussian spherical laser beam will lead us at the same time, to the understanding of focussing and collimating such laser beams and designing spatial filters etc.

§5.1 **Definition:** Spherical wave

It is a propagating wave whose phase (or wave front) in a complex amplitude description is a spherical surface emanating from a point source in space and whose amplitude decreases inversely as the distance from the source

$$\text{i.e.} \qquad \begin{matrix}\text{Spherical}\\ \text{wave}\end{matrix} \propto \frac{f(kr \pm \omega t)}{r} = \frac{e^{-i(kr \pm \omega t)}}{r} \tag{5-1}$$

where r is the radius of the spherical wave. $\frac{1}{r}$ represents the attenuation of the wave. In spherical coordinates (r, Θ, ϕ) with the point source as origin, $f(kr \pm \omega t)$ represents the propagating wave function either diverging from the point source $[f(kr - \omega t)]$ or converging towards the point source $[f(kr + \omega t)]$. It depends only on r (the radius) and independent of direction.

We now consider the wave front in the vicinity of the direction of observation, i.e. the z - direction, as shown in Fig. 5-1. We assume a paraxial laser beam whose wavefront is a spherical wave. We look at its propagation in the vicinity of the z - direction. Hence, at time t, when the wavefront reaches the (x - y) plane at position z,

$$\psi(r) \sim \frac{e^{-i(kr - \omega t)}}{r} \tag{5-2}$$

$$\sim \frac{e^{-ikr}}{r} \quad \text{at a fixed time t.} \tag{5-3}$$

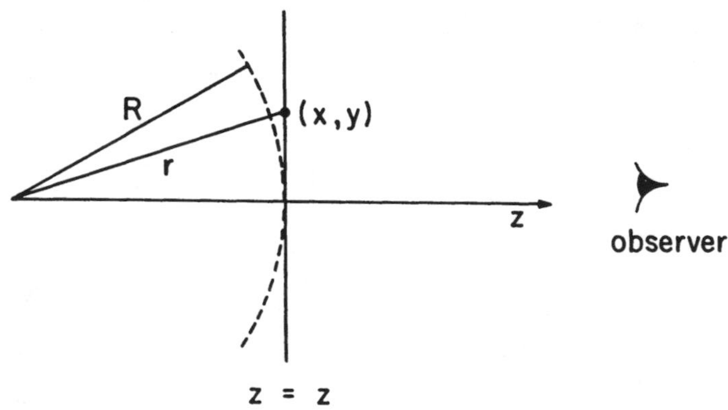

Fig. 5-1. Paraxial approximation of a spherical wavefront.

We now ask what the wave function is at a point (x, y, z) on the plane z = z. We have,

$$r^2 = x^2 + y^2 + z^2 \tag{5-4}$$

Since we are concerned only with paraxial rays, x « z, y « z, and eq. (5-4) becomes

$$r^2 = z^2 \left(1 + \frac{x^2}{z^2} + \frac{y^2}{z^2}\right) \tag{5-5}$$

$$r = z\left[1 + \frac{x^2}{z^2} + \frac{y^2}{z^2}\right]^{1/2}$$

$$\simeq z\left[1 + \frac{x^2}{2z^2} + \frac{y^2}{2z^2}\right] \tag{5-6}$$

where we have expanded the square root in a Taylor series and kept only the first two terms, because x « z, y « z. Eq. (5-3) becomes

$$\psi\ (r) \sim \frac{e^{-ikz\left[1 + \frac{x^2}{2z^2} + \frac{y^2}{2z^2}\right]}}{z\left[1 + \frac{x^2}{2z^2} + \frac{y^2}{2z^2}\right]} \tag{5-7}$$

This is the part of the wavefront propagation in the direction $r = (x^2 + y^2 + z^2)^{1/2}$; not in the z - direction. However, because of the assumption of paraxial ray, the transverse extent of the wave front is small (x « z, y « z) and around the z - axis. We thus make the approximation that r is almost in the z - direction so that r = z = R, where R is the radius of curvature of the wavefront at the z = z plane. Hence, eq. (5-7) becomes

$$\psi \sim \frac{e^{-ikR\left(1 + \frac{x^2}{2R^2} + \frac{y^2}{2R^2}\right)}}{R\left(1 + \frac{x^2}{2R^2} + \frac{y^2}{2R^2}\right)} \tag{5-8}$$

The nominator in eq. (5-8),

i.e.
$$e^{-ikR\left(1 + \frac{x^2}{2R^2} + \frac{y^2}{2R^2}\right)} \tag{5-9}$$

represents an approximated propagating wavefront of a paraxial spherical wave at the point (x, y) on the plane z = z. The denominator is simply an attenuation factor. The transverse phase variation of eq. (5-9) is

$$e^{-ikR\left(x^2 + y^2\right)/2R^2} = e^{-ik\left(x^2 + y^2\right)/2R} \tag{5-10}$$

§5.2 **Definition:** Gaussian amplitude variation of a wavefront

A general wavefront on the plane z = z may vary in amplitude from point to point on the plane. In the present case, we assume the variation follows a Gaussian distribution,

i.e. $$\text{Wave amplitude} \sim e^{-\left(x^2 + y^2\right)/w^2} \tag{5-11}$$

where w ≡ spot size of the Gaussian distribution.

§5.3 **Definition:** Gaussian Spherical laser beam

Using the definitions in the previous two sections, a paraxial laser beam having a spherical wavefront and a Gaussian amplitude distribution across a plane at z = z can thus be represented by combining eq. (5-10) and (5-11). We define the transverse part of the wavefront as

$$u(x, y) = \sqrt{\frac{2}{\pi}}\frac{1}{w}\, e^{-ikR\left(\frac{x^2}{2R^2} + \frac{y^2}{2R^2}\right)}\, e^{-\left(x^2 + y^2\right)/w^2} \tag{5-12}$$

The factor $\sqrt{\frac{2}{\pi}}\frac{1}{w}$ is a normalizing factor so that

$$\int |u(x, y)|^2 dxdy = 1 \tag{5-13}$$

u(x, y) represents the electric field strength of the laser beam.

Eq. (5-12) can be made more compact by grouping the exponentials together:

$$-ikR\left(\frac{x^2+y^2}{2R^2}\right) - \frac{x^2+y^2}{w^2}$$

$$= \left[-i\frac{2\pi}{\lambda} R \bullet \frac{1}{2R^2} - \frac{1}{w^2}\right] \left(x^2+y^2\right)$$

$$= -i\frac{\pi}{\lambda}\left[\frac{1}{R} - \frac{i\lambda}{\pi w^2}\right] \left(x^2+y^2\right)$$

$$\equiv -i\frac{\pi}{\lambda}\frac{x^2+y^2}{q} \tag{5-14}$$

Where $$\frac{1}{q} \equiv \frac{1}{R} - i\frac{\lambda}{\pi w^2} \tag{5-15}$$

and λ is the wavelength.

Substituting eq. (5-14) into (5-12),

$$u(x, y) = \sqrt{\frac{2}{\pi}}\frac{1}{w} \bullet e^{-i\frac{\pi}{\lambda}\frac{x^2+y^2}{q}} \tag{5-16}$$

This can be recast in a similar form as eq. (5-10)

i.e. $$u(x, y) = \sqrt{\frac{2}{\pi}}\frac{1}{w} e^{-ik\frac{x^2+y^2}{2q}} \tag{5-17}$$

Comparing (5-17) and (5-10), we can say that q is a complex radius of curvature of the Gaussian spherical laser beam.

§5.4 Huygen-Fresnel's diffraction approach to the propagation of a wavefront

We shall make use of the consequence of Huygen-Fresnel diffraction and examine how a given wavefront propagates through a homogeneous isotropic linear medium. The medium is usually characterized by the index of refraction. In the present discussion, the index n is incorporated in the notation of the wavelength λ such that

$$\lambda_o = n\lambda \tag{5-18}$$

where λ_o is the wavelength in vacuum. (Eq. (5-18) can be verified by noting that the frequency ν of the E-M wave is the same in both the vacuum and the homogeneous isotropic linear medium, so that $\nu = c/\lambda_o = v/\lambda$ where v is the speed of light in the medium. Since $n = c/v$, we obtain eq. (5-18).) Thus, all formulation will look as if the propagation is in the "vacuum" with wavelength λ.

We consider a wavefront at the plane $z = z_o$. The transverse part of the wavefront is $u_o(x_o, y_o)$. Referring to Fig. 5-2, we like to know what this wavefront becomes after propagating to the plane $z = z'$. According to Huygen-Fresnel's principle of superposition of secondary waves, each point on the wavefront at $z = z_o$ can be considered as a point source of secondary spherical waves so that the wavefront at $z = z'$ could be considered as the superposition of these secondary waves. The result of such a consideration gives (see any optics text on diffraction)

$$u(x, y, z) = \frac{i}{\lambda} \iint\limits_{\substack{\text{input}\\ \text{plane}}} u_o(x_o, y_o) \, \frac{1 + \cos\alpha}{2} \, \frac{e^{-ik|\vec{r} - \vec{r}_o|}}{|\vec{r} - \vec{r}_o|} \, dx_o \, dy_o \tag{5-19}$$

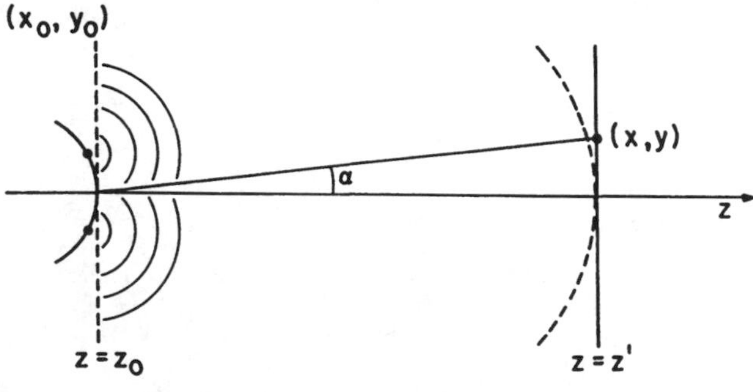

Fig. 5-2. Diffraction of a spherical paraxial wavefront after propagation through a distance in space.

where $\vec{r}_o \equiv (x_o, y_o, z_o)$

$\vec{r} = (x, y, z')$

Eq. (5-19) is valid under the following assumptions:

(a) all dimensions $\gg \lambda$

(b) $\Delta u(x, y) \equiv$ variation of $u(x, y) \ll \lambda$

(c) $\Delta u_o(x_o, y_o) \ll \lambda$

If we make some further assumptions:

(d) the wavefront is that of a paraxial ray;

i.e. $\alpha \approx 0$ and $x, x_o \ll z' - z_o$

$y, y_o \ll z' - z_o$

These lead to

$$\frac{1 + \cos\alpha}{2} \approx 0 \tag{5-20}$$

and

$$|r - r_o| \approx (z' - z_o) \equiv z \tag{5-21}$$

and

$$k|\vec{r} - \vec{r}_o| = k\left[(z - z_o)^2 + (y - y_o)^2 + (x - x_o)^2\right]^{1/2}$$

$$\approx kz\left[1 + \frac{(x - x_o)^2}{2z^2} + \frac{(y - y_o)^2}{2z^2}\right] \tag{5-22}$$

(see eq. 5-6)

Inspecting eq. (5-19), we see that it represents a superposition of spherical waves

$$\frac{e^{-ik|\vec{r} - \vec{r}_o|}}{|\vec{r} - \vec{r}_o|}$$

with the appropriate weighting factor $u_o(x_o, y_o)$ and a geometrical factor $\frac{1 + \cos\alpha}{2}$. Using eqs. (5-20), (5-21) and (5-22), eq. (5-19) becomes

$$u(x, y, z) = \frac{i}{\lambda} \iint_{\substack{\text{input}\\ \text{plane}}} u_o(x_o, y_o) \frac{e^{-ikz\left[1 + \frac{(x - x_o)^2}{2z^2} + \frac{(y - y_o)^2}{2z^2}\right]}}{z} dx_o\, dy_o$$

$$= \frac{ie^{-ikz}}{\lambda z} \iint_{\substack{\text{input}\\ \text{plane}}} u_o(x_o, y_o) e^{-ik\left[\frac{(x - x_o)^2}{2z} + \frac{(y - y_o)^2}{2z}\right]} dx_o\, dy_o \qquad (5\text{-}23)$$

Eq. (5-23) is the general integral equation of the propagation of a paraxial wavefront a long distance z away.

§5.5 Propagation of a Gaussian plane wave

We start to apply eq. (5-23) to different special cases. The present one is a Gaussian plane wave. According to eq. (5-3), a spherical wave becomes a plane wave when $r \to \infty$. Hence, for a Gaussian plane wave, using eq. (5-16), (5-15) with $R \to \infty$, we have

$$u_o(x_o, y_o) = \sqrt{\frac{2}{\pi}} \frac{1}{w_o} e^{-i \frac{\pi}{\lambda} \frac{x_o^2 + y_o^2}{q_o}} \qquad (5\text{-}24)$$

where $\quad q_o^{-1} = 0 - i \dfrac{\lambda}{\pi w_o^2} \quad$ from eq. (5-15)

$$q_o = \frac{\pi w_o^2}{-i\lambda} = \frac{i\pi w_o^2}{\lambda} \qquad (5\text{-}25)$$

where w_o is the spot size at (x_o, y_o). Substituting into eq. (5-24), we have

$$u_o(x_o, y_o) = \sqrt{\frac{2}{\pi}} \frac{1}{w_o} e^{-i \frac{\pi}{\lambda} \cdot \frac{x_o^2 + y_o^2}{\pi w_o^2/(-i\lambda)}}$$

$$= \sqrt{\frac{2}{\pi}}\frac{1}{w_o}\, e^{-\frac{x_o^2 + y_o^2}{w_o^2}} \tag{5-26}$$

which is a Gaussian distribution. This spot size of a Gaussian plane wave is called the beam waist. Substituting eq. (5-26) into (5-23), we get

$$u(x, y, z) = \frac{ie^{-ikz}}{\lambda z}\int\int_{-\infty}^{\infty} \sqrt{\frac{2}{\pi}}\frac{1}{w_o}\, e^{-\frac{x_o^2 + y_o^2}{w_o^2}}\, e^{-ik\left[\frac{(x - x_o)^2}{2z} + \frac{(y - y_o)^2}{2z}\right]} dx_o dy_o \tag{5-26}$$

Simplifying and integrating, we obtain

$$u(x, y, z) = \sqrt{\frac{2}{\pi}}\frac{1}{w(z)}\exp\{-i[(kz - \psi(z)]\}\exp\left[-i\frac{k}{2}\frac{x^2 + y^2}{q(z)}\right] \tag{5-27}$$

where

$$q(z) \equiv q_o + z = z + i\frac{\pi w_o^2}{\lambda} \tag{5-28}$$

$$\frac{1}{q(z)} = \frac{1}{R(z)} - i\frac{\lambda}{\pi w^2(z)} = \frac{1}{z + i\pi w_o^2/\lambda} \tag{5-29}$$

$$w(z) = w_o\sqrt{1 + \left(\frac{\lambda z}{\pi w_o^2}\right)^2} \tag{5-30}$$

$$\psi(z) = \tan^{-1}\left(\frac{\lambda z}{\pi w_o^2}\right) \tag{5-31}$$

$$R(z) = z + \left(\frac{\pi w_o^2}{\lambda}\right)^2\frac{1}{z} \tag{5-32}$$

For the sake of comparison, we rewrite eq. (5-24) as follows:

$$u_o(x_o, y_o) = \sqrt{\frac{2}{\pi}}\frac{1}{w_o}\exp\left\{-i\frac{\pi}{\lambda}\frac{x_o^2 + y_o^2}{q_o}\right\} \tag{5-33}$$

and compare with eq. (5-27). One sees that after propagation through a long distance z (z » (x_o, y_o) » λ) a Gaussian plane wave at the plane $z = z_o$, given by eq. (5-33), with spot size (waist) w_o and complex radius of curvature q_o (eq. 5-25) is transformed into a diverging Gaussian spherical wavefront $u(x, y, z)$ due to diffraction, whose spot size is $w(z)$ (eq. 5-30), complex radius of curvature $q(z)$ (eq. 5-28) and the real radius of curvature $R(z)$ (eq. 5-32), plus a phase increase of $[kz - \psi(z)]$ (eq. 5-31). This is a very important result because it says that any Gaussian plane wave, after propagation and diffraction, becomes a diverging Gaussian spherical wave. This means that (see below) the plane $z = z_o$ is the origin of the diverging wave; i.e. it is the "focal plane" of the diverging wave. Since such kind of propagation is reversible, a converging Gaussian spherical wavefront will become a Gaussian plane wave at the focal plane. These are shown in Fig. 5-3.

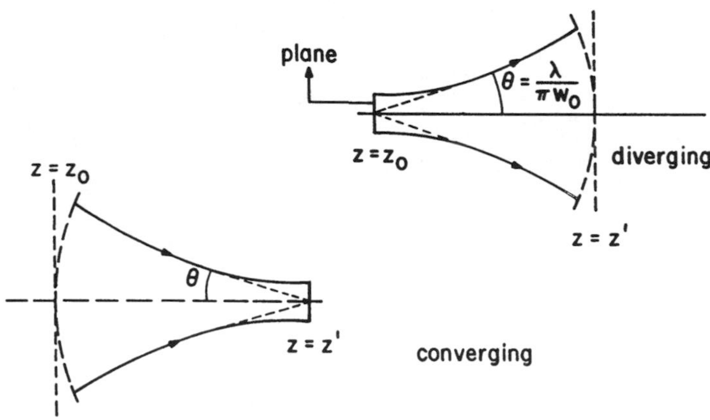

Fig. 5-3. Converging and diverging Gaussian spherical wave propagation.

<u>Consequence (1)</u>

At large $z = z' - z_o$, eq. (5-32) becomes

$$R(z) = z + \left(\frac{\pi w_o^2}{\lambda}\right)^2 \frac{1}{z}$$

$$\approx z + 0 \qquad \left(if\ z \gg \frac{\pi w_o^2}{\lambda}\right)$$

$$= z \tag{5-34}$$

This shows that the real radius of curvature R of the diverging spherical wavefront at z = z' is equal to z. This means that the plane at $z = z_o$ is indeed at the center of curvature of the outgoing spherical wave at a long distance; i.e. far field (Fig. 5-3).

Consequence (2)

Again, for large $z \gg \frac{\pi w_o^2}{\lambda}$

$$w(z) = w_o \sqrt{1 + \left(\frac{\lambda z}{\pi w_o^2}\right)^2} \qquad \text{(from eq. 5-30)}$$

$$\approx w_o \frac{\lambda z}{\pi w_o^2}$$

$$= z \bullet \frac{\lambda}{\pi w_o} \tag{5-35}$$

$$\equiv z\, \Theta \tag{5-36}$$

where $$\Theta \equiv \frac{\lambda}{\pi w_o} \tag{5-37}$$

is the (cone) angle of divergence (or simply divergence) at the far field (Fig. 5-3). Thus, looking at the problem in reverse, a converging Gaussian spherical wavefront will converge at the focal plane with a spot size (waist) given by

$$w(z) = z\, \Theta$$

or $$w(z) \equiv f\, \Theta \tag{5-38}$$

where f is the focal distance ; i.e. the distance between the spherical surface and its focus. Eq. (5-38) give a convenient general estimate of the spot size at the focus of a lens (see below).

§5.6 Propagation of a general Gaussian spherical wavefront

From eq. (5-17), the transverse part of the general Gaussian spherical wavefront is

$$u(x, y) = \sqrt{\frac{2}{\pi}}\frac{1}{w}\, e^{-k\frac{x^2+y^2}{2q}} \tag{5-17}$$

and one can substitute this into eq. (5-23) to obtain the wavefront at a distance z away. However, there is a better and physically clearer way to obtain the result.

Referring back to the previous section, any diverging Gaussian spherical wavefront must have come from a focal plane where the wavefront is a Gaussian plane wave. Similarly, any converging Gaussian spherical wavefront will converge to a focal plane where the wavefront is a Gaussian plane wave. Thus, if we start with a general diverging Gaussian spherical wavefront, we should first ask where its focal plane was and then work from the focal plane, and calculate what it should be at a distance z from the initial Gaussian spherical wavefront. The case of a converging wavefront is similar. Let us now do a calculation for a diverging wavefront at $z = z_1$ from the beam waist at $z = 0$. We need to find what z_1 is (Fig.5-4). The complex radius of curvature is, from eq. (5-29),

$$\frac{1}{q(z_1)} = \frac{1}{R(z_1)} - i\,\frac{\lambda}{\pi w^2(z)}$$

or

$$q\,(z_1) = \left[\frac{1}{R(z_1)} - i\,\frac{\lambda}{\pi w^2(z_1)}\right]^{-1} \tag{5-39}$$

But

$$q\,(z_1) = q_o + z_1 = i\,\frac{\pi w_o^2}{\lambda} + z_1 \qquad \text{(from eq. 5-28)} \tag{5-40}$$

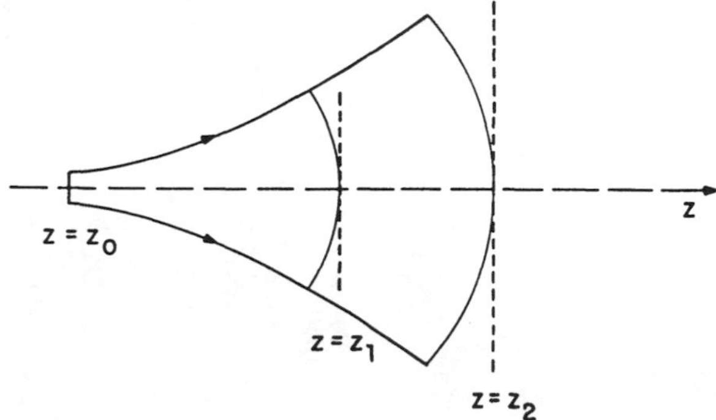

Fig. 5-4. Determination of the wavefront of a diverging Gaussian spherical wave after propagation through a distance in space.

Equating the real and imaginary parts of the right hand sides of eq. (5-39) and (5-40). we have:

real part:
$$z_1 = \mathrm{Re}\left[\frac{1}{R(z_1)} - i\frac{\lambda}{\pi w^2(z_1)}\right]^{-1}$$

$$= \frac{\dfrac{1}{R(z_1)}}{R^2(z_1) + \dfrac{\lambda^2}{\pi^2 w^4(z_1)}} \qquad (5\text{-}41)$$

imaginary part:
$$\frac{\pi w_o^2}{\lambda} = \frac{\dfrac{\lambda}{\pi w^2(z_1)}}{R^2(z_1) + \dfrac{\lambda^2}{\pi^2 w^4(z_1)}}$$

or
$$w_o = \frac{\lambda/[\pi w(z_1)]}{\left\{R^2(z_1) + \dfrac{\lambda^2}{\pi^2 w^4(z_1)}\right\}^{1/2}} \qquad (5\text{-}42)$$

Let
$$w(z_1) \equiv w_1$$
$$R(z_1) \equiv R_1$$

Eq. (5-42) becomes

$$w_o = \frac{w_1}{\left\{1 + \frac{\pi^2 w_1^4}{\lambda^2 R_1^2}\right\}^{1/2}} \tag{5-43}$$

Now knowing w_o, we find the wavefront at $z = z_2$ from the beam waist, (Fig. 5-4). Using eq. (5-28), the complex radius of curvature is

$$q(z_2) = q_o + z_2 = z_2 + i\,\frac{\pi w_o^2}{\lambda} \tag{5-44}$$

But from eq. (5-40), $$i\frac{\pi w_o^2}{\lambda} = q(z_1) - z_1 \tag{5-45}$$

Substituting (5-45) into 5-44), we have

$$\begin{aligned} q(z_2) &= z_2 + q(z_1) - z_1 \\ &= q(z_1) + (z_2 - z_1) \end{aligned} \tag{5-46}$$

and the wavefront at $z = z_2$ is given by eq. (5-27), with $z \to z_2$. We note here that eq. (5-46) gives the transformation of $q(z_1)$ into $q(z_2)$ for the propagation of a Gaussian spherical wave in a homogeneous medium. The case of a converging wavefront is left as a straightforward exercise for the reader.

§5.7 Propagation of a Gaussian spherical wavefront through a thin lens

In geometrical optics, diverging spherical wavefront is transformed into a converging spherical wavefront by a thin spherical lens, Fig. (5-5). We can use the matrix transformation of a thin lens in chapter IV to calculate R_2. Note that we are always dealing with paraxial rays.

$$\begin{pmatrix} r_o \\ r_o' \end{pmatrix} = \begin{pmatrix} 1 & 0 \\ -\frac{1}{f} & 1 \end{pmatrix} \begin{pmatrix} r_i \\ r_i' \end{pmatrix} \qquad \text{(from eq. (4-3)}$$

i.e. $$r_o = r_i \tag{5-47}$$

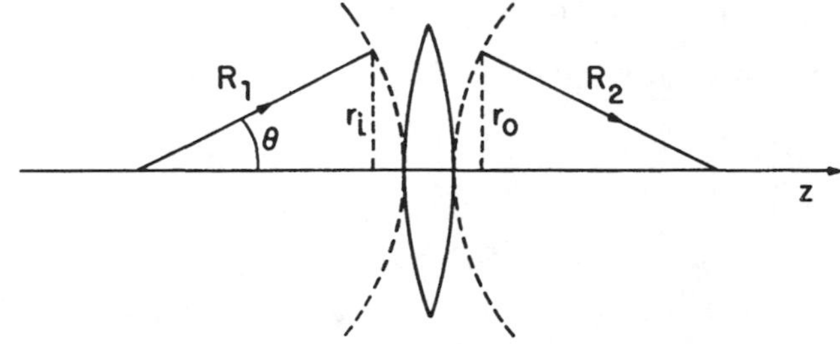

Fig. 5-5. Propagation of a Gaussian spherical wave through a thin lens.

and $$r_o' = r_i' - {}^{r_i}/_f \tag{5-48}$$

For paraxial rays, $r_o' \approx r_o/R_2$ and

$r_i' \approx r_i/R$; eq. (5-48) ⇒

$$\frac{r_o}{R_2} = \frac{r_i}{R_1} - \frac{r_i}{f}$$

Using eq. (5-47), this becomes

$$\frac{1}{R_2} = \frac{1}{R_1} - \frac{1}{f} \tag{5-49}$$

Now, if the incident wavefront is a Gaussian spherical type, the wavefront is still spherical, while the amplitude distribution is Gaussian. Assuming no loss through the lens, the amplitude distribution at the output side of the lens should be unchanged. Hence, adding $\left(-\frac{i\lambda}{\pi w_2^2}\right)$ to both sides of eq. (5-49) gives

$$\frac{1}{R_2} - i\frac{\lambda}{\pi w_2^2} = \frac{1}{R_1} - i\frac{\lambda}{\pi w_2^2} - \frac{1}{f} \tag{5-50}$$

But, $w_2 = w_1$ because the amplitude distribution is unchanged for a thin lens. We thus have, using eq. (5-29),

$$\frac{1}{q_2} = \frac{1}{q_1} - \frac{1}{f} \tag{5-51}$$

i.e. a thin lens transforms a Gaussian spherical wave in a similar way as it tranforms an ordinary spherical wave.

§5.8 Focal spot size

We can now make use of the result of section 5.7 to calculate the transverse diameter d_o at the focus of a thin lens when a Gaussian spherical wave is focussed down to a "spot". This is shown in Fig. 5-6. We already know that in the propagation of a Gaussian plane wave into a Gaussian spherical wavefront, (Fig. 5-7), the beam radius at z, i.e. w(z), is related to the beam waist w_o by eq. (5-30).

i.e. $$w(z) = w_o\left[1 + \left(\frac{z}{z_R}\right)^2\right]^{1/2} \qquad z_R \equiv \frac{\pi w_o^2}{\lambda}$$

$$\simeq \frac{w_o z}{z_R} \qquad \text{if } z \gg z_R$$

$$= \frac{\lambda z}{\pi w_o}$$

Hence, regrouping the left and right hand sides,

$$w_o w(z) \simeq \frac{\lambda z}{\pi} \tag{5-52}$$

Because of the reversibility of optical rays (waves), the propagation in the reverse direction is also true and eq. (5-52) is still valid. That is,

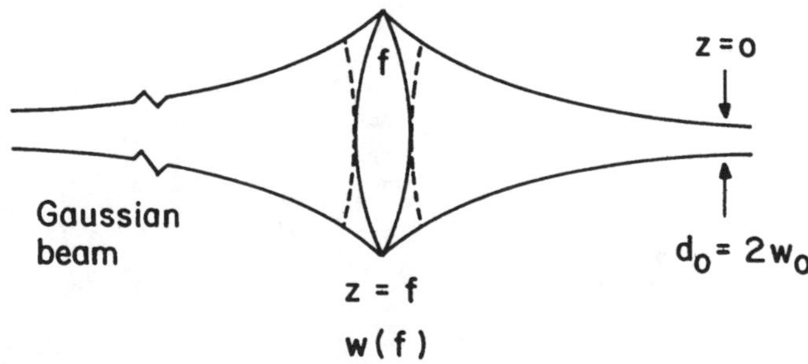

Fig. 5-6. Gaussian spherical wave focussing by a lens.

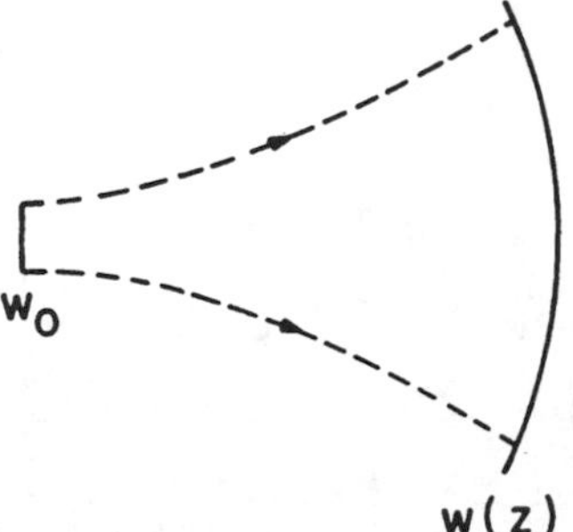

Fig. 5-7. Gaussian spherical wave.

(Fig. 5-6) if the lens transforms the incident Gaussian spherical wave into one that has a waist w(z) ($z = f$) at the output side of the lens, its relationship with w_o at the focus ($z = o$) is also

$$w_o w(z) \simeq \frac{\lambda z}{\pi}$$

$$w_o w(f) \simeq \frac{\lambda f}{\pi}$$

$$w_o = \frac{\lambda f}{\pi w(f)}$$

or

$$d_o = 2w_o \simeq \frac{2\lambda f}{\pi w(f)} \tag{5-53}$$

Definition: <u>99% criterion</u>.

It just happens that for a Gaussian spherical laser beam, 99% of its power (assuming a CW laser) is concentrated within the diameter (<u>exercise</u>)

$$D = \pi w(f) \tag{5-54}$$

Using this criterion, eq. (5-53) becomes

$$d_o = \frac{2\lambda f}{D} \tag{5-55}$$

$$= 2\lambda f^{\#} \tag{5-56}$$

where

$$f^{\#} \equiv f/D \tag{5-57}$$

Eq. (5-55) says that if we can measure experimentally at the input side of the thin lens the diameter D of the laser beam containing 99% of its power, and if f and λ is known, we can calculate the focal diameter d_o through which 99% of the laser power passes.

It should be noted that even if the laser beam is a purely Gaussian spherical wave and is paraxial, the calculation of d_o is still not exact because we have neglected the aberration of the lens. Moreover, most laser beams are not perfectly Gaussian spherical. Hence, eq. (5-53) to (5-57) represent only some practical estimates. Any better knowledge of the real d_o has to be measured directly at the focus. This poses a challenge in the measurement of the focal spot size of very intense laser pulses. Direct measurement is impossible because any measuring device (film, detector arrays, translating pin-holes, etc.) set at the focus will be severely damaged. Using beam splitting technique requires very good beam splitting optical surfaces so that the wavefront is not disturbed. Even more difficult is the measurement of i.r. laser focal diameter because of the lack of sensitive material. Although all these difficulties have been overcome, the price is high and there is still a lot of room for improvement.

Often, experimental measurement of the focal spot size is represented by an integrated energy distribution across the focal diameter d_o as shown in Fig. 5-8. The diameter or full width at half the maximum value of the distribution is defined as the width, or FWHM. For a more detail discussion, see appendix I.

There is another practical way of estimating the focal spot size. This is by way of geometrical optics (Fig. 5-9). The laser beam divergence angle

$$\Theta = 2\Theta' \tag{5-58}$$

can be calculated as follows.

$$\frac{d_o/2}{f} = \tan \Theta' \approx \Theta' \text{ for small } \Theta'$$

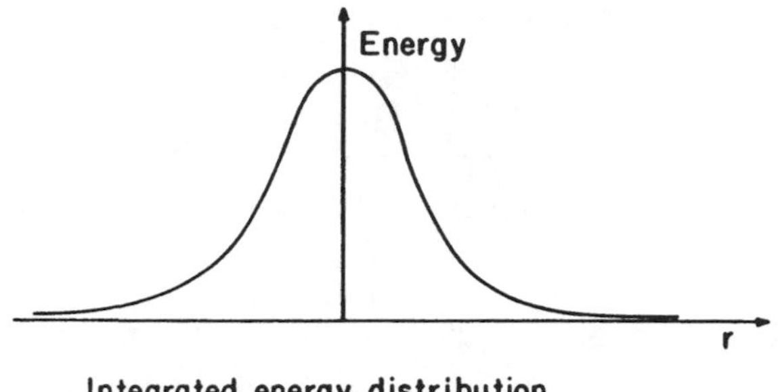

Fig. 5-8. Energy distribution of a laser beam.

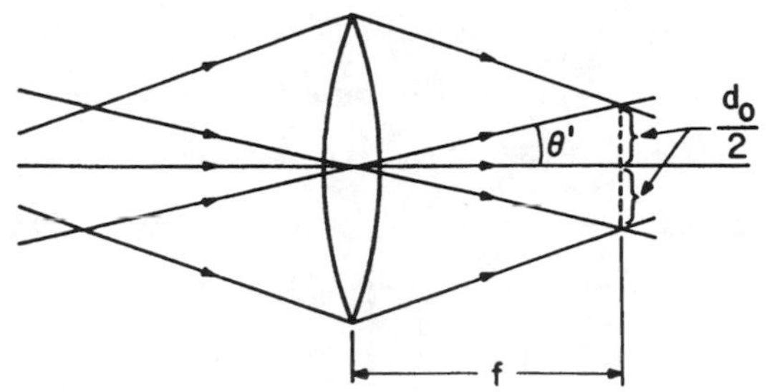

Fig. 5-9. Quick estimation of the focal spot size.

Hence $$d_o = f \cdot 2\theta'$$

or $$d_o = f\,\theta \qquad (5\text{-}59)$$

Normally, one can measure the divergence angle at the input side of the lens and apply eq. (5-59) to calculate d_o. It should be warned again that such measurements, as well as those using eq. (5-53) to (5-57), are estimates. More precise measurements have to be made directly.

§5.9 Modes

In a laser cavity, the laser radiation "oscillates" at different possible modes. Each oscillating mode is a wave of a fixed frequency with a certain transverse energy (field amplitude) distribution satisfying both the

Maxwell's equations and the boundary conditions for a stable oscillation, i.e. $0 \le g_1 g_2 \le 1$ (see chapter IV). In practice one loosely distinguishes these modes into two types, spatial or transverse modes and axial or temporal modes, although they are interwoven together (see below).

Definition: Spatial (transverse mode)

At a fixed laser frequency, the cavity can in general sustain stable waves with different kinds of transverse amplitude distributions. An example is the Gaussian spherical wave which is the lowest order transverse mode in a laser cavity with spherical mirrors. (This mode is sometimes called the fundamental transverse mode.) Higher order modes can exist whose angle of divergence is larger than that of the fundamental transverse mode. Hence, very often, one puts a diaphragm inside a laser cavity and make the hole as small as possible so that only the fundamental transverse mode can pass through the hole. The laser output is thus Gaussian spherical wave. However, the frequency at which this fundamental transverse mode oscillates is not unique. Other allowed axial frequencies can also oscillate with the same transverse distribution. This leads to the following definition.

Definition: Axial or temporal modes

At a fixed transverse field distribution in a laser cavity, the stable oscillation demands that there is a standing wave along the axis of the cavity, as shown in Fig. (5-10). i.e. the cavity length L should be:

$$L = n \left(\frac{\lambda}{2}\right) \qquad n = \text{very large integer} \tag{5-60}$$

Since $\lambda = \frac{c}{\nu}$

where c is the appropriate speed of light in the media inside the cavity and ν is the frequency of the laser; eq. (5-60) becomes

$$L = n \cdot \frac{c}{2\nu} \tag{5-61}$$

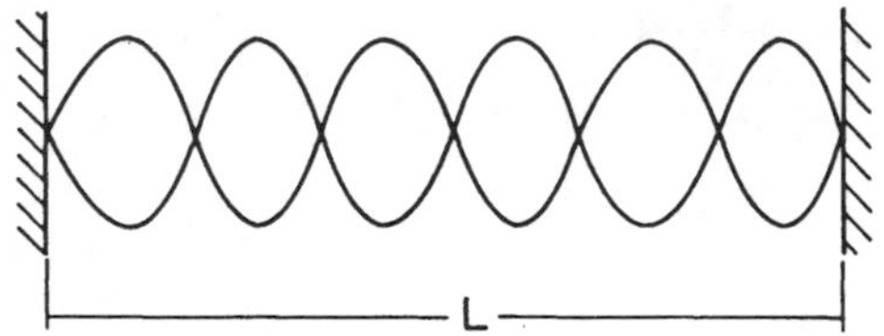

Fig. 5-10. Standing wave in a laser cavity.

we define ν_n as the frequency corresponding to a large integer n. Eq. (5-61) becomes

$$\nu_n = n\left(\frac{c}{2L}\right) \tag{5-62}$$

We call ν_n an axial mode (frequency). In a laser cavity, many such axial modes can be sustained, the separation between an adjacent pair being

$$\begin{aligned} \nu_{n+1} - \nu_n &= \left(n + 1\right)\left(\frac{c}{2L}\right) - n\left(\frac{c}{2L}\right) \\ &= \frac{c}{2L} \end{aligned} \tag{5-63}$$

which is a constant. Thus, under the gain curve of the laser (Fig. 5-11), there can be several axial modes which experience the gain and hence will oscillate. The simultaneous operation of many axial modes results in mode beating (interference effect) among themselves so that a temporal modulation

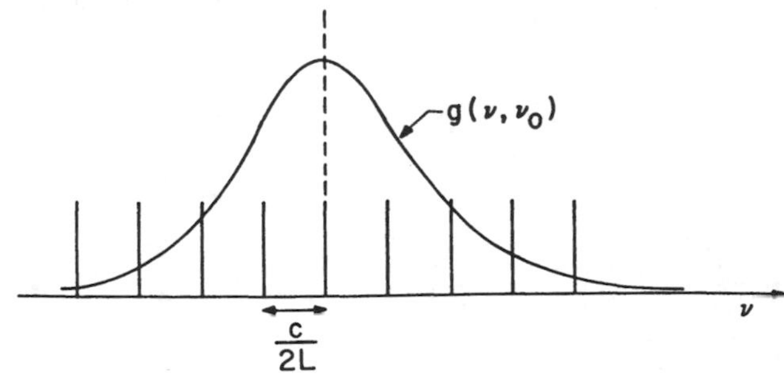

Fig. 5-11. Gain curve of a laser oscillator and the axial modes of the cavity.

of the wave exists. If we detect the laser output at a fixed position z along the laser axis, the power will be modulated in time (Fig. 5-12).

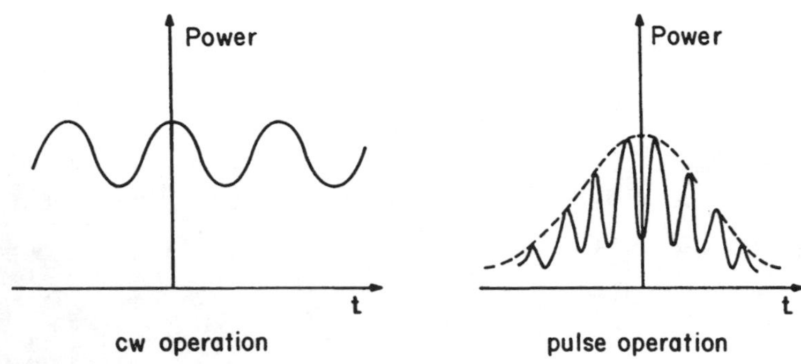

Fig. 5-12. Temporal mode beatings.

§5.10 Spatial-temporal modes

Any appropriate combination of a frequency and a single transverse amplitude distribution satisfying the lasing conditions in a cavity constitutes a spatial-temporal mode. A general statement of the lasing condition from the point of view of wave propagation is the following. The laser field's amplitude distribution and its phase should reproduce themselves after one round trip inside the cavity, i.e. gain = loss and the total phase change after one round trip is $2m\pi$ ($m = 0,1,2$). It is the phase change condition that leads to the allowed frequency (mode) oscillation.

For example, for pedagogical reason, we assume that a plane wave oscillates in the cavity. i.e. apart from the amplitude, the propagating part of the wave is $\exp\{-i(kz - \omega t)\}$ where z is the laser axis. After one round trip, we should have (phase condition: the wave reproduces itself)

$$\exp\{-i(kz - \omega t)\} = \exp\{-i[k(z+2L) - \omega t]\}$$

i.e. $\exp[-i(k2L)] = 1$

$$k\,2\,L = 2\,m\,\pi \qquad (m = 0,1,2\ldots)$$

$$\frac{2\pi}{\lambda} \bullet 2\,L = 2\,m\,\pi$$

$$L = m\,(\lambda/2)$$

which is eq. (5-60), the condition for stable oscillation.

A more realistic example is a Gaussian spherical wave which can be a fundamental mode of the cavity. Referring to eq. (5-27), the phase part due to propagation is $\exp\{-i(kz - \psi(z))\}$ with $\psi(z) = \tan^{-1}\left(\left|\frac{\lambda z}{\pi w_o^2}\right|\right)$. Round trip reproduction of the wave means

$$\exp\{-i[kz - \psi(z)]\} = \exp\{-i[k(z + 2L) - \psi(z + 2L)]\}$$

or $\quad k(z + 2L) - \psi(z + 2L) - (kz - \psi(z)) = 2\,m\,\pi$

i.e. $$k \bullet 2L - [\psi(z + 2L) - \psi(z)] = 2\,m\,\pi \tag{5-64}$$

eq. (5-64) is the condition that governs the laser frequencies through

$$k = \frac{2\pi}{\lambda} = \frac{2\pi\nu}{c} \quad .$$

A general example is a general high order Gaussian mode given by

$$u_{mn}(x, y, z) = \sqrt{\frac{1}{2^{m+n} m!n!}}\,\frac{1}{w(z)}\,H_m\left(\frac{\sqrt{2}\,x}{w(z)}\right)\,H_n\left(\frac{\sqrt{2}\,y}{w(z)}\right)$$

$$\bullet \exp\left(-i\,\frac{k}{2}\,\frac{x^2 + y^2}{q(z)}\right)\,\exp\left[-ikz + i(m + n + 1)\psi(z)\right] \tag{5-65}$$

where $q(z)$, $w(z)$ and $\psi(z)$ are given by eq. 5-29, 30 and 31 respectively, and where $H_n(x)$ is the Hermite polynomial. This mode becomes the fundamental Gaussian spherical mode when $m = n = 0$ and one often refers to the latter case as a <u>TEM_{oo} mode</u>. Round trip reproduction of wave means

$$\exp[-ikz + i(m + n + 1)\psi(z)] = \exp\,[-ik(z + 2L) + i(m + n + 1)\,\psi(z + 2L)]$$

i.e. $$k(z + 2L) - (m + n + 1)\,\psi(z + 2L)$$

$$-\,[kz - (m + n + 1)\,\psi(z)] = 2q\pi \quad (q = 0,1,2,\ldots)$$

or $$k \bullet 2L - (m + n + 1)\,[\psi(z + 2L) - \psi(z)] = 2q\pi \tag{5-66}$$

Using $$k = \frac{2\pi\nu}{c} \equiv \frac{2\pi}{c}\,\nu_{qmn} \tag{5-67}$$

and eq. 5-29 to 31, and after some lengthy calculation (left as an <u>exercise</u> for the reader), one obtains the general expression for the frequency of a general stable cavity Gaussian mode.

$$\nu_{qmn} = \left[q + (m + n + 1) \frac{\cos^{-1}\sqrt{g_1 g_2}}{\pi} \right] \frac{c}{2L} \tag{5-68}$$

where $$g_i = 1 - \frac{L}{R_i} \qquad i = 1,2$$

and R_i is the radius of curvature of the ith mirror.

In general, on the frequency scale, the modes look like the sketch shown in Fig. 5-13. Because the spacing between the qth and $(q + 1)^{th}$ group of lines is c/2L, q labels the axial modes. For each axial mode, there are many transverse modes possible, labelled by (mn). The spacing between adjacent transverse modes depends on $g_1\, g_2$.

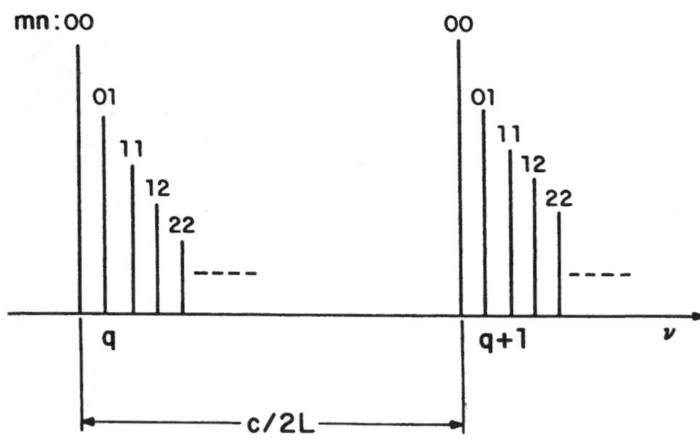

Fig. 5-13. Spatial-temporal modes of a cavity.

Closing comment:

No attempt is made here to give a detailed description of the modes inside a cavity based on the solution of the Maxwell's equations. It is long and tedious and is beyond the scope of this book. Readers who are interested in the details can consult any advanced laser textbook.

CHAPTER VI

OPTICAL ANISOTROPY IN A LOSSLESS MEDIUM

We now come to a basic physical phenomenon, namely, optical anisotropy, that underlies a large number of passive as well as active optical elements in modern optics. These comprise elements such as: electro-optical modulators, acousto-optical modulator, nonlinear optical elements, wave plates (quarter, half, full waves etc.), Faraday rotators, birefringent filters etc. Because this basic phenomenon is so important for the understanding of so many modern optical operations, it is necessary to have a thorough discussion of the physical process. The discussion would involve quite a bit of tedious calculation. Quite often, at the end of a set of calculations, the reader is lost in the mathematics and has to review what he is up to. As such, the author, for pedagogical reasons tries to pursue the discussion in a way which is not so systematic in the expert's eyes. The reader will be reminded again and again at what he is looking for and where he is going.

When we send an electromagnetic wave into a medium, the electric response of the material is fundamentally an induced polarization (dipole moment/volume) generated by the electric field of the E-M wave. Assuming an isotropic medium,

$$\vec{P} = \epsilon_o \chi \vec{E}$$

where χ is the susceptibility, $\vec{P}$ and $\vec{E}$ are the polarization and electric field respectively, in a very small region of the medium. The usual definition of the field $\vec{D}$ is

$$\vec{D} \equiv \vec{P} + \epsilon_o \vec{E}$$
$$\equiv \epsilon \vec{E}$$

where ϵ is the dielectric constant. It is related to the refractive index n and the E-M wave's propagation velocity v by

$$n^2 = \epsilon/\epsilon_o$$
$$v = \frac{c}{n} = \frac{c}{\sqrt{\epsilon/\epsilon_o}}$$

Thus, once we know v and $\vec{E}$, we can calculate $\vec{P}$. Because we usually measure n and/or v, but not $\vec{P}$, we would say that the response of the medium is known if we know n and/or v.

In this chapter, the medium is assumed anisotropic. Thus the induced polarization is also anisotropic. We shall ask what the response of the anisotropic medium to an E-M wave is and analyse this response in terms of, not $\vec{P}$, but the various n's and/or v's together with the associated fields $\vec{D}$'s and $\vec{E}$'s.

§6.1 Optical anisotropy

In the interaction between radiation and matter, the response of the material medium is not always isotropic. That is to say, the response depends on the direction of propagation as well as the polarization of the radiation. We assume that only the electric response of the medium is anisotropic while the magnetic response is always isotropic. This is reasonable (apart from some exceptions, see §10.1, Optical Activity) even for laser fields strong enough to damage the material, because

$$E = cB \tag{6-1}$$

where E is the electric field; B, the associated magnetic field of the radiation and c, the speed of light; i.e. E » B. (If the laser field is very strong, molecular dissociation, vaporization and plasma formation, etc. (damage) will take place. Even though the magnetic field of the laser may now become significant, its effect on the initial material will not be seen any more because the material is already damaged.) With this in mind, we consider an electrically anisotropic medium and make the following statement.

The dielectric constant of the anisotropic medium is not a constant any more, but a tensor. This dielectric tensor is symmetric. In mathematical form, it is

$$\epsilon = \begin{pmatrix} \epsilon_{11} & \epsilon_{12} & \epsilon_{13} \\ \epsilon_{21} & \epsilon_{22} & \epsilon_{23} \\ \epsilon_{31} & \epsilon_{32} & \epsilon_{33} \end{pmatrix} \tag{6-2}$$

and

$$\epsilon_{ij} = \epsilon_{ji} \qquad \text{(symmetric condition)} \tag{6-3}$$

Confirmation

The dielectric constant describes the response of the material to the electric field of the radiation. Under the influence of an electric field, the medium will be polarized. The induced polarization $\vec{P}$ (dipole moment per unit volume), the displacement $\vec{D}$ and the electric field $\vec{E}$ are related by

$$\vec{D} = \epsilon_o \vec{E} + \vec{P} \tag{6-4}$$

In an isotropic medium, $\vec{D}$, $\vec{E}$ and $\vec{P}$ are parallel to one another while in an anisotropic medium, they are not. Also,

$$\vec{P} = \epsilon_o \chi \vec{E} \tag{6-5}$$

where χ is the electric susceptibility tensor. In matrix form, eq. (6-5) becomes

$$\begin{pmatrix} P_1 \\ P_2 \\ P_3 \end{pmatrix} = \epsilon_o \begin{pmatrix} \chi_{11} & \chi_{12} & \chi_{13} \\ \chi_{21} & \chi_{22} & \chi_{23} \\ \chi_{31} & \chi_{32} & \chi_{33} \end{pmatrix} \begin{pmatrix} E_1 \\ E_2 \\ E_3 \end{pmatrix} \tag{6-6}$$

Substituting into eq. (6-4), we have

$$\begin{pmatrix} D_1 \\ D_2 \\ D_3 \end{pmatrix} = \epsilon_o \begin{pmatrix} E_1 \\ E_2 \\ E_3 \end{pmatrix} + \epsilon_o \begin{pmatrix} \chi_{11} & \chi_{12} & \chi_{13} \\ \chi_{21} & \chi_{22} & \chi_{23} \\ \chi_{31} & \chi_{32} & \chi_{33} \end{pmatrix} \begin{pmatrix} E_1 \\ E_2 \\ E_3 \end{pmatrix}$$

$$= \epsilon_o \begin{pmatrix} 1+\chi_{11} & \chi_{12} & \chi_{13} \\ \chi_{21} & 1+\chi_{22} & \chi_{23} \\ \chi_{31} & \chi_{32} & 1+\chi_{33} \end{pmatrix} \begin{pmatrix} E_1 \\ E_2 \\ E_3 \end{pmatrix} \tag{6-7}$$

Definition: <u>dielectric tensor</u> ϵ

$$\epsilon \equiv \epsilon_o \begin{pmatrix} 1+\chi_{11} & \chi_{12} & \chi_{13} \\ \chi_{21} & 1+\chi_{22} & \chi_{23} \\ \chi_{31} & \chi_{32} & 1+\chi_{33} \end{pmatrix} \tag{6-8}$$

$$\equiv \begin{pmatrix} \epsilon_{11} & \epsilon_{12} & \epsilon_{13} \\ \epsilon_{21} & \epsilon_{22} & \epsilon_{23} \\ \epsilon_{31} & \epsilon_{32} & \epsilon_{33} \end{pmatrix} \tag{6-9}$$

Substituting eq. (6-8) and (6-9) into (6-7),

$$\begin{pmatrix} D_1 \\ D_2 \\ D_3 \end{pmatrix} = \begin{pmatrix} \epsilon_{11} & \epsilon_{12} & \epsilon_{13} \\ \epsilon_{21} & \epsilon_{22} & \epsilon_{23} \\ \epsilon_{31} & \epsilon_{32} & \epsilon_{33} \end{pmatrix} \begin{pmatrix} E_1 \\ E_2 \\ E_3 \end{pmatrix} \tag{6-10}$$

or

$$D_i = \sum_{i=j}^{3} \epsilon_{ij} E_j$$

This explains eq. (6-2). **We still need to prove** $\epsilon_{ij} = \epsilon_{ji}$.

Assumptions:

a) ϵ_{ij} $(i,j = 1,2,3)$ **are real.** (Complex values of ϵ_{ij} are possible. This would lead to complex indices of refraction resulting in the absorption by the medium. Thus, this assumption of real ϵ_{ij} is equivalent to assuming a lossless or non-absorbing medium.)

b) The medium is homogeneous and magnetically isotropic.

c) The medium is linear, i.e. ϵ_{ij}'s do not depend on $\vec{E}$.

Now, two of the Maxwell's equations in the medium without a current are

$$\nabla \times \vec{H} = \frac{\partial \vec{D}}{\partial t} \tag{6-12}$$

$$\nabla \times \vec{E} = -\frac{\partial \vec{B}}{\partial t} \tag{6-13}$$

where

$$\vec{B} = \mu \vec{H} \tag{6-14}$$

$$\mu \approx \mu_o$$

Because of the assumption that the medium is isotropic magnetically, μ is a constant real number.

$\vec{E} \bullet$ [eq. (6-12)] gives

$$\vec{E} \bullet (\nabla \times \vec{H}) = \vec{E} \bullet \left(\frac{\partial \vec{D}}{\partial t} \right) \tag{6-15}$$

NOW, $$\nabla \bullet (\vec{E} \times \vec{H}) = \vec{H} \bullet (\nabla \times \vec{E}) - \vec{E} \bullet (\nabla \times \vec{H}) \tag{6-16}$$

which is a vector identity. Substituting eq. (6-16) into eq. (6-15), we have

$$\vec{H} \bullet (\nabla \times \vec{E}) - \nabla \bullet (\vec{E} \times \vec{H}) = \vec{E} \bullet \left(\frac{\partial \vec{D}}{\partial t} \right) \tag{6-17}$$

Since $\vec{H} \bullet (\nabla \times \vec{E}) = \vec{H} \bullet \left(-\frac{\partial \vec{B}}{\partial t} \right)$ (From eq. (6-13)

eq. (6-17) becomes

$$-\nabla \bullet (\vec{E} \times \vec{H}) = \vec{H} \bullet \left(\frac{\partial \vec{B}}{\partial t} \right) + \vec{E} \bullet \left(\frac{\partial \vec{D}}{\partial t} \right) \tag{6-18}$$

and since $\vec{S} \equiv$ Poynting vector $\equiv \vec{E} \times \vec{H}$

eq. (6-18) becomes $$-\nabla \bullet \vec{S} = \mu \vec{H} \bullet \left(\frac{\partial \vec{H}}{\partial t} \right) + \vec{E} \bullet \left(\frac{\partial \vec{D}}{\partial t} \right) \tag{6-19}$$

But $$-\nabla \bullet \vec{S} = \frac{dW}{dt} \equiv \text{rate of change of EM energy density} \tag{6-20}$$

which is simply a continuity equation for energy flow.

i.e. the rate of change of electromagnetic energy density is equal to the out flow of energy flux $(-\nabla \bullet \vec{S})$.

Since
$$W = W_{electric} + W_{magnetic} \equiv W_e + W_m$$
$$= \frac{1}{2}\vec{E} \bullet \vec{D} + \frac{1}{2}\vec{H} \bullet \vec{B} \tag{6-21}$$

Note that $\vec{E}$ is not parallel to $\vec{D}$ while $\vec{B} = \mu\vec{H}$ (μ = constant)

i.e. $\vec{B}$ is parallel to $\vec{H}$

From eq. (6-21),

$$\therefore \quad \frac{dW}{dt} = \frac{d}{dt}\left\{ \frac{1}{2} \sum_{i,j=1}^{3} E_i \,\epsilon_{ij}\, E_j \right\} + \frac{1}{2}\frac{d\vec{H}}{dt} \bullet (\mu\vec{H}) + \frac{1}{2}\vec{H} \bullet \frac{d}{dt}(\mu\vec{H})$$

$$\frac{dW}{dt} = \frac{1}{2}\left[\sum_{i,j} \dot{E}_i \,\epsilon_{ij}\, E_j + \sum_{i,j} E_i \,\epsilon_{ij}\, \dot{E}_j \right] + \mu\vec{H} \bullet \left(\frac{d\vec{H}}{dt} \right) \tag{6-22}$$

Comparing eq. (6-19), (20) and (22), we have, by equating the right hand side of eq. (6-19) and that of (6-22),

$$\frac{1}{2}\left[\sum_{ij} \dot{E}_i \,\epsilon_{ij}\, E_j + \sum_{ij} E_i \,\epsilon_{ij}\, \dot{E}_j \right] = \vec{E} \bullet \frac{\partial \vec{D}}{\partial t}$$
$$= \sum_i E_i \frac{\partial}{\partial t}\left(\sum_j \epsilon_{ij}\, E_j \right)$$
$$= \sum_{ij} E_i \,\epsilon_{ij}\, \dot{E}_j$$
$$\frac{1}{2} \sum_{ij} \dot{E}_i \,\epsilon_{ij}\, E_j = \frac{1}{2} \sum_{ij} E_i \,\epsilon_{ij}\, \dot{E}_j \tag{6-23}$$

Exchanging i and j on the right hand side of eq. (6-23) doesn't change its value. Hence, eq. (6-23) becomes

$$\frac{1}{2} \sum_{ij} \dot{E}_i \,\epsilon_{ij}\, E_j = \frac{1}{2} \sum_{ij} E_j \,\epsilon_{ji}\, \dot{E}_i$$

Thus
$$\epsilon_{ij} = \epsilon_{ji} \tag{6-24}$$

which proves the symmetric condition of eq. (6-3).

§6.2 Electromagnetic wave interaction with an anisotropic medium (general considerations)

We assume that the anisotropic medium is lossless (hence transparent) to the electromagnetic wave in question. Thus, ϵ_{ij}'s are real. Let a linearly polarized, monochromatic, plane EM wave propagate in the direction $\hat{k}$ inside the medium. The transverse electric field of the EM wave will induce a polarization $\vec{P}$ given by eq. (6-6) which re-radiates.

Statement (1): Given a propagation direction the re-radiated EM waves consist of two monochromatic plane waves with two different phase velocities governed by the surface of wave normals. i.e. there is a unique index of refraction for each of the two waves. When $\hat{k}$ is along some specific (generally two) directions, (called optic axes) only one secondary wave is re-radiated. These are shown schematically in Fig. 6-1. The phenomenon is called birefringence.

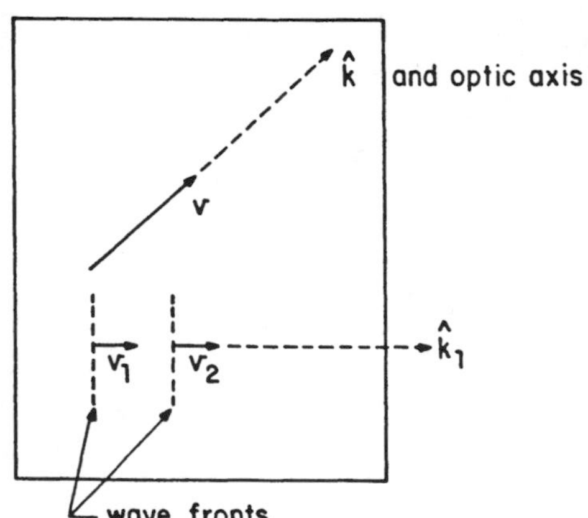

Fig. 6-1. An E-M wave entering an anisotropic medium in an arbitrary direction $\hat{k}_1$ induces two waves propagating at two different velocities v_1 and v_2, whereas in the direction of the optic axis, only one wave propagates at a velocity v.

Confirmation

(1) Dielectric tensor

By a suitable transformation, the dielectric tensor can be transformed into a diagonal form (in its so-called principal coordinate system, c.f. any 3D analytical geometry or matrix algebra),

i.e.
$$\epsilon = \begin{pmatrix} \epsilon_x & 0 & 0 \\ 0 & \epsilon_y & 0 \\ 0 & 0 & \epsilon_z \end{pmatrix} \quad (6\text{-}25)$$

Henceforth, we name the x - y - z co-ordinates the principal dielectric axes.

(2) Wave representation We assume that the EM wave is a plane wave interacting with the material inside the anisotropic medium.

i.e.
$$\vec{E}(\vec{r}, t) = \vec{E} \exp[i(\omega t - \vec{k} \bullet \vec{r})] \quad (6\text{-}26)$$

$$\vec{H}(\vec{r}, t) = \vec{H} \exp[i(\omega t - \vec{k} \bullet \vec{r})] \quad (6\text{-}27)$$

(3) Wave equation We rewrite two of the four Maxwell's equations:

$$\nabla \times \vec{E} + \frac{\partial \vec{B}}{\partial t} = 0 \qquad \text{Faraday's law} \quad (6\text{-}28)$$

$$\nabla \times \vec{H} - \frac{\partial \vec{D}}{\partial t} = 0 \qquad \text{Modified Ampere's law} \quad (6\text{-}29)$$

Substituting eq. (6-26) and (6-27) into (6-28) and (6-29), and using the following consequences when operating on plane waves:

$$\nabla = - i\vec{k} \quad (6\text{-}30)$$

$$\frac{\partial}{\partial t} = i\,\omega \quad (6\text{-}31)$$

(Exercise: the reader should try to prove eq. (6-30) and (6-31))

we have

$$\vec{k} \times \vec{E} = \omega\mu\vec{H} \quad (6\text{-}32)$$

$$\vec{k} \times \vec{H} = - \omega\epsilon\vec{E} \quad (6\text{-}33)$$

$\vec{k}$ x eq. (6-32) gives

$$\vec{k} \times \vec{k} \times \vec{E} = \omega\mu\vec{k} \times \vec{H}$$

$$= \omega\mu(-\ \omega\epsilon)\vec{E} \quad \text{using eq. (6-33)}$$

$$\vec{k} \times \vec{k} \times \vec{E} + \omega^2\mu\epsilon\vec{E} = 0 \tag{6-34}$$

Equation (6-34) is the equivalent of the wave equation. Substituting eq. (6-25) into (6-34), after some lengthy calculation, one obtains an equation $\omega(\vec{k})$ = constant in the k-space (i.e. wave vector space). This equation is called Fresnel's equation of wave normals. In the 3-D k-Space it represents two sheets of connected surfaces (see Fig. 6-2), for a constant ω. Along a general direction of propagation $\hat{k}$, the wave vector $\vec{k}$ will intersect the two surfaces, giving two solutions of k. Since $k = \frac{n\omega}{c} = \frac{\omega}{v}$, two solutions of k means two values of v, the phase velocity, or two indices of refraction. This is what has just been stated.

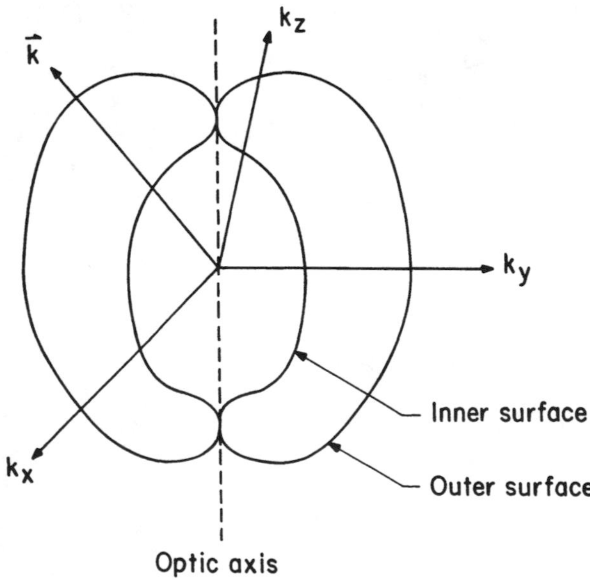

Fig. 6-2. A cross section of the surface of wave normals.

(4) Fresnel's equation of wave normals We now demonstrate two methods of calculation that lead to the Fresnel's equation of wave normals (i.e. $\omega(\vec{k})$ = constant).

Method (a)

Since $\vec{a} \times (\vec{b} \times \vec{c}) = (\vec{a} \cdot \vec{c})\vec{b} - (\vec{a} \cdot \vec{b})\vec{c}$

which is a vector identity, eq. (6-34) becomes

$$(\vec{k} \cdot \vec{E})\vec{k} - k^2 \vec{E} + \omega^2 \mu\epsilon\vec{E} = 0 \qquad (6\text{-}35)$$

In component form,

$$(\vec{k} \cdot \vec{E})k_i - k^2 E_i + \omega^2 \mu\epsilon_i E_i = 0 \qquad (6\text{-}36)$$

$$(i - 1,2,3 \text{ or } x,y,z)$$

$$E_i = \frac{k_i (\vec{k} \cdot \vec{E})}{k^2 - \omega^2 \mu\epsilon_i} \qquad (6\text{-}37)$$

Now, k_i x (eq. 6-37), and using eq. (6-25) gives

$$k_x E_x = \frac{k_x^2 (\vec{k} \cdot \vec{E})}{k^2 - \omega^2 \mu\epsilon_x} \qquad (6\text{-}38)$$

$$k_y E_y = \frac{k_y^2 (\vec{k} \cdot \vec{E})}{k^2 - \omega^2 \mu\epsilon_y} \qquad (6\text{-}39)$$

$$k_z E_z = \frac{k_z^2 (\vec{k} \cdot E)}{k^2 - \omega^2 \mu\epsilon_z} \qquad (6\text{-}40)$$

Adding eq. (6-38), (6-39) and (6-40),

$$k_x E_x + k_y E_y + k_z E_z = (k \cdot \vec{E}) \left\{ \frac{k_x^2}{k^2 - \omega^2 \mu\epsilon_x} + \frac{k_y^2}{k^2 - \omega^2 \mu\epsilon_y} + \frac{k_y^2}{k^2 - \omega^2 \mu\epsilon_z} \right\}$$

i.e. $\vec{k} \cdot \vec{E} = (\vec{k} \cdot \vec{E}) \{ \quad \}$

or $1 = \{ \quad \}$

Exchanging left and right sides,

$$\frac{k_x^2}{k^2 - \omega^2\mu\epsilon_x} + \frac{k_x^2}{k^2 - \omega^2\mu\epsilon_y} + \frac{k_x^2}{k^2 - \omega^2\mu\epsilon_z} = 1 \tag{6-41}$$

This is one form of the Fresnel's equation of wave normals, representing two sheets of connected surfaces in the 3-D k-space. Fig. 6-2 shows schematically a section of the surfaces with two symmetrical points of connection. In general, there are two more such points. The origin is the point where the material re-radiates. Fig. 6-3 shows a 3-D version that is cut into two halves. If the $\vec{k}$ vector (direction of propagation) is along the line joining a pair of connected points, there is only one solution of k; hence one phase velocity or one index of refraction. This direction is called the <u>optic axis</u>. There are two such axes in general. Now, if we define a unit vector, $\vec{s}$, in the k-space such that its components are the direction cosines of $\hat{k}$;

i.e.
$$\vec{s} \equiv \hat{k} = s_x\,\hat{x} + s_y\,\hat{y} + s_z\,\hat{z} \tag{6-42}$$

where s_x, s_y and s_z are the direction cosines; using

$$\vec{k} = \frac{2\pi}{\lambda}\vec{s} = \frac{2\pi}{v/\nu}\vec{s} = \frac{2\pi\nu}{c/n}\vec{s} = \frac{\omega}{c}n\vec{s} \tag{6-43}$$

we shall transform eq. (6-41) into another form.

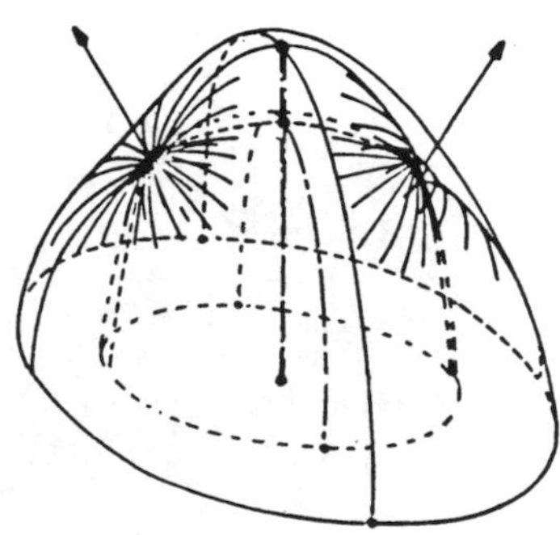

Fig. 6-3. A 3-D view of half the surface of wave normals. The two arrows indicate the optic axes.

Since
$$\vec{k}\cdot\vec{k} = k^2 = \frac{\omega^2}{c^2}\, n^2\, \vec{s}\cdot\vec{s} = \frac{\omega^2 n^2}{c^2} \tag{6-44}$$

$$k_i = \frac{\omega}{c}\, n\, s_i \quad (i = x,y,z) \tag{6-45}$$

the first term on the left hand side of eq. (6-41) becomes

$$\frac{k_x^2}{k^2 - \omega^2\mu\epsilon_x} = \frac{\frac{\omega}{c}\, n\, s_x \bullet \frac{\omega}{c}\, n\, s_x}{\frac{\omega^2}{c^2}\, n^2 - \omega^2\mu\epsilon_x}$$

$$= \frac{\frac{\omega^2 n^2}{c^2}\, s_x^2}{\frac{\omega^2}{c^2}\, n^2 - \frac{\omega^2\mu\epsilon_o}{\epsilon_o}\,\epsilon_x}$$

$$= \frac{\omega^2 n^2 s_x^2 \,/\, c^2}{\frac{\omega^2}{c^2}\, n^2 - \frac{\omega^2}{c^2}\,\frac{\epsilon_x}{\epsilon_o}}$$

$$= \frac{n^2\, s_x^2}{n^2 - \frac{\epsilon_x}{\epsilon_o}} \tag{6-46}$$

Similar expressions can be derived for the other two terms at the left hand side of eq. (6-41) which now becomes

$$\frac{s_x^2}{n^2 - \epsilon_x/\epsilon_o} + \frac{s_y^2}{n^2 - \epsilon_y/\epsilon_o} + \frac{s_z^2}{n^2 - \epsilon_z/\epsilon_o} = \frac{1}{n^2} \tag{6-47}$$

This is another form of the Fresnel's equation of wave normals showing the explicit dependence on the index n. It describes the same type of surfaces of two connected sheets in the s-(or k-) space, giving two solutions of n in a general direction $\vec{s}$, (cf. Fig. 6-2 and 6-3 with $\vec{k}$ changed into $\vec{s}$.) confirming again part of our statement at the beginning of this section.

Method (b)

This method is the direct substitution of eq. (6-25) into (6-34), writing the latter in the matrix form and solving the resulting equation. The latter requires that the determinant of the matrix be zero. Such a condition leads to, after simplification, eq. (6-41). The calculation is left as an exercise for the reader.

Statement 2: The direction of the electric field vector associated with the $\vec{k}$ vector is

$$\begin{pmatrix} \dfrac{k_x}{k^2 - \omega^2\mu\epsilon_x} \\ \dfrac{k_y}{k^2 - \omega^2\mu\epsilon_y} \\ \dfrac{k_z}{k^2 - \omega^2\mu\epsilon_z} \end{pmatrix}$$

Confirmation:

Eq. (6-37) can be rewritten as

$$\frac{E_i}{\vec{k}\cdot\vec{E}} = \frac{k_i}{k^2 - \omega^2\mu\epsilon_i} \quad (i = x,y,z) \qquad (6\text{-}48)$$

Hence,

$$E_x : E_y : E_z = \frac{E_x}{\vec{k}\cdot\vec{E}} : \frac{E_y}{\vec{k}\cdot\vec{E}} : \frac{E_z}{\vec{k}\cdot\vec{E}}$$

$$= \frac{k_x}{k^2 - \omega^2\mu\epsilon_x} : \frac{k_y}{k^2 - \omega^2\mu\epsilon_y} : \frac{k_z}{k^2 - \omega^2\mu\epsilon_z} \qquad \text{(using eq. 6-48)}$$

i.e. the direction of the vector

$$\vec{V} \equiv \begin{pmatrix} \dfrac{k_x}{k^2 - \omega^2 \mu \epsilon_x} \\ \dfrac{k_y}{k^2 - \omega^2 \mu \epsilon_y} \\ \dfrac{k_z}{k^2 - \omega^2 \mu \epsilon_z} \end{pmatrix} \equiv \begin{pmatrix} V_x \\ V_y \\ V_z \end{pmatrix} \tag{6-49}$$

is parallel to the direction of the electric field

$$\vec{E} \equiv \begin{pmatrix} E_x \\ E_y \\ E_z \end{pmatrix}$$

as shown in Fig. 6-4. Since there are two solutions of k (or n) for a fixed frequency ω and a fixed direction of propagation, $\vec{V}$ will have two solutions, $\vec{V}_1$, $\vec{V}_2$, say, and the corresponding electric field $\vec{E}_1$ and $\vec{E}_2$ will be parallel to $\vec{V}_1$ and $\vec{V}_2$, respectively.

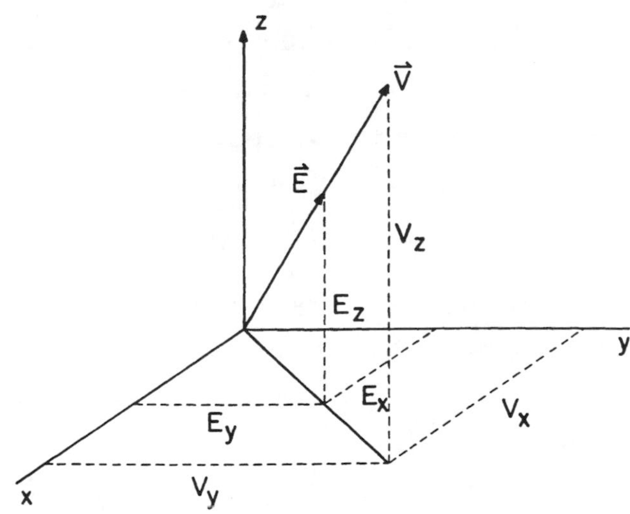

Fig. 6-4. Vector relationship showing the direction of the electric vector.

Statement 3: The electric field $\vec{E}$, the field $\vec{D}$ and the wave vector $\vec{k}$ (or $\vec{s}$) lie in the same plane and $\vec{D} \perp \hat{k}$.

Confirmation

$$\because \quad \nabla \cdot \vec{D} = 0 \quad \text{(One of Maxwell's equations)}$$

$$\therefore \quad - ik \cdot \vec{D} = 0 \quad (\because \nabla = - i\vec{k} \text{ for plane waves see eq. 6-30.})$$

$$\text{or} \quad \vec{D} \perp \hat{k} \tag{6-50}$$

This proves the last part of the statement.

Because of eq. (6-34) which we rewrite as follows:

$$\vec{k} \times \vec{k} \times \vec{E} + \omega^2\mu\epsilon\vec{E} = 0 \tag{6-51}$$

$$\hat{k} \times \hat{k} \times \vec{E} \;\; \propto - \epsilon\vec{E} = - \vec{D}$$

$$\therefore \quad - \vec{D} \parallel \hat{k} \times (\hat{k} \times \vec{E}) \tag{6-52}$$

Referring to Fig. 6-5(a), it shows pictorially the combination of eq. (6-50) and (6-52) in the horizontal plane.

Now $\qquad (\hat{k} \times \vec{E}) \;\perp\; \hat{k} \times (\hat{k} \times \vec{E})$

by inspection, i.e. $(\hat{k} \times \vec{E})$ is a vertical vector perpendicular to the horizontal plane, (Fig. 6-5(b)). But $\vec{E} \perp (\hat{k} \times \vec{E})$ and $\hat{k} \perp (\hat{k} \times \vec{E})$ by inspection. Hence $\vec{E}$ has to be in the same plane as $\hat{k}$ is, i.e. the horizontal plane (Fig. 6-5(c)). Hence, $\vec{D}$, $\vec{E}$, $\hat{k}$ lie in the same plane and $\vec{D} \perp \hat{k}$.

Statement 4: The electromagnetic energy propagates in a different (ray) direction than $\hat{k}$; i.e. the Poynting vector $\vec{S}$ doesn't coincide with the wave vector $\vec{k}$, but $\vec{S}$ is in the same plane as $\vec{k}$, $\vec{E}$ and $\vec{D}$.

Confirmation

We just remind ourselves that $\hat{k}$ is the direction of propagation of the wavefront (surface of constant phase) in the anisotropic medium.

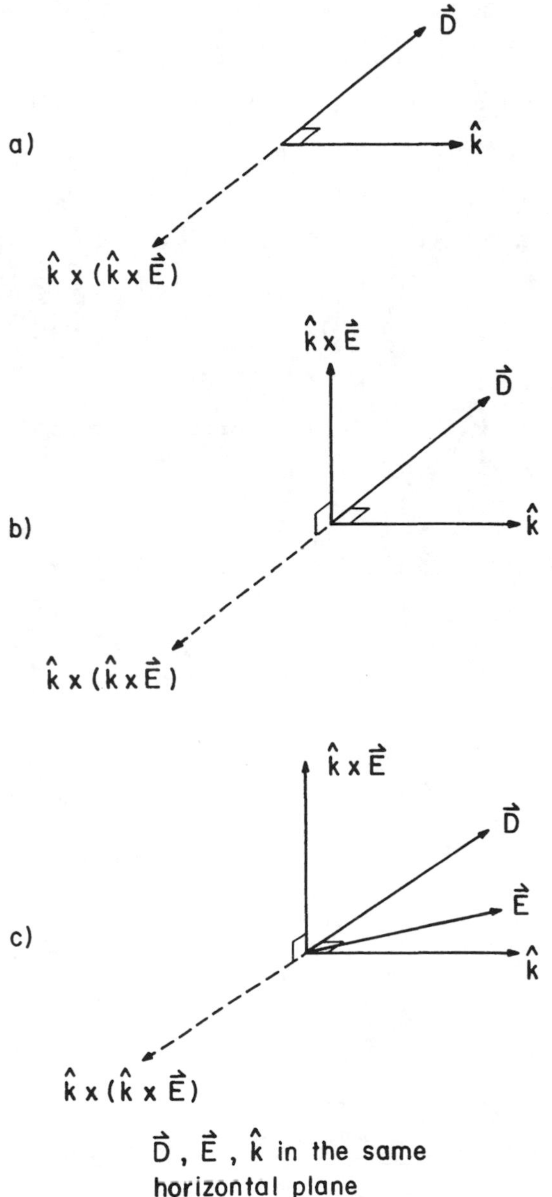

Fig. 6-5. Relationship between $\vec{D}$, $\hat{k}$, $\vec{E}$ and $\hat{k} \times \vec{E}$.

Normally, in an isotropic medium, $\vec{S} \parallel \vec{k}$. However, because $\vec{E}$ is not parallel to $\vec{D}$ in an anisotropic medium, $\vec{S}$ is not parallel to $\vec{k}$ either, as shown in Fig. 6-6. Fig. 6-6(a) shows the result of statement 3, i.e. they are drawn in the same horizontal plane.

Now,
$$H \perp \hat{k} \tag{6-53}$$

$$\text{(because } \nabla \cdot \vec{B} = 0$$

$$-i\vec{k} \cdot \vec{B} = 0 \qquad \text{(by eq. 6-30.)}$$

$$\vec{k} \perp \vec{B}$$

$$\text{or} \quad \vec{k} \perp \vec{H}$$

as shown in Fig. (6-6(a)) as the vertical vector. Since $\vec{S} = \vec{E} \times \vec{H}$, (definition of the Poynting vector)

$$\vec{S} \perp \vec{H}$$

Hence $\vec{S}$ has to be in the horizontal plane (Fig. 6-6(b)). In addition, $\vec{S} \perp \vec{E}$ from the definition of the Poynting vector. We can thus see from Fig. (6-6(b)) that

$$\alpha = \beta \ . \tag{6-54}$$

Conclusion: $\hat{S}$, $\hat{k}$, $\vec{E}$, $\vec{D}$ are in the same plane and the angle $(\hat{S}, \hat{k})$ is equal to the angle $(\vec{E}, \vec{D})$.

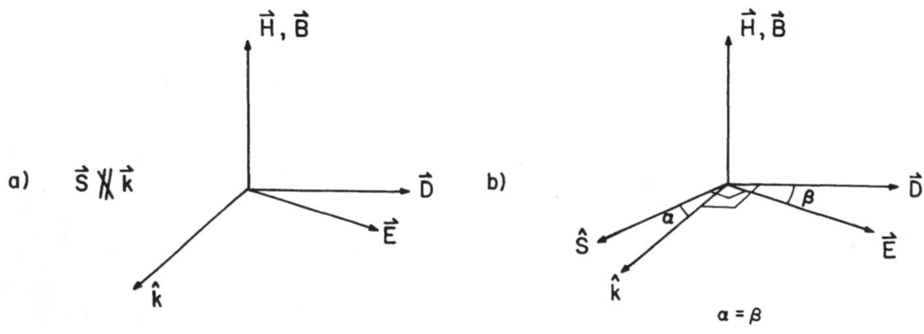

Fig. 6-6. Relationship between $\vec{H}$, $\vec{B}$, $\vec{D}$, $\vec{E}$, $\hat{k}$ and $\hat{S}$.

<u>Statement 5</u> The energy propagation velocity, usually referred to as the group or <u>ray velocity</u> ($\vec{v}_r$), makes an angle α with the wavefront propagation velocity or phase velocity ($\vec{v}_p$), and

$$\cos\alpha = \frac{v_p}{v_r} \tag{6-55}$$

<u>Confirmation</u>

We start with eq. (6-32) and (6-33) which are rewritten as follows:

$$\vec{k} \times \vec{E} = \omega\mu\vec{H} \tag{6-56}$$

$$\vec{k} \times \vec{H} = -\omega\epsilon\vec{E} = -\omega\vec{D} \tag{6-57}$$

We shall derive a relationship between the total energy density in the E-M field and the Poynting vector $\vec{S}$. This is then used in the definition of $\vec{v}_r$ leading to the desired result. The total energy density in the E-M field is the sum of the electric and magnetic energy densities:

$$W = W_e + W_m \qquad \text{(e: electric, m: magnetic)}$$

$$= \frac{1}{2}\vec{E}\cdot\vec{D} + \frac{1}{2}\vec{B}\cdot\vec{H}$$

$$= \frac{1}{2}\vec{E}\cdot\left(-\frac{\vec{k}}{\omega}\times\vec{H}\right) + \frac{1}{2}\mu\vec{H}\cdot\left(\frac{\vec{k}}{\omega\mu}\times\vec{E}\right) \quad \text{(using eq. (6-56) and (6-57)}$$

$$= -\frac{1}{2\omega}\vec{k}\cdot(\vec{H}\times\vec{E}) + \frac{1}{2\omega}\vec{k}\cdot(\vec{E}\times\vec{H}) \quad \text{(where use has been made of}$$

the vector identity: $\vec{a}\cdot(\vec{b}\times\vec{c}) = \vec{b}\cdot(\vec{c}\times\vec{a}) = \vec{c}\cdot(\vec{a}\times\vec{b})$). Continuing,

$$W = -\frac{1}{2\omega}\vec{k}\cdot(-\vec{S}) + \frac{1}{2\omega}\vec{k}\cdot\vec{S} \qquad (\because \vec{S} = \vec{E}\times\vec{H}) \tag{6-58'}$$

$$= \frac{\vec{k}\cdot\vec{S}}{\omega} \tag{6-58}$$

Now, the ray velocity (or group velocity) $\vec{v}_r$ is defined as the velocity of the transport of the total E-M energy;

$$\vec{v}_r \equiv \frac{\vec{S}}{W}$$

$$\equiv \frac{\text{(energy/sec) crossing a unit area } (\perp \vec{S})}{\text{W (total energy/vol.)}}$$

Hence
$$v_r = \frac{|\vec{S}|}{W}$$
$$= \frac{|\vec{S}|}{\frac{\vec{k}\cdot\vec{S}}{\omega}} \qquad \text{(by eq. 6-58)}$$
$$= \frac{\omega}{\vec{k}\cdot\hat{S}}$$
$$= \frac{\omega}{k\cos\alpha} \qquad \text{(see Fig. 6-6(b))}$$
$$= \frac{v_p}{\cos\alpha} \qquad \left(\frac{\omega}{k} = \frac{\omega}{\frac{\omega}{c}n} = \frac{c}{n} = v_p\right)$$

Hence
$$\cos\alpha = \frac{v_p}{v_r} \qquad \text{confirming eq. (6-55)}$$

Statement 6 In the $\vec{S}$ - space, there is another surface $\omega(\vec{S})$ = constant giving two group or ray velocities (v_r) or ray indices (n_r). This surface (again of two sheets) is called the ray surface and there is a rule of duality between the ray surface and the surface of wave normals (Statement 1). The origin of the ray surface is the re-radiating point.

Confirmation

We notice that eq. (6-35), namely
$$(\vec{k}\cdot\vec{E})\vec{k} - k^2\vec{E} + \omega^2\mu\epsilon\vec{E} = 0 \tag{6-59}$$
has led to the Fresnel equation of wave normals. We slightly transform eq. (6-59) into
$$\vec{D} = \epsilon\vec{E} = \frac{k^2}{\omega^2\mu}\left\{\vec{E} - \hat{k}\,(\hat{k}\cdot\vec{E})\right\} \tag{6-60}$$

We shall derive an analogous equation with $\vec{D}$ and $\vec{E}$ interchanged so as to obtain the rule of duality. This rule will "logically" give us the ray surface. We define the unit vector
$$\vec{t} \equiv \hat{S} \tag{6-61}$$

Using eq. (6-60), we have

$$\vec{D} \cdot \vec{t} = \frac{k^2}{\omega^2 \mu} \left\{ \vec{E} \cdot \vec{t} - (\hat{k} \cdot \vec{t})\ (\hat{k} \cdot \vec{E}) \right\}$$

$$= \frac{-k^2}{\omega^2 \mu} (\hat{k} \cdot \vec{t})\ (\hat{k} \cdot \vec{E})\quad (\because \vec{E} \perp \vec{t}) \tag{6-62}$$

Using again eq. (6-60) and (6-62),

$$\vec{D} - (\vec{D} \cdot \vec{t})\vec{t} = \frac{k^2}{\omega^2 \mu} \left\{ \vec{E} - \hat{k}\ (\hat{k} \cdot \vec{E}) + (\hat{k} \cdot \vec{t})\ (\hat{k} \cdot \vec{E})\vec{t} \right\}$$

$$= \frac{k^2}{\omega^2 \mu} \left\{ (E \cos \alpha)\vec{D} + (E \sin \alpha)\hat{k} \right.$$

$$\left. -(E \sin \alpha)\hat{k} + (\cos \alpha)\ (E \sin \alpha)\vec{t} \right\} \text{ (see Fig. (6-6(b))}$$

$$= \frac{k^2}{\omega^2 \mu} \left\{ (E\cos\alpha)\ (\cos\alpha)\hat{E} + (E\cos\alpha)\ (\sin\alpha)\ (-\vec{t}) + (E\cos\alpha\ \sin\alpha)\vec{t} \right\}$$

(see again Fig. 6-6(b))

$$= \frac{k^2}{\omega^2 \mu} \vec{E} \cos^2\alpha \tag{6-63}$$

But
$$\cos^2\alpha = \left(\frac{v_p}{v_r} \right)^2 = \left(\frac{(c/n)^2}{(c/n_r)^2} \right) = \frac{n_r^2}{n^2} \tag{6-64}$$

where n_r is the <u>ray index</u>

$$n_r = \frac{c}{n_r}$$

Hence, eq. (6-63) becomes

$$\vec{D} - (\vec{D} \cdot \vec{t})\vec{t} = \frac{k^2}{\omega^2 \mu} \vec{E} \left(\frac{n_r^2}{n^2} \right)$$

$$= \frac{\omega^2\ n^2/c^2}{\omega^2 \mu} \vec{E} \left(\frac{n_r^2}{n^2} \right)$$

i.e. $$\vec{E} = \frac{\mu c^2}{n_r^2}\left\{ \vec{D} - (\vec{D}\cdot\vec{t})\vec{t} \right\} \tag{6-65}$$

Comparing this equation with eq. (6-60), which is rewritten in the following form:

$$\vec{D} = \frac{n^2}{c^2\mu}\left\{ \vec{E} - (\vec{E}\cdot\hat{k})\hat{k} \right\} \tag{6-66}$$

we can write down the following <u>rule of duality</u> to transform from the $\vec{k}$ - space to the $\vec{S}$ - (or $\vec{t}$ -) space:

$$\begin{aligned} \vec{k}\text{ - space} &\rightarrow \vec{S}\text{ (or }\vec{t}\text{) space} \\ \vec{D} &\rightarrow \vec{E} \\ \frac{1}{\mu} &\rightarrow \mu \\ \frac{1}{c} &\rightarrow c \\ n &\rightarrow \frac{1}{n_r} \\ \hat{k} &\rightarrow -\vec{t} \end{aligned}$$

Since eq. (6-66) has led to the Fresnel equation of wave normals, the ray surface can be obtained by simply applying the above transformation rule to the equation of wave normals. For example, eq. (6-47):

$$\frac{s_x^2}{n^2 - \epsilon_x/\epsilon_o} + \frac{s_y^2}{n^2 - \epsilon_y/\epsilon_o} + \frac{s_z^2}{n^2 - \epsilon_z/\epsilon_o} = \frac{1}{n^2}$$

becomes

$$\frac{t_x^2}{\frac{1}{n_r^2} - \frac{1}{\epsilon_x\epsilon_o}} + \frac{t_y^2}{\frac{1}{n_r^2} - \frac{1}{\epsilon_y\epsilon_o}} + \frac{t_y^2}{\frac{1}{n_r^2} - \frac{1}{\epsilon_z\epsilon_o}} = n_r^2 \tag{6-68}$$

Eq. (6-68) is the equation of the ray surface. It again is a surface of two shells similar to Fig. (6-2) and (6-3), with $\vec{k}$ (or $\vec{s}$) replaced by $\vec{t}$. Thus for a general direction $\vec{t}$ emanating from the origin where the material re-radiates there are two solutions for n_r. (Note: there is a relationship between the ray surface and the surface of the wave normals namely that the latter is the pedal surface of the former. The reader is referred to Born and Wolf.)

<u>Statement 7</u> Using the index ellipsoid and the inverse (or Fresnel) ellipsoid, one can show that $\vec{D}_1 \perp \vec{D}_2$ and $\vec{E}_1 \perp \vec{E}_2$ respectively, where 1 and 2 refer to the two solutions of the E-M wave propagation in the anisotropic medium (governed by the surface of wave normals and the ray surface).

<u>Confirmation</u>

We shall first define the index ellipsoid and the inverse (or Fresnel) ellipsoid. The electric part of the energy density of the E-M wave is

$$W_e = \frac{1}{2}\vec{E}\cdot\vec{D}$$

$$2W_e = (E_x\ ,\ E_y\ ,\ E_z)\begin{pmatrix}\epsilon_x & & 0\\ & \epsilon_y & \\ 0 & & \epsilon_z\end{pmatrix}\begin{pmatrix}E_x\\ E_y\\ E_z\end{pmatrix}$$

$$= \epsilon_x E_x^2 + \epsilon_y E_y^2 + \epsilon_z E_z^2 \quad \text{(\underline{inverse} or \underline{Fresnel Ellipsoid})} \tag{6-69}$$

$$= \frac{D_x^2}{\epsilon_x} + \frac{D_y^2}{\epsilon_y} + \frac{D_z^2}{\epsilon_z} \qquad \text{(\underline{index ellipsoid})} \tag{6-70}$$

(a) <u>Index ellipsoid</u> or <u>optical indicatrix</u>)

Using the definitions:

$$\vec{r} \equiv (x,y,z)$$

$$\equiv \frac{\vec{D}}{\sqrt{2W_e\epsilon_o}} \tag{6-71}$$

$$n_i^2 = \frac{\epsilon_i}{\epsilon_o} \qquad (i = x,y,z) \tag{6-72}$$

eq. (6-70) becomes

$$\frac{x^2}{n_x^2} + \frac{y^2}{n_y^2} + \frac{z^2}{n_z^2} = 1 \tag{6-73}$$

which is the <u>index ellipsoid</u> with major axes $2n_x$, $2n_y$ and $2n_z$ along the x,y,z axes. In general, any point on the surface of the index ellipsoid is a solution of the index n or the field $\vec{D}$ because x, y, z are proportional to D_x , D_y , D_z (eq. (6-71)). However, as shown in Fig. 6-7(a) for a given direction of propagation $\hat{k}$ of the wavefront, the field $\vec{D}$ associated with the wave vector $\vec{k}$ should lie in the plane perpendicular to $\vec{k}$, because $\vec{D} \perp \vec{k}$ (eq. 6-50). This plane intersects the surface of the index ellipsoid in an ellipse, A B C. This ellipse is redrawn in Fig. (6-7(b)). Any vector

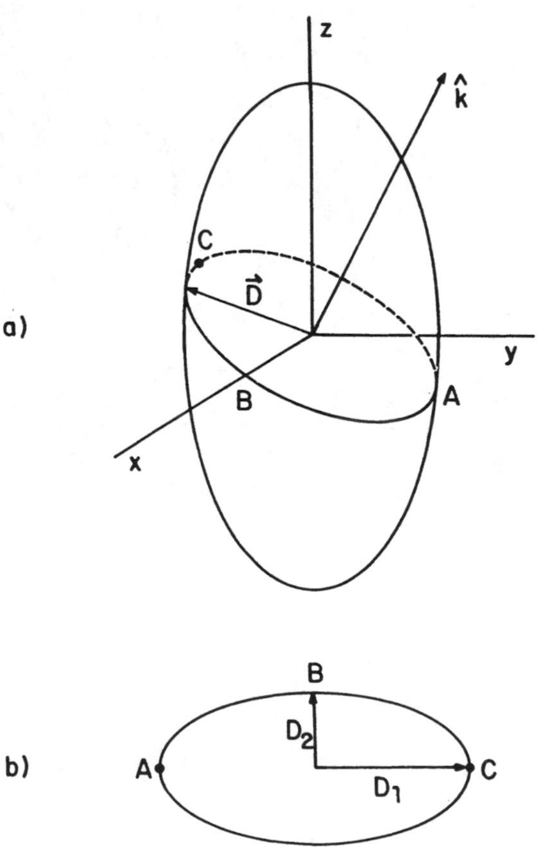

Fig. 6-7. Index ellipsoid.

drawn from the origin of the ellipsoid to the ellipse is a solution of $\vec{D}$ in the material in the infinitesimal region surrounding the origin. Because $\vec{D}$ is the response of the material to the travelling plane E-M wave, it is also a travelling wave of the same frequency (i.e. linear response), and the above solution means that the tip of the vector $\vec{D}$ traces out the ellipse; in other words, $\vec{D}$ is elliptically polarized. As will be seen in chapter VII, any elliptically polarized vector can be described by the vector combination of two orthogonal vector waves with two wave vectors $\vec{k}_1$ and $\vec{k}_2$ both propagating in the same direction of $\hat{k}$ (or $\hat{s}$) which we denote by the z' - direction here:

$$\vec{D}_1 = \vec{D}_{10} \cos (\omega t - k_1 z')$$

$$\vec{D}_2 = \vec{D}_{20} \cos (\omega t - k_2 z')$$

where z' is the position of the point of response (origin of the index ellipsoid) with respect to an arbitrary coordinate system. The above two equations can be rewritten in the more familiar form (used in chapter VII):

$$\vec{D}_1 = \vec{D}_{10} \cos (\omega t - k_1 z')$$

$$\vec{D}_2 = \vec{D}_{20} \cos (\omega t - k_1 z' + \epsilon)$$

where $\epsilon \equiv$ relative phase difference between $\vec{D}_1$ and $\vec{D}_2$

$$\equiv (k_1 - k_2) z'$$

$2D_{10}$ and $2D_{20}$ are the principal axes of the ellipse. Mathematically, it can indeed be rigorously proved that the lengths of the principal axes of the ellipse are the two solutions of the index n (or v, or k, or D) given by the surface of wave normals, for a propagation direction $\hat{k}$. The fields $\vec{D}$ corresponding to these two solutions are along the principal axes, and thus, are orthogonal (see Born and Wolf for detail.) We use the convention of right hand rotation so that $(\vec{D}_1 \times \vec{D}_2)$ points in the direction of $\hat{k}$. These two waves $\vec{D}_1$ and $\vec{D}_2$ are sometimes called the <u>normal</u> <u>modes</u> in the medium for a particular propagation direction $\hat{k}$. In all, one can thus say that for a given direction $\hat{k}$, there are two solutions of $\vec{D}$, namely $\vec{D}_1$, $\vec{D}_2$; and $\vec{D}_1 \perp \vec{D}_2$. Each of these fields ($\vec{D}_1$ and $\vec{D}_2$) propagates with a different velocity which is governed by the surface of wave normals

(b) Fresnel or inverse ellipsoid

Eq. (6-69) can be rewritten as

$$\frac{E_x^2}{\left(\sqrt{\frac{2W_e}{\epsilon_x}}\right)^2} + \frac{E_y^2}{\left(\sqrt{\frac{2W_e}{\epsilon_y}}\right)^2} + \frac{E_z^2}{\left(\sqrt{\frac{2W_e}{\epsilon_z}}\right)^2} = 1 \qquad (6\text{-}74)$$

Eq. (6-74) is an ellipsoid whose surface points give solutions to the electric field $\vec{E}$. However, as in the case of the index ellipsoid, for a given direction $\vec{t}$ (or $\hat{S}$), (see Fig. 6-8a), the solution of $\vec{E}$ should lie in the plane perpendicular to $\vec{t}$, since $\vec{E} \perp \hat{S}$ (or $\vec{t}$) by the definition of the Poynting vector. The solution becomes a vector from 0 terminating at the intersection ellipse, which is redrawn in Fig. 6-8(b). Similar to the case of $\vec{D}$, this electric field can be "decomposed" into two orthogonal fields $\vec{E}_1$ and $\vec{E}_2$ so that $\vec{E}_1 \times \vec{E}_2$ is in the direction of $\vec{t}$ (using the right hand convention). We again conclude that for a given direction $\vec{t}$, there are two solutions of $\vec{E}$, namely $\vec{E}_1$ and $\vec{E}_2$; and $\vec{E}_1 \perp \vec{E}_2$. Each of these fields ($\vec{E}_1$ and $\vec{E}_2$) propagates with a different velocity governed by the ray surface.

Statement 8 Given a direction $\hat{k}$ of the wave vector, the fields associated with $\vec{k}$ are such that ($\vec{D}_1$,$\vec{E}_1$) and ($\vec{D}_2$,$\vec{E}_2$) lie in two perpendicular planes with $\vec{k}$ as the intersection line between the planes.

Confirmation

This is easily seen in Fig. (6-9). According to statement 3, $\vec{D}$, $\vec{E}$, $\vec{k}$ lie the same plane. Hence. $\vec{D}_1$,$\vec{E}_1$,$\vec{k}$ lie in one plane, and $\vec{D}_2$,$\vec{E}_2$,$\vec{k}$ lie in another plane so that $\vec{k}$ is the intersection line of the two

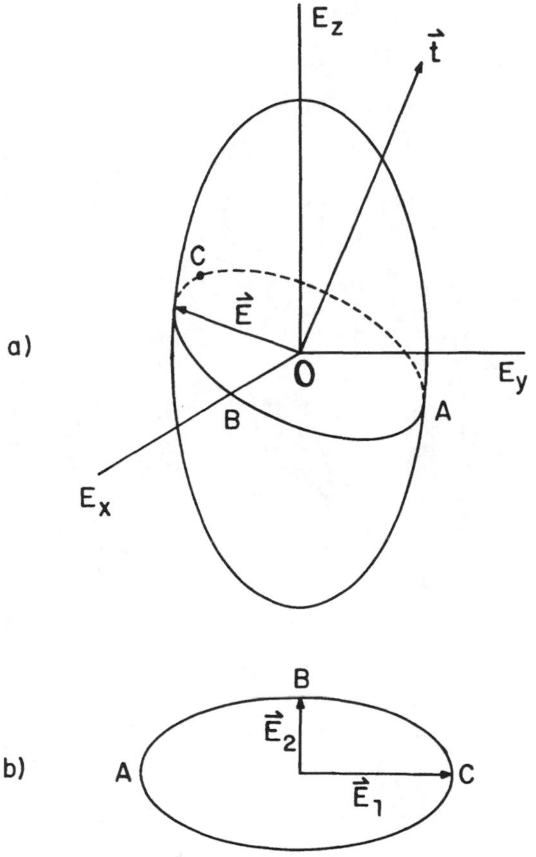

Fig. 6-8. Fresnel or inverse ellipsoid.

planes. Since $\vec{D}_1 \perp \vec{D}_2$ (statement 7, first part), the two planes must be perpendicular to each other.

Statement 9 Given a direction $\hat{S}$ (or $\vec{t}$) of the Poynting vector, the fields associated with $\vec{S}$ are such that $(\vec{E}_1\ ,\ \vec{D}_1)$ and $(\vec{E}_2\ ,\ \vec{D}_2)$ lie in two perpendicular planes with $\vec{t}$ (or $\hat{S}$) as the intersection line between the two planes.

<u>Confirmation</u>

As shown in Fig. (6-9(b)), since in general $\vec{t},\vec{D},\vec{E},\vec{k}$ lie in the same plane (statement 4), we should have $\vec{t},\vec{E}_1,\vec{D}_1$ lying in one plane and $\vec{t},\vec{E}_2,\vec{D}_2$ lying in another. Thus, $\vec{t}$ forms the intersection line between the two planes. Since $\vec{E}_1 \perp \vec{E}_2$ (statement 7), the two planes should be perpendicular.

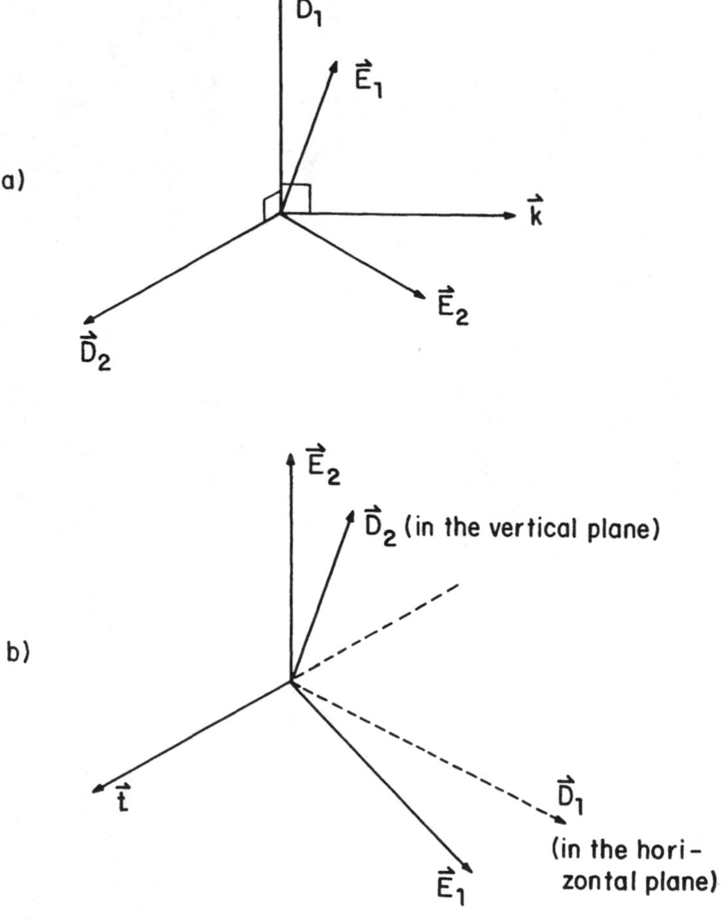

Fig. 6-9. Relationship between $\vec{E}_1$, $\vec{E}_2$, $\vec{D}_1$, $\vec{D}_2$, $\hat{k}$ and $\vec{t}$.

<u>Statement 10</u> $\vec{D}_1 \perp \vec{E}_2$

$\vec{D}_2 \perp \vec{E}_1$

$\hat{k} \cdot (\vec{E}_1 \times \vec{H}_2) = 0$

$\hat{k} \cdot (\vec{E}_2 \times \vec{H}_1) = 0$

<u>Confirmation</u>

This is left as an <u>exercise</u> for the reader. He merely has to draw the appropriate $\vec{H}$ vectors onto Fig. (6-9a) and the resultant figure will prove the above relationship.

<u>Statement 11</u> Given a wave vector $\hat{k}$, there are in general two $\vec{t}$'s lying in two planes perpendicular to each other with $\vec{k}$ as the intersection line of the two planes.

<u>Confirmation</u>

According to statement 8, given a wave vector $\vec{k}$, there are two pairs of (E_1 , D_1) and (E_2 , D_2) lying in two perpendicular planes with $\vec{k}$ as the intersection line. Since $\vec{t}$, $\vec{k}$, $\vec{E}$, $\vec{D}$ all lie in the same plane (statement 4), $(\vec{t}_1 , \vec{E}_1 , \vec{D}_1)$ and $(\vec{t}_2 , \vec{E}_2 , \vec{D}_2)$ should lie in two perpendicular planes with $\vec{k}$ as the intersection line. (Fig. 6-10(a)).

<u>Statement 12</u> Given a Poynting vector's direction $\vec{t}$, there are in general two $\vec{k}$'s lying in two planes perpendicular to each other with $\vec{t}$ as the intersection line of the two planes.

<u>Confirmation</u>

As in statement 11, using statement 9 and 4, the above statement can be seen easily (Fig. 6-10(b)).

<u>Summary</u>

Because of the anisotropy, the dielectric constant ϵ becomes a tensor (matrix) which gives different values of ϵ in different directions. The resultant fields $\vec{D}$ and $\vec{E}$ generated by the interaction of a plane E-M wave with the medium are not parallel. Maxwell's equations lead to the conclusion that when a plane E-M wave propagates in an anisotropic medium, it

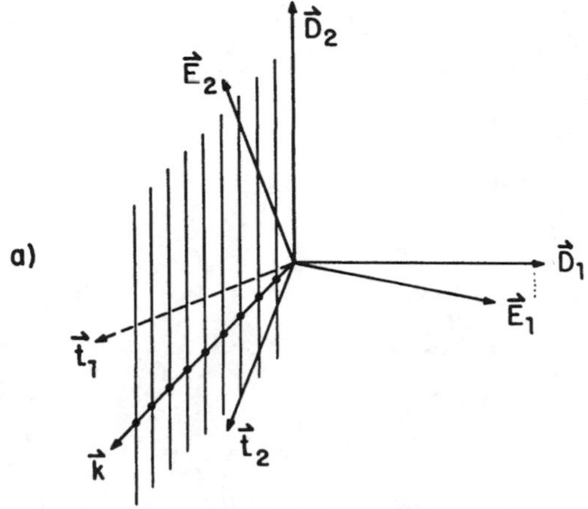

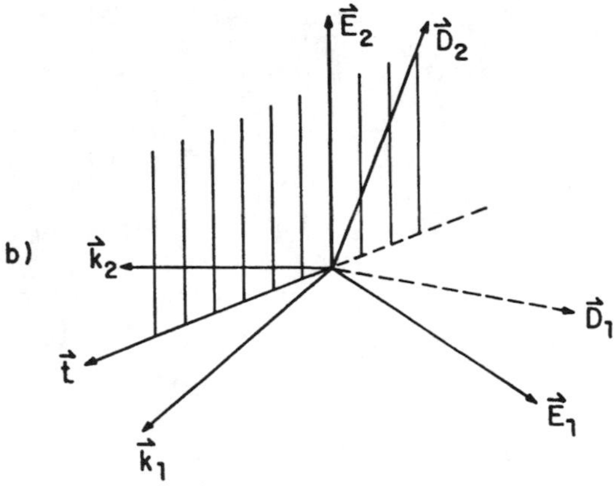

Fig. 6-10. Relationship between $\vec{D}$, $\vec{E}$, $\hat{k}$ and $\vec{t}$.

decomposes into two plane waves of different propagation velocities, each of them linearly polarized, and the fields perpendicular to each other. ($\vec{D}_1 \perp \vec{D}_2$ in the direction of $\vec{k}$ and/or $\vec{E}_1 \perp \vec{E}_2$ in the direction of $\vec{t}$ or $\hat{S}$).

§6.3 Classification of anisotropic material optically

Most anisotropic materials are crystals. It is the different types of crystal symmetries that give rise to the anisotropy. A full understanding of the matter requires a good knowledge of crystallography; we cannot go into such details because it would be outside the scope of this book. Rather, we assume that all crystals of interest have already been studied and classified according to the crystal specialists. The following optical classification of anisotropic materials thus ignores the crystal symmetry aspect.

Classification	ϵ_{ij}
isotropic	$\begin{pmatrix} \epsilon & & 0 \\ & \epsilon & \\ 0 & & \epsilon \end{pmatrix}$
uniaxial	$\begin{pmatrix} \epsilon_x & & 0 \\ & \epsilon_x & \\ 0 & & \epsilon_z \end{pmatrix}$
biaxial	$\begin{pmatrix} \epsilon_x & & 0 \\ & \epsilon_y & \\ 0 & & \epsilon_z \end{pmatrix}$

(1) Isotropic material

This is the normal material (dielectrics) one assumes in any book on electromagnetism. Its interaction with the E-M wave is well known.

(2) Uniaxial material

According to our definition,

$$\epsilon_y = \epsilon_x \tag{6-75}$$

We define

$$n_o^2 \equiv \frac{\epsilon_x}{\epsilon_o} \tag{6-76}$$

$$n_e^2 \equiv \epsilon_z/\epsilon_o \tag{6-77}$$

Substituting eq. (6-75 , 76, 77) into the equation of wave normals, eq. (6-41), and using

$$k^2 = k_x^2 + k_y^2 + k_z^2 \tag{6-78}$$

$$= \frac{\omega^2}{c^2} n^2 \tag{6-79}$$

One obtains (the detailed calculation is left as an exercise):

$$(n^2 - n_o^2) \left\{ (k_x^2 + k_y^2)\, n_o^2\, n^2 + k_z^2\, n_e^2\, n^2 - n_o^2\, n_e^2 \frac{\omega^2}{c^2} n^2 \right\} = 0 \tag{6-80}$$

This leads to two simultaneous solutions:

$$n^2 - n_o^2 = 0 \tag{6-81}$$

and

$$\frac{k_x^2 + k_y^2}{n_e^2} + \frac{k_z^2}{n_o^2} = \frac{\omega^2}{c^2} \tag{6-82}$$

Eq. (6-81) can be transformed into

$$k_x^2 + k_y^2 + k_z^2 \equiv k^2 = \frac{\omega^2 n_o^2}{c^2} \qquad \text{using eq. (6-79)}$$

or

$$\frac{k_x^2}{\frac{\omega^2 n_o^2}{c^2}} + \frac{k_y^2}{\frac{\omega^2 n_o^2}{c^2}} + \frac{k_z^2}{\frac{\omega^2 n_o^2}{c^2}} = 1 \tag{6-83}$$

And eq. (6-82) is rewritten:

$$\frac{k_x^2}{\frac{\omega^2 n_e^2}{c^2}} + \frac{k_y^2}{\frac{\omega^2 n_e^2}{c^2}} + \frac{k_z^2}{\frac{\omega^2 n_o^2}{c^2}} = 1 \tag{6-83'}$$

The simultaneous solutions in the $\vec{k}$ - space (wave normals) are thus a spherical surface (eq. (6-83)) and an ellipsoid (eq. 6-83')) whose axis of rotational symmetry is around the z - axis. Note that the length of the principal axis of the ellipsoid along the z - direction is equal to the radius of the sphere ($\omega n_o/c$). Thus, the ellipsoid and the sphere should touch. Depending on whether n_e is larger or smaller than n_o, we have the surface of the wave normals shown schematically in Fig. (6-11). We define the anisotropy as positive (negative) uniaxial if $n_e > (<) n_o$. n_o and n_e are defined as the indices of the ordinary (o) and extraordinary (e) waves. (We notice again that the origin is the re-radiating point.)

The polarizations of the ordinary and extraordinary waves can be investigated by using the index ellipsoid. Because it is uniaxial, we define that the indices in the x - and y - directions to be equal to n_o. We thus have the ellipsoid with the z - axis as the rotational symmetric axis shown in Fig. 6-12. Its equation is

$$\frac{D_x^2}{2\,\epsilon_x W_e} + \frac{D_y^2}{2\,\epsilon_x W_e} + \frac{D_z^2}{2\,\epsilon_z W_e} = 1 \quad \text{(From eq. 6-70)} \tag{6-84}$$

Because of the rotational symmetry, any propagation direction $\hat{k}$ of the wave vector is symmetric around z; i.e. for an arbitrary $\hat{k}$, one can choose the y - z plane to coincide with $\hat{k}$ without loss of generality. The y - z plane (containing $\hat{k}$ and the z - axis of symmetry) is called the principal plane. Any change in direction of $\hat{k}$ becomes the change in the angle Θ in the y - z plane. The ellipse formed by the intersection of the plane perpendicular to $\hat{k}$ and the index ellipsoid defines the field $\vec{D}$ having two components along the principal axes of the ellipse. Because this ellipse is symmetrical about the principal plane (y - z plane), one of the principal axes of the ellipse is always perpendicular to the principal plane and the other lies in the plane. We denote them by $\vec{D}_o$ and $\vec{D}_e$, respectively. Whatever the direction of $\hat{k}$ is (i.e. changing θ), $\vec{D}_o$ which is now along the x - axis in Fig. 6-12, is unchanged; hence, it is the isotropic ordinary wave. $\vec{D}_e$ is perpendicular to $\vec{D}_o$. Since $\vec{D}_o$ is isotropic,

$$\vec{D}_o \parallel \vec{E}_o$$

i.e. $\vec{E}_o$ is in the x - direction. By statement 3 of the previous section, $\hat{k}$, $\vec{D}_e$, $\vec{E}_e$ should lie in the same y - z plane. Hence $\vec{E}_o \perp \vec{E}_e$. That is to say,

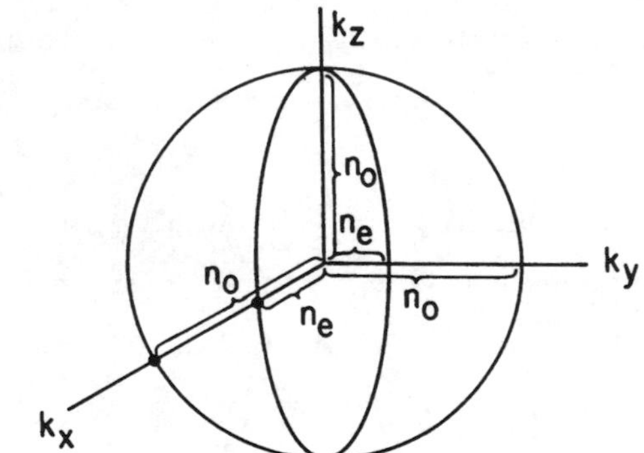

$n_e < n_o$ = negative uniaxial

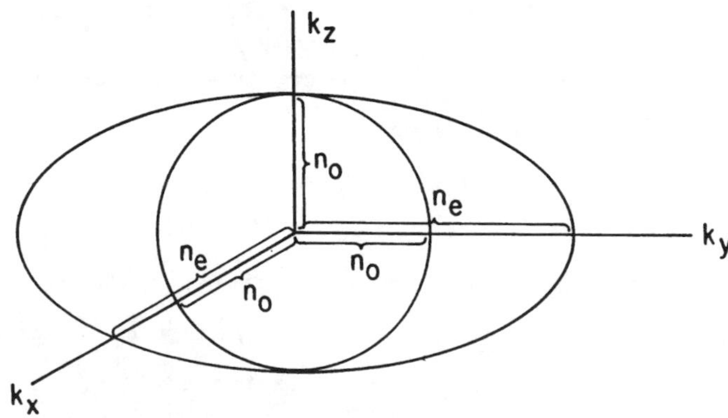

$n_e > n_o$ = positive uniaxial

Fig. 6-11. Surfaces of wave normals of uniaxial materials.

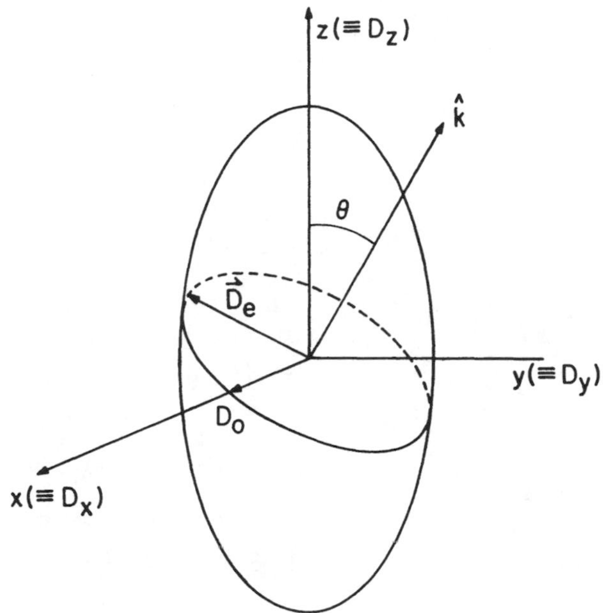

Fig. 6-12. Index ellipsoid of a uniaxial material.

when an E-M wave interacts with a uniaxial medium, it generates two waves with two different propagation velocities. Both the ordinary (isotropic) wave and the extraordinary waves are linearly polarized and the two polarizations $\vec{E}_o$ and $\vec{E}_e$ are orthogonal.

At $\Theta = 0$, ($\hat{k}$ coincides with the z - axis), $\vec{D}_e = \vec{D}_o$, because the ellipse becomes a circle; i.e. when the wave vector is in the z - direction, there is only one wave velocity. This means that the z - axis is the optic axis. <u>It also means that $\vec{E}_o$, the polarization of the ordinary wave, is always perpendicular to the optic axis</u>. The electric vector $\vec{E}_e$ lies in the y - z plane; i.e. in the principal plane containing $\vec{k}$ and the optic axis.

Because $\vec{D}_o \parallel \vec{E}_o$ in the ordinary wave, the angle between $\vec{D}_o$ and $\vec{E}_o$ is zero. This means that $\vec{k}_o \parallel \vec{S}_o$ (statements 4 and 5). However, the extraordinary wave will always have $\vec{D}_e \nparallel \vec{E}_e$; and hence $\vec{k}_e \nparallel \vec{S}_e$; i.e. the wave vector $\vec{k}_e$ is in a different direction than the Poynting vector $\vec{S}_e$.

(3) Biaxial material

This is the general case for optical anisotropy. We assume

$$\epsilon_x < \epsilon_y < \epsilon_z \tag{6-85}$$

$$(\text{or } v_x > v_y > v_z), \text{ since } \frac{\epsilon_x}{\epsilon_o} \equiv n_x^2 \text{ , and } v_x = \frac{c}{n_x} \tag{6-85}$$

In statement 1, we have mentioned a second method of deriving the Fresnel equation of wave normals, and have left it as an exercise for the readers. Those who have done this will first of all arrive at the following equation:

$$\begin{vmatrix} \omega^2\mu\epsilon_x - k_y^2 - k_z^2 & k_x k_y & k_x k_z \\ k_y k_x & \omega^2\mu\epsilon_y - k_x^2 - k_z^2 & k_y k_z \\ k_z k_x & k_z k_y & \omega^2\mu\epsilon_z - k_x^2 - k_y^2 \end{vmatrix} = 0 \tag{6-86}$$

Eq. (6-86) is essentially an equation of the wave normals in a "confused" form. We could examine it more closely by looking at the projections of the equation on the $k_x = 0$, $k_y = 0$, and $k_z = 0$ planes successively (i.e. on the planes perpendicular to the principal dielectric axes; see §6.2, statement 1.) We again leave it as an <u>exercise</u> to the reader to simplify eq. (6-86) under the above three separate conditions, thus obtaining the projections on the $k_x = 0$, $k_y = 0$ and $k_z = 0$ planes as shown in Fig. 6-13 a, b and c respectively. In (a), a circle lies inside an oval, in (b), a circle intersects an oval and in (c), an oval is inside a circle. We consider only the case in Fig. 6-13(b). Simplification of eq. (6-86) gives simultaneously two solutions:

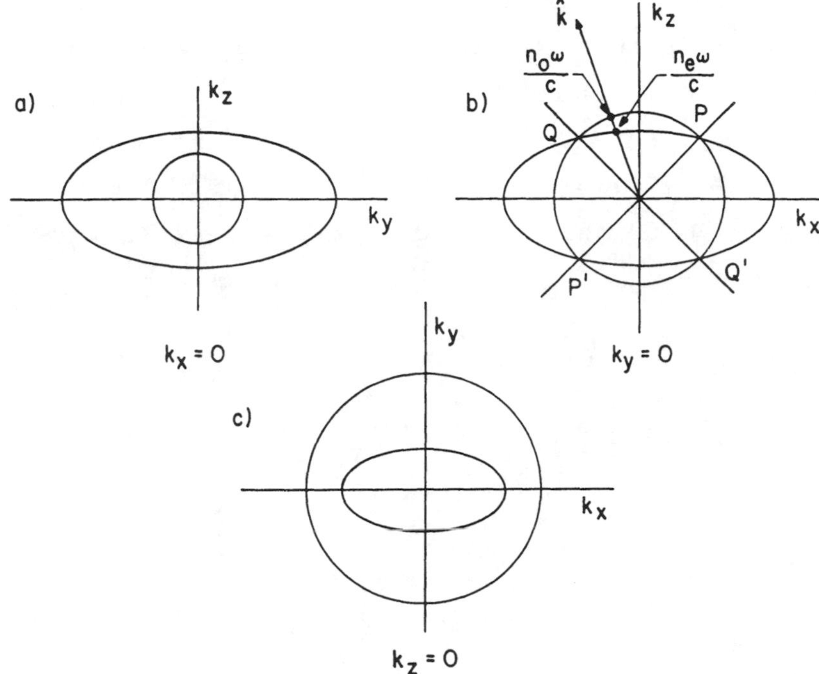

Fig. 6-13. Projections of the surface of wave normals of a biaxial material on the x-z, y-z and x-z planes.

$$\frac{k_x^2}{n_y^2} + \frac{k_z^2}{n_y^2} = \frac{\omega^2}{c^2} \tag{6-87}$$

and

$$\frac{k_x^2}{n_z^2} + \frac{k_z^2}{n_x^2} = \frac{\omega^2}{c^2} \tag{6-88}$$

Eq. (6-87) and (6-88) are represented in Fig. 6-13(b) by the circle and the ellipse respectively. (Note: an ellipse is a special case of an oval.) The two intersect because of our assumption that $\epsilon_x < \epsilon_y < \epsilon_z$ which means

$$n_x < n_y < n_z \quad \left(\because n_i^2 = \frac{\epsilon_i}{\epsilon_o}, \qquad i = x,y,z \right).$$

In a general direction of propagation $\hat{k}$, the two intersection points with the ellipse and with the circle give the values of the two wave vectors. For the intersection point with the circle, the direction of the polarization (direction of $\vec{E}_o$) of the wave associated with the wave vector $(n_o\omega/c)$ can be obtained from eq. (6-49) of statement 2 of the previous section, using $k_y = 0$. We see that the x - and z - components have finite values under the condition $k_y = 0$ while the y - component becomes

$$\frac{k_y}{k^2 - \omega^2\mu\epsilon_y} = \frac{k_y}{k^2 - \omega\ \mu\epsilon_o \frac{\epsilon_y}{\epsilon_o}}$$

$$= \frac{k_y}{k_x^2 + k_z^2 + k_y^2 - \frac{\omega^2}{c^2} n_y^2}$$

$$= \frac{k_y}{\frac{\omega^2}{c^2} n_y^2 + k_y^2 - \frac{\omega^2}{c^2} n_y^2} \qquad \text{(using eq. 6-87 for the circle)}$$

$$= \frac{k_y}{k_y^2}$$

$$= \frac{1}{k_y} \quad \rightarrow \infty \text{ as } k_y \rightarrow 0$$

Hence, the direction of $\vec{E}_o$ (or polarization) becomes

$$\begin{pmatrix} \text{finite} \\ \infty \\ \text{finite} \end{pmatrix}$$

i.e. the polarization for the isotropic wave $\vec{E}_o$ is in the y - directon; or perpendicular to the x - z plane. This also means that $\vec{E}_o$ is always perpendicular to the optic axes POP' and QOQ'. For the extraordinary wave, i.e. the intersection point of $\hat{k}$ and the ellipse (eq. 6-88), the same consideration leads to the following direction of $\vec{E}_e$: (using eq. (6-88) and (6-49))

$$\begin{pmatrix} \text{finite} \\ 0 \\ \text{finite} \end{pmatrix}$$

Hence, the polarization $\vec{E}_e$ of the extraordinary wave lies in the x - z plane. Since $\vec{E}_o$ is perpendicular to the x - z plane,

$$\vec{E}_o \perp \vec{E}_e \tag{6-89}$$

Similarly one can show that for propagation directions in the $k_x = 0$ and $k_z = 0$ planes, eq. (6-89) is always valid. (This is again left as an exercise.) We thus conclude that in a general direction of propagation $\vec{k}$ in a biaxial material, there are two waves that propagate with two different velocities. Their polarizations are always linear and orthogonal.

§6.4 Double refraction at a boundary

When a plane wave of wave vector $\vec{k}_i$ propagates from an isotropic medium into an anisotropic medium, it will generate two waves with different velocities of propagation (or different wave vectors). The reason why this happens can be seen from the following analysis.

Consider the point 0 in Fig. 6-14(a) where the incident wave vector $\vec{k}_i$ intersects the boundary separating the isotropic and anisotropic media. When interacting with the incident E-M wave, the material at the anisotropic side of the boundary will re-radiate according to the "rules" (or statements) discussed in §6.2. That is to say, the re-radiated wave vectors $\vec{k}'$ will terminate on the surface of wave normals governed by eq. (6-41), which is rewritten as follows:

$$\frac{k_x^2}{k'^2 - \omega^2\mu\epsilon_x} + \frac{k_y^2}{k'^2 - \omega^2\mu\epsilon_y} + \frac{k_z^2}{k'^2 - \omega^2\mu\epsilon_z} = 1$$

We use this particular form of the wave normals because we shall be dealing with the wave vectors rather than refractive indices or velocities. Fig. 6-14(a) shows half of the surface of wave normals inside the anisotropic medium. The surface of wave normals is of two sheets, of course.

Now, referring to Fig. 6-14(b), from an arbitrary reference frame 0', the point 0 is at the position $\vec{r}$. The boundary conditions at 0 require that

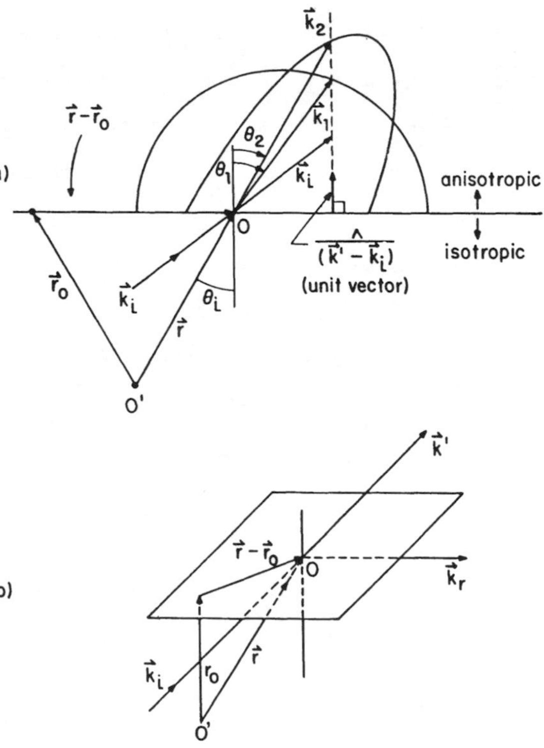

Fig. 6-14. Double refraction. The point 0 which should be at the common center of the half-ellipse and half-circle is purposely displaced leaving more space for the drawing.

the tangential components of $\vec{E}$ be continuous across the boundary;

i.e. $$(\vec{E}_i)_t \exp[j(\omega t - \vec{k}_i \cdot \vec{r})] + (\vec{E}_r)_t \exp[j(\omega t - \vec{k}_r \cdot \vec{r} + \epsilon_r)]$$
$$= (\vec{E}')_t \exp[j(\omega t - \vec{k}' \cdot \vec{r} + \epsilon')] \qquad 6\text{-}90)$$

where the subscript "t" means "tangential"; "i" and "r" mean "incident" and "reflected"; "'" (prime) means transmitted; $j = \sqrt{-1}$; ϵ is the relative phase difference of the reflected (ϵ_r) or transmitted (ϵ') wave with respect to the incident wave. Eq. (6-90) is valid only if all the phases are equal. In particular,

$$\omega t - \vec{k}_i \cdot \vec{r} = \omega t - \vec{k}' \cdot \vec{r} + \epsilon' \qquad (6\text{-}91)$$

i.e. $$\vec{r} \cdot (\vec{k}_i - \vec{k}') = \epsilon' = \text{constant} \equiv \vec{r}_o \cdot (\vec{k}_i - \vec{k}') \qquad (6\text{-}92)$$

where $\vec{r}_o$ is a constant vector terminating on the boundary plane at an arbitrary point. Eq. (6-92) which is the equation of a plane becomes

$$(\vec{r} - \vec{r}_o) \cdot (\vec{k}' - \vec{k}_i) = 0 \qquad (6\text{-}93)$$

Since $\vec{r}$ and $\vec{r}_o$ both terminate on the boundary plane, and $\vec{r}$ is a variable vector (varies as the position 0 varies), $(\vec{r} - \vec{r}_o)$ is a general vector in this plane. Eq. (6-93) means that the vector $(\vec{k}' - \vec{k}_i)$ is perpendicular to the boundary plane spanned by $(\vec{r} - \vec{r}_o)$. But the vector $(\vec{k}' - \vec{k}_i)$ intersects the surface of wave normals at two points (Fig. 6-14a) because $\vec{k}'$, the refracted wave vector, should terminate at the surface of wave normals. Hence, there will be two transmitted wave vectors $\vec{k}_1$, $\vec{k}_2$.

Again, according to eq. (6-92), we should have (Fig. 6-14a)

$$k_i \sin\Theta_i = k' \sin\Theta' \qquad (6\text{-}94)$$

where $$\Theta' = \begin{Bmatrix} \Theta_1 \\ \Theta_2 \end{Bmatrix} \qquad (6\text{-}95)$$

Rewriting (6-94):

$$k_i \sin\Theta_i = \begin{cases} k_1 \sin\Theta_1 \\ k_2 \sin\Theta_2 \end{cases} \tag{6-96}$$

which is the Snell's law for refraction at an anisotropic boundary. It should be noted that $|\vec{k}_2|$ depends on Θ_2 while $|\vec{k}_1|$, being isotropic, is independant of Θ_1. We stress that $\hat{k}_1$ and $\hat{k}_2$ denote the propagation direction of two wave fronts. The ray or the Poynting vector, namely the E - M energy transport directions, are those associated with $\hat{k}_1$ and $\hat{k}_2$ according to statements 4, 5 and 6 of section 6.2. We denote the two ray directions as $\vec{t}_1$ and $\vec{t}_2$, each having a fixed angular relationship with $\hat{k}_1$ and $\hat{k}_2$, respectively according to statement 5. Then, if $\vec{k}_i$ represents the direction of an incident laser beam, the two refracted beams that will be observed (i.e. energy transport) will be along $\vec{t}_1$ and $\vec{t}_2$, not $\hat{k}_1$ and $\hat{k}_2$. In the particular drawing in Fig. 6-14a, since $\vec{k}_1$ terminates on the spherical surface, it represents the isotropic wave. Hence $\vec{k}_1 \parallel \vec{t}_1$.

One can also show that $\vec{k}_1$, $\vec{k}_2$, $\vec{k}_i$ and $\vec{k}_r$ all lie in the same plane of incidence using the argument that is usually used in analyzing refraction across the interface of two isotropic media. We leave this as an exercise to the reader.

Finally, according to the discussion in section 6.3, the electric fields $\vec{E}_1$, $\vec{E}_2$ associated with $\vec{k}_1$, $\vec{k}_2$ are perpendicular, because E_1 belongs to the ordinary (isotropic) wave while $\vec{E}_2$ belongs to the extraordinary wave.

§6.5 Conical emission from a biaxial crystal

When a wave propagates along the optic axis of a biaxial crystal, the wave degenerates into a cone of rays. Fig. 6-15 shows the same cross section as

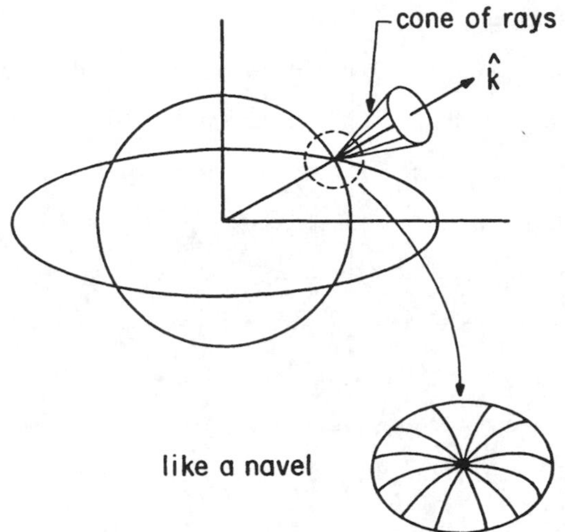

Fig. 6-15. Conical refraction.

Fig. 6-13(b). When the wave vector $\vec{k}$ is in the direction of the optic axis, the ray or Poynting vector will be in the direction of $\vec{v}_r$ or group velocity.

Since
$$\vec{v}_r = \nabla_k \omega \quad \left(\text{see chap. 1, } v = \frac{d\omega}{dk} \right) \tag{6-97}$$

and the wave surface is in fact

$$\omega(\vec{k}) = \text{constant} \tag{6-98}$$

$\vec{v}_r$ is normal to the wave normals. In particular, in the vicinity of the optic axis, the surface of the wave normals becomes a singular point resembling a navel (see Fig. (6-15), inset). $\vec{v}_r$ or the directions of the associated Poynting vectors are thus radiating out in the form of a cone from around the "navel". Since the Poynting vectors represent the true E-M energy propagation, the detector will see a cone of light. More details can be found in Born and Wolf.

§6.6 Physical discussion

It is easy to incorrectly make the following statements. Since for each direction $\hat{k}$ inside an anisotropic medium, there are in general two solutions $\vec{D}_1$ and $\vec{D}_2$, these two fields should each be associated with a wave vector $\vec{k}_1$ and $\vec{k}_2$. Thus one has a pair of solutions $(\vec{k}_1, \vec{D}_1)$ and $(\vec{k}_2, \vec{D}_2)$. Because of this, one would deduce that along $\vec{k}_1$, there can again be a pair of solutions and so is the case for $\vec{k}_2$. And this would propagate indefinitely so that there would be many $\vec{k}_n$'s $(n = 1,2,3,\ldots)$ branching out from the original $\vec{k}$. The same question applies to the Poynting vector $\vec{S}$ (or $\vec{t}$) which has two solutions $(\vec{t}_1, \vec{E}_1)$ and $(\vec{t}_2, \vec{E}_2)$. One would then say that $\vec{t}_1$ and $\vec{t}_2$ would branch out in the same way and so on until there are many $\vec{t}_n$'s $(n = 1,2,3,\ldots)$. We explain why such statements are incorrect.

When one says that the equation of the wave normals gives two solutions $(\vec{k}_1, \vec{D}_1)$ and $(\vec{k}_2, \vec{D}_2)$ for a given $\hat{k}$, it means that the anisotropic medium can support two plane waves of wave vectors $\vec{k}_1$ and $\vec{k}_2$ in the same direction as $\hat{k}$. Each of the plane wave fronts contains $\vec{D}_1$ and $\vec{D}_2$ respectively.

In other words, when a plane EM wave spanning the entire space of the anisotropic medium interacts with the material inside it the electric field $\vec{E}(\vec{r}, t)$ at every point of the wave front will induce in the material a polarization which re-radiates according to the Maxwell's equations. The solution of Maxwell's equations at each such point gives rise to a surface of wave normals. In a given direction $\hat{k}$, the combination of all the re-radiated wave fronts from all the re-radiating "point" sources on the initial wave front gives rise to two total re-radiated wave fronts of wave vectors $\vec{k}_1$, and $\vec{k}_2$, i.e. each wave front represents one of the following solutions:

$$\vec{D}_i = \vec{D}_{io} \exp[i(\omega t - \vec{k}_i \cdot \vec{r})]; \quad i = 1,2 \tag{6-99}$$

Thus, for each direction $\hat{k}$, the medium supports only two waves spanning the whole medium.

Similar argument can be applied to the ray surface, the ray direction $\vec{t}$ and the two wave fronts

$$\vec{E}_i = \vec{E}_{io} \exp[i(\omega t - \vec{k}_i \cdot \vec{r})]; \quad i = 1,2 \tag{6-100}$$

The above discussion applies to the hypothetical situation in which a plane wave with an arbitrary wave vector $\vec{k}$ is assumed to exist inside the anisotropic medium. In practice, we need to send such a wave from outside the medium. This coupling from outside encounters a restriction to the propagation, namely the boundary condition at the interface between the incident and transmitted media. If the incident medium is an isotropic medium, the transmission into the anisotropic medium leads to the phenomenon of double refraction. This means that the incident plane wave excites the material across the whole interface. The radiation re-radiated by all the "point" sources at the anisotropic side of the interface satisfies both the solution given by the surfaces of wave normals and the boundary condition (the tangential $\vec{E}$'s and $\vec{H}$'s are continuous). The result of this is that the anisotropic medium supports in general two plane waves spanning the whole medium; the wave vectors are $\vec{k}_1$ and $\vec{k}_2$. But now $\vec{k}_1$ and $\vec{k}_2$ each satisfies a different Snell's law. That means each of the two waves will propagate with a different velocity in different directions. The electric fields $\vec{E}_1$ and $\vec{E}_2$ associated with the two refracted waves are orthogonal according to section 6.3.

CHAPTER VII

POLARIZATION, ITS MANIPULATION AND JONES VECTORS

So far, we have only talked about plane E-M waves which have linear (or plane) polarizations. That is to say, the electric vector has a fixed orientation in space as the wave propagates. In fact, ideal natural sources are assumed to randomly emit E-M wave trains polarized linearly in all possible directions transverse to the direction of propagation.

It is only after the emission that different waves combine in different ways so that the resultant polarization becomes complicated (circular or elliptic). Such combination is based on the principle of superposition of waves without which the nature of polarization would certainly be completely different. We shall review in this chapter all states of polarization before talking about the application of optical anisotropy to produce different states of polarization and the matrix mathematics for the propagation of polarized E-M waves in anisotropic material.

§7.1 Superposition of E-M waves

In what follows, all waves are assumed travelling in the z - direction. For any plane or linearly polarized monochromatic E-M wave, one can choose a laboratory coordinate system to describe the electric field vector $\vec{E}(\vec{r}, t)$ associated with the wave. Thus, in this coordinate system, one can decompose $\vec{E}(\vec{r}, t)$ into $\vec{E}_x(\vec{r}, t)$ and $\vec{E}_y(\vec{r}, t)$ in the x - and y - directions of the laboratory coordinate system. The x and y directions are arbitrary; say, horizontal and vertical. If there is another E-M wave of the same frequency propagating in the same laboratory coordinate system, it too can be decomposed in the same way so that the x - and y - components can be superimposed giving a resultant component in the x - direction and a resultant component in the y - direction. In general, there can be many monochromatic plane waves of the same frequency each of which being decomposed into an x- and a y-component. One can then superimpose all the x - components and all the y - components separately resulting in two resultant components of travelling waves:

$$\vec{E}_x = \hat{i}\, E_{ox} \cos(\omega t - kz + \epsilon_x) \tag{7-1}$$

$$\vec{E}_y = \hat{j}\, E_{oy} \cos(\omega t - kz + \epsilon_y) \tag{7-2}$$

ϵ_x , ϵ_y are the phases. We rewrite these two expressions as:

$$\vec{E}_x = \hat{i}\, E_{ox} \cos(\omega t - kz) \tag{7-3}$$

$$\vec{E}_y = \hat{j}\, E_{oy} \cos(\omega t - kz + \epsilon) \tag{7-4}$$

where
$$\epsilon = \epsilon_y - \epsilon_x \tag{7-5}$$

This is valid because we will be dealing with the phase difference ϵ only.

<u>Definition</u>: E_y leads E_x by ϵ if $\epsilon > 0$

and E_y lags behind E_x by ϵ if $\epsilon < 0$.

The above definition can be seen by inspecting eq. (7-3) and (7-4). If $\epsilon > 0$ and at a fixed position in space, E_y will become zero (for example) at $\frac{\epsilon}{\omega}$ seconds earlier than E_x (Fig. 7-1a). A similar explanation applies to phase lag if $\epsilon < 0$. Such phase lag or lead is easier to see by considering the complex representation (Fig. 7-1b and c). Eq. (7-3) and (7-4) can now be vectorially added so as to obtain a resultant electric field. The

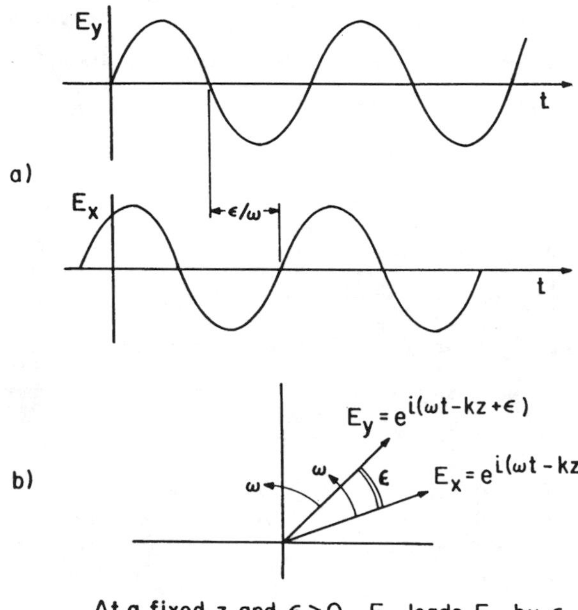

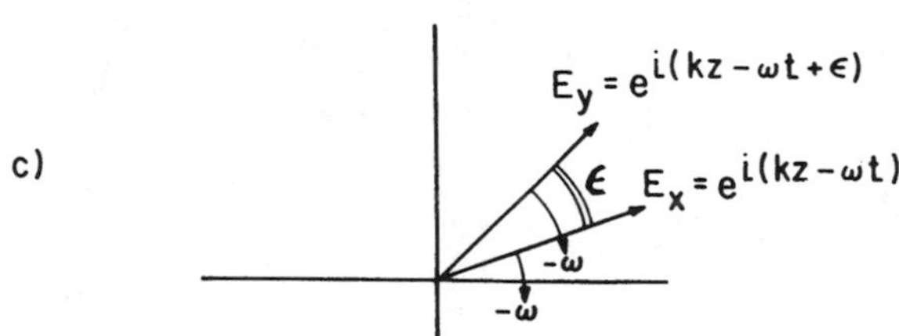

Fig. 7-1. Two sinusoidal oscillations with a phase difference: definition of phase lead and phase lag.

state (of polarization) of the resultant electric field depends on ϵ. We discuss this in the following sections.

§7.2 Linear polarization

Condition:
$$\epsilon = \begin{cases} 2n\pi \\ (2n+1)\pi \end{cases} \quad (n = 0,1,2,\ldots) \tag{7-6}$$

$$E_{ox} \neq E_{oy}$$

Thus,
$$E_y = E_{oy} \begin{cases} \cos(\omega t - kz + 2n\pi) \\ \cos(\omega t - kz + (2n+1)\pi) \end{cases}$$

$$= \pm E_{oy} \cos(\omega t - kz) \tag{7-7}$$

$$\therefore \vec{E} = \vec{E}_x + \vec{E}_y$$

$$= \vec{E}_{ox} \cos(\omega t - kz) \pm \vec{E}_{oy} \cos(\omega t - kz)$$

$$= \left(\vec{E}_{ox} \pm \vec{E}_{oy}\right) \cos(\omega t - kz) \tag{7-8}$$

Each of the vector amplitudes ($\vec{E}_{ox} \pm \vec{E}_{oy}$) has a fixed orientation in the laboratory coordinate system (Fig. 7-2) and they are thus linearly polarized.

Traditionally, one defines linear polarization as $\epsilon = \pi$; i.e. setting $n = 0$ in eq. (7-6). This is because $2n\pi$ doesn't change the cosine or exponential function.

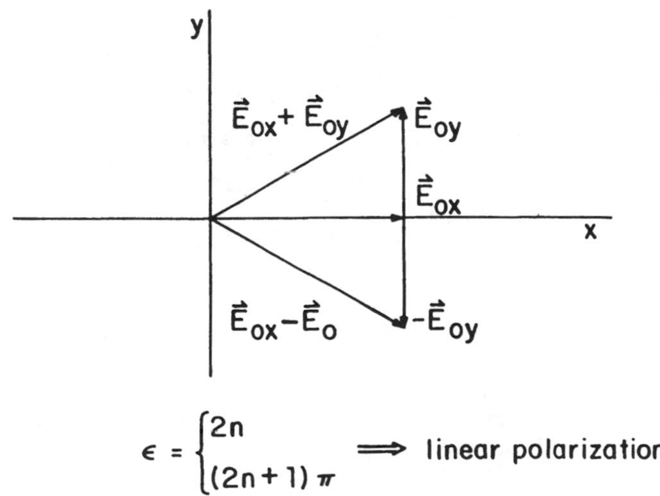

$$\epsilon = \begin{cases} 2n \\ (2n+1)\pi \end{cases} \Longrightarrow \text{linear polarization}$$

Fig. 7-2. Definition of linear polarization.

§7.3 Circular polarization

Conditions: $\epsilon = 2n\pi \pm \pi/2 \quad (n = 0,1,2,\ldots)$ (7-9)

and $E_{ox} = E_{oy} = E_o$

Thus,
$$E_y = E_o \cos(\omega t - kz + 2n\pi \pm \pi/2)$$
$$= \pm E_o \sin(\omega t - kz) \quad (7\text{-}10)$$

Also,
$$E_x = E_o \cos(\omega t - kz) \quad (7\text{-}11)$$

Now, the resultant vector traces out a circle in the x - y plane. To see this, we add:

$$E_x^2 + E_y^2 = E_o^2 [\sin^2(\omega t - kz) + \cos^2(\omega t - kz)] = E_o^2 \quad (7\text{-}12)$$

i.e. the x and y components together form a circle of radius E_o; or the tip of $\vec{E}_o$ traces out a circle in the x - y plane. Now at a certain time t, the x and y components are given by eqs. (7-10) and (7-11) (see Fig. 7-3).

Thus,
$$\tan\beta \equiv \frac{E_y}{E_x} = \pm \tan(\omega t - kz) \quad (7\text{-}12)$$

or
$$\beta = \pm (\omega t - kz) \quad (7\text{-}13)$$

$$\therefore \frac{d\beta}{dt} = \pm \omega \qquad \text{for } \epsilon = 2n\pi \pm \pi/2 \text{ and } z = \text{constant} \quad (7\text{-}14)$$

This means that β changes in time at a fixed position z (Fig. 7-3); i.e. the resultant electric field $\vec{E}_o$ rotates at the rate of $\pm\,\omega$ in the counter-clockwise or clockwise direction for ($\epsilon = 2n\pi + \pi/2$) and ($\epsilon = 2n\pi - \pi/2$) respectively, if one looks into the direction of the light beam at a fixed position z. The above senses of rotation can also be called left-handed (counter-clockwise) and right-handed (clockwise) because they follow the left and right hand "rules" of rotation respectively. All these are confusing. Worse is that the above definitions of the senses of rotation will be reversed if one changes the phase of the wave from (ωt - kz) to (kz - ωt). (A more detailed comment follows the section on elliptic polarization). Also, because the words clockwise and counter-clockwise will

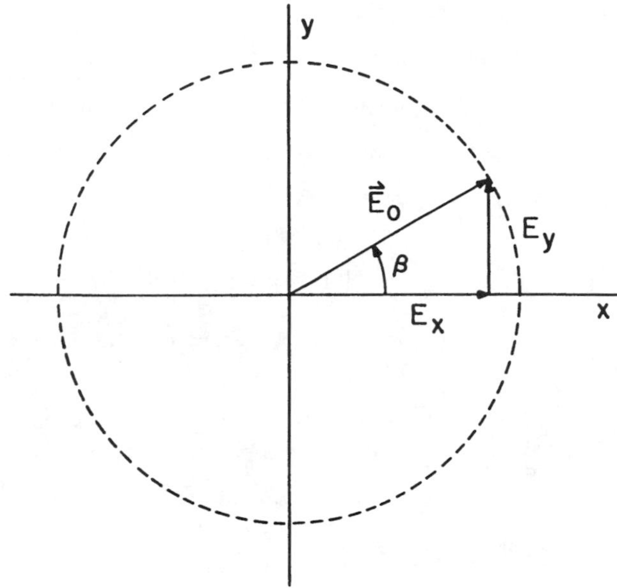

Fig. 7-3. Definition of circular polarization.

sooner or later become antique words since most mechanical clocks and watches will most probably be replaced by digital clocks or watches, younger generations will probably find it difficult to use such words. As such, a symbol will be used to represent the sense of rotation and state of polarization. They are defined as follows, assuming we always look into the light beam.

Circular polarizations are simply represented by two circles with opposite senses of rotation: ↻ and ↺

Again, traditionally, as in the case of linear polarization one defines circular polarization as when $\epsilon = \pm \pi/2$; i.e. $n = 0$.

§7.4 General or elliptic polarization

For all other values of ϵ (other than $2n\pi$, $(2n+1)\pi$, $2n\pi \pm \pi/2$) and arbitrary values of E_{ox} and E_{oy}, the state of polarization of the resultant electric field is elliptical. This can be seen by calculating the following.

$$\left(\frac{E_x}{E_{ox}}\right)^2 + \left(\frac{E_y}{E_{oy}}\right)^2 - 2\left(\frac{E_x}{E_{ox}}\right)\left(\frac{E_y}{E_{oy}}\right)\cos\epsilon$$

= ...(exercise for the reader using eq. (7-3 and 4))

$$= \sin^2\epsilon \qquad (7\text{-}15)$$

Eq. (7-15) is an ellipse. One can make a proper rotation of the x - y axes such that eq. (7-15) becomes a familiar equation of an ellipse.

$$\left(\frac{E'_x}{E'_{ox}}\right)^2 + \left(\frac{E'_y}{E'_{oy}}\right)^2 = 1 \qquad (7\text{-}16)$$

The " ' " (prime) means the fields are now components in the rotated x' y' coordinate system. The details of the rotation of coordinates are given below.

Set:

$$\begin{aligned} E_{ox} &= x_o & E'_{ox} &= x'_o \\ E_{oy} &\equiv y_o & E'_{oy} &= y'_o \\ E_x &= x & E'_x &= x' \\ E_y &= y & E'_y &= y' \end{aligned} \qquad (7\text{-}17)$$

We do the coordinate rotation:

$$x = x'\cos\alpha - y'\sin\alpha \qquad (7\text{-}18)$$

$$y = x'\sin\alpha + y'\cos\alpha \qquad (7\text{-}19)$$

Substitute eq. (7-18) and (7-19) into (7-15) and requiring that eq. (7-16) be valid, (Fig. 7-4), one obtains (exercise again for the reader)

$$\tan 2\alpha = \frac{2x_o\, y_o}{x_o^2 - y_o^2}\cos^2\epsilon \qquad (7\text{-}20)$$

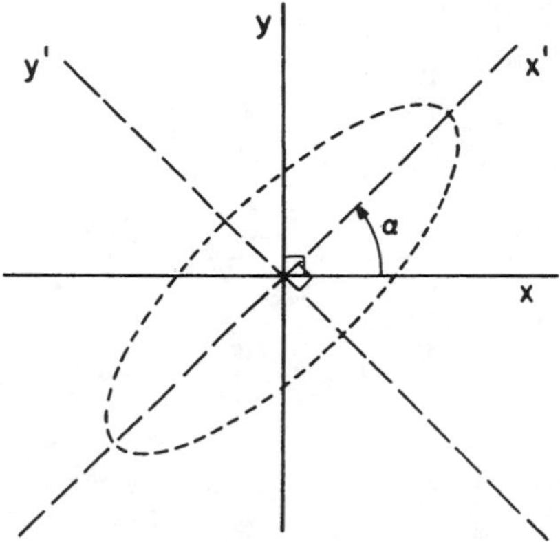

Fig. 7-4. Definition of elliptic polarization.

and

$$x_o'^2 = \frac{x_o^2\, y_o^2\, \sin^2\epsilon}{y_o^2\cos^2\alpha + x_o^2\sin^2\alpha - 2x_o y_o \sin\alpha\cos\alpha\cos\epsilon} \tag{7-21}$$

$$y_o'^2 = \frac{x_o^2\, y_o^2\, \sin^2\epsilon}{y_o^2\cos^2\alpha + x_o^2\sin^2\alpha + 2x_o y_o \sin\alpha\cos\alpha\cos\epsilon} \tag{7-22}$$

After this rotation, the "normal" look of the ellipse in the $x'y'$ coordinates is shown in Fig. 7-4. The sense of rotation of the tip of the resultant electric field around the ellipse can be determined as follows. In the x'y' frame, we define β as (see Fig. 7-5)

$$\tan\beta = \frac{y'}{x'} \equiv \frac{E_y'}{E_x'} \qquad \text{(by eq. 7-17)}$$

$$= \frac{E_{oy}'\cos(\omega t - kz + \epsilon)}{E_{ox}'\cos(\omega t - kz)} \tag{7-23}$$

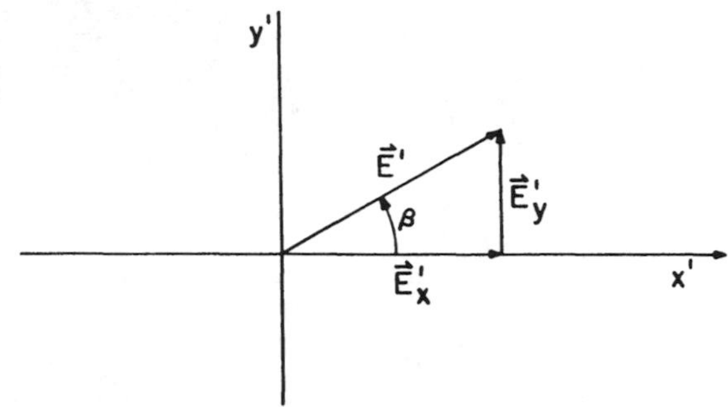

Fig. 7-5. Relationship between the orthogonal electric fields of an elliptically polarized wave.

Note: Since from eq. (7-20), $\alpha = \alpha(x_o , y_o , \epsilon)$ only, the rotation of the z - y coordinates by α doesn't affect $(\omega t - kz)$. Hence, we use the initial cosine terms in eq.(7-23). After some simplification,

$$\tan\beta = \frac{E'_{oy}}{E'_{ox}} \left\{ \cos\epsilon - \tan(\omega t - kz)\sin\epsilon \right\} \tag{7-24}$$

Differentiate with respect to time t at a fixed position z,

$$\sec^2\beta \frac{d\beta}{dt} = \frac{E'_{oy}}{E'_{ox}} \left\{ -\left[\sin \epsilon \sec^2(\omega t - kz)\right] \omega \right\}$$

Hence

$$\frac{d\beta}{dt} = - \frac{E'_{oy}}{E'_{ox}} \omega \frac{\sec^2(\omega t - kz)}{\sec^2\beta} \sin\epsilon \tag{7-25}$$

$\because$ the first three parts on the right hand side are positive, we have

$$\frac{d\beta}{dt} \propto - \sin\epsilon \tag{7-26}$$

We have the following results:

For an observer at a fixed position z looking into the light beam,

$$(-\sin\epsilon) > 0 \text{ means } \frac{d\beta}{dt} > 0 \text{ leading to}$$

$$(-\sin\epsilon) < 0 \text{ means } \frac{d\epsilon}{dt} < 0 \text{ leading to} \qquad (7\text{-}27a)$$

$$(-\sin\epsilon) = 0 \text{ means } \frac{d\beta}{dt} = 0 \text{ leading to}$$

The symbol of an ellipse represents the state of elliptic polarization whose sense of rotation is indicated. The last line for linear polarization (double ended arrow inclined at an arbitrary angle) is true because $\left(\frac{d\beta}{dt} = 0\right)$ leads to β = constant, i.e. the resultant electric vector has a fixed orientation in space as the wave propagates. Furthermore, we add, for completeness, that for $E_{ox} = E_{oy}$ and $-\sin\epsilon = \pm 1$, one obtains circular polarizations (<u>Exercise</u> for the reader who will prove that $(E_{oy}/E_{ox} = E'_{oy}/E'_{ox})$.)

To summarize, one can determine the orientation of the ellipse using eq.(7-20), the major and minor axes of the ellipse by eq. (7-21) and (7-22) and the sense of rotation by eq. (7-26). All these four equations depend only on ϵ and $r \equiv y_o/x_o$. Thus, the state of polarization, its size, its orientation and its sense of rotation are completely determined if one knows r and ϵ. Because of this, one can represent <u>all</u> the states, their sizes and their senses of rotation in a complex plane using the vector (Fig. 7-6)

$$re^{i\epsilon} \qquad (7\text{-}27b)$$

each point in this complex plane corresponds to one polarization. We shall analyze this.

From eq. (7-27a) and Fig. (7-6), we see that the upper half of the complex plane contains all polarization states whose sense of rotation is (sinε > 0) while the the lower half is (sinε < 0). The horizontal axis (sinε = 0) contains all the states of linear polarization with different

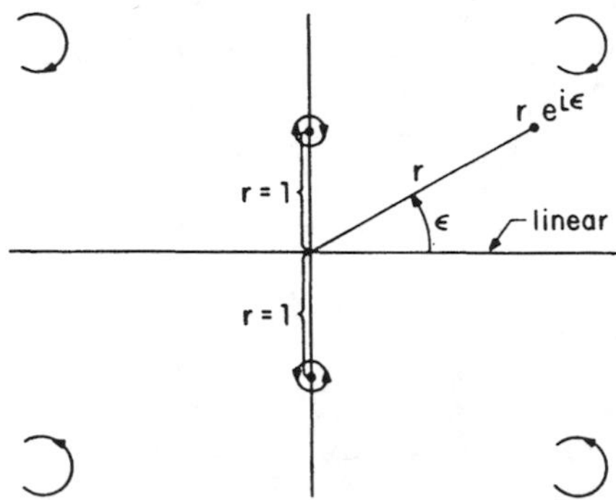

Fig. 7-6. Complex plane defining different polarization states of an E-M wave.

orientations. There are only two points (r = 1) on the vertical axis that give the two states of circular polarization (along the vertical axis, $\sin\epsilon = \pm 1$; plus the condition r = 1; one obtains circular polarization.). Besides the above special cases, all the other points in the complex plane give elliptic polarizations.

Fig. 7-7 shows polarization states in the complex plane around three circles, $r < 1$, $r = 1$ and $r > 1$. The reader should try (exercise) to prove that the orientations of the elliptic polarizations and linear polarizations are indeed what are shown in Fig. 7-7.

How to use the polarization chart of Fig. 7-7

To use the chart, one needs to know r and ϵ. We can then locate the point $re^{i\epsilon}$ in the complex plane and determine at least quickly and qualitatively what the orientation and sense of rotation of the polarization state are. Only when the points lie along the horizontal axis will the polarization be linear, and circular polarizations occur only at two points on the vertical axis ($\epsilon = \pm \pi/2$), namely $r = \pm 1$ (i.e. $E_{ox} = E_{oy}$).

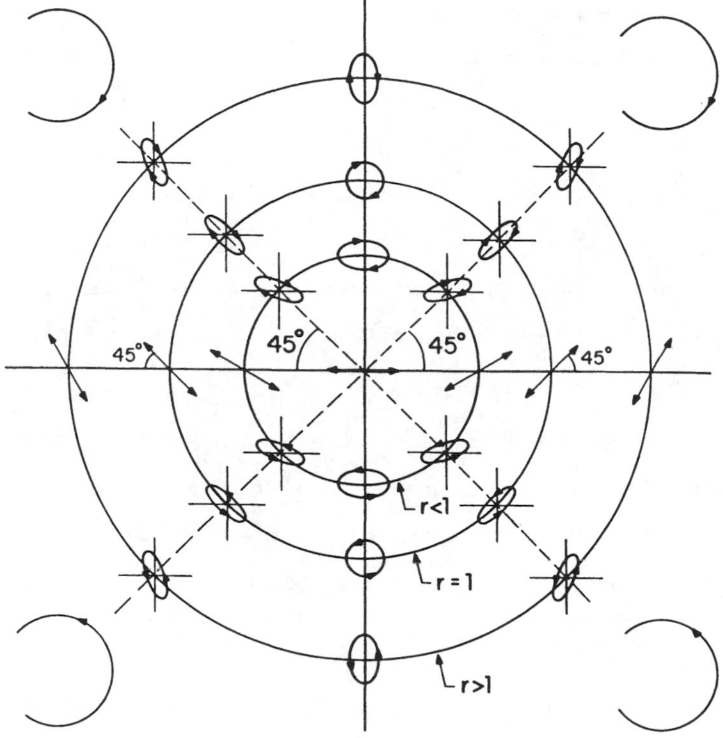

Fig. 7-7. Polarization chart in the complex plane of polarization.

§7.5 Some comments on the sense of rotation of circular and elliptical polarization

The sense of rotation (left-handed or right-handed; clockwise or counter-clockwise) of circular and elliptical polarizations depends on two definitions.

(a) the position of the observer: whether he looks into the light beam or he looks in the beam's direction of propagation. (Most people including the author choose the former. If one chooses the latter definition, all the senses of rotation will be reversed.)

(b) the mathematical description of the wave propagation: whether the propagating wave function is described by $f(\omega t - kz)$ or by $f(kz - \omega t)$. (For example, $\cos(\omega t - kz)$ or $\cos(kz - \omega t)$.) (Note: kz can be replaced by $\vec{k} \cdot \vec{r}$). Both usages are equally popular. The sense of rotation of the polarization resulting from one definition of the wave function is opposite to that resulting from the other. This can be seen in eq. (7-13) easily by changing $(\omega t - kz)$ to $(kz - \omega t)$ and the sign of $d\beta/dt$ in eq. (7-14) is reversed.

Such relative sense of rotation of the polarization (circular or elliptical) might lead to confusion. Thus, when talking about the sense of rotation, it is wise to first define the two "frames of reference" mentioned in (a) and (b). Comparison between results will then be consistent.

One might thus ask the following question. If the sense of rotation is so relative, are there really circular and elliptical polarizations that rotate in definite senses. The answer is yes. Circular and elliptical polarizations do have definite senses of rotation. Whether we call one right-handed or left-handed (counter-clockwise or clockwise) depends strictly on our choice of definition. It is similar to watching a spinning top. The sense of rotation is definitely there but whether it is described as right or left, clockwise or anticlockwise etc. is a matter of choice.

§7.6 Anisotropic material as polarizer

Using the phenomenon of double refraction in an anisotropic material described in chapter VI, one can devise ways to isolate linearly polarized light from an E-M radiation with random polarizations. One almost historical way is to use a Nicol prism (Fig. 7-8(a)). The Nicol prism is essentially an assembly of two anisotropic prisms sticking together by a transparent cement. Usually, the prisms are calcite while the cement is Canada balsam. The optic axes of the two prisms are parallel to each other as shown in Fig. 7-8a . A randomly polarized E-M wave enters the prism at non-zero angle of incidence. Double refraction produces two beams of orthogonal linear polarizations, one (ordinary (O) wave) perpendicular to the principal plane containing the optic axis, the other (extraordinary (e) wave) in the principal plane. (Thus, the polarization of the O-wave is always perpendicular to the optic axes.) When these two beams (waves) of different velocities (hence different indices of refraction n_o , n_e) arrive at the interface of the two prisms, the O-wave is totally internally reflected and the extraordinary wave passes through into the second prism. This requires that the cement's index of refraction n_c be between n_o and n_e; i.c. $n_e < n_c < n_o$. No further double refraction nor double wavefronts are generated in the second medium because the polarization of the extra-ordinary wave is now in the principal plane containing the optic axis (see §6.3(2)).

Different variations of such types of prism exist. They are used as beam splitters of laser beams inside and outside laser cavities (<u>Glan</u> <u>air</u> or <u>Glan Foucault</u> <u>prisms</u>, <u>Glan-Thomson</u> prisms, again made of calcite Fig. 7-8b). A laser beam of arbitrary polarization enters the surface normally. The two waves (O and e) thus created inside the prism propagate in the same direction but with different velocities and orthogonal polarizations. As in the case of the Nicol prism, one (O-wave) is rejected and one (e-wave) is transmitted. If the laser beam's polarization is parallel to that of the O-wave, it will be rejected; if it is parallel to that of the e-wave, it will be transmitted.

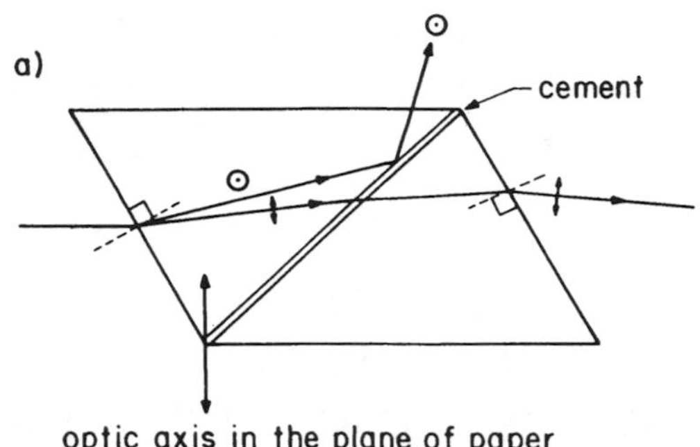

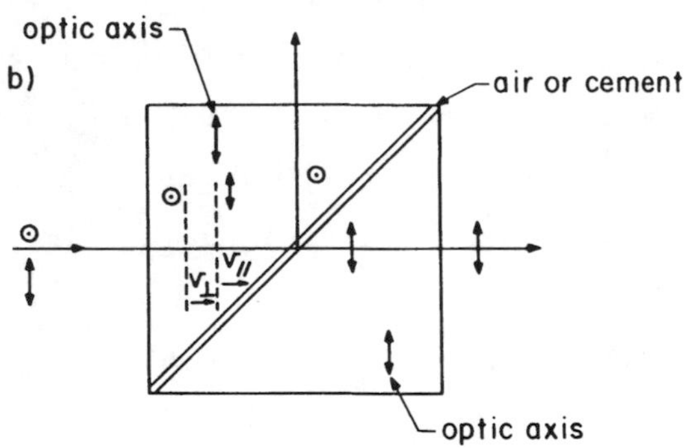

Fig. 7-8. (a) A Nicol prism.
(b) A Glan prism.

§7.7 Wave plates and wave retarders

Using an anisotropic material, one can transform one state of polarization of an E-M wave into another. This is usually accomplished by a wave plate with parallel incident and exit surfaces. As shown in Fig. (7-9), a plane monochromatic E-M wave of any polarization enters a plate of anisotropic material (usually uniaxial crystals) at normal incidence. The optic axis is perpendicular to z as shown, assuming the material to be a uniaxial crystal. Thus, the two waves (ordinary and extraordinary) generated in the medium will both travel in the same direction but at different velocities v_o and v_e

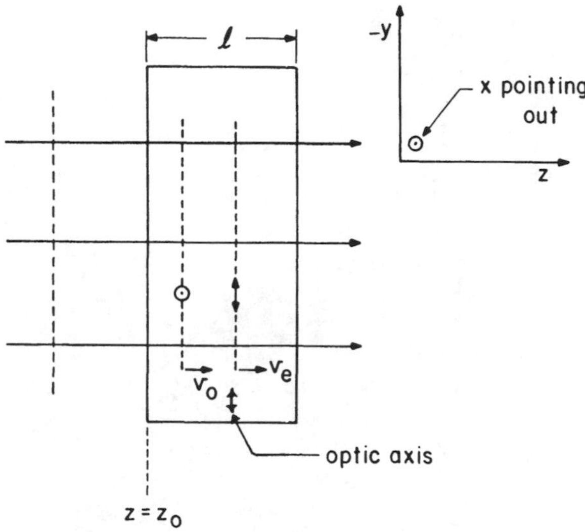

Fig. 7-9. A general wave plate.

corresponding to two indices n_o and n_e. These two waves are initially in phase at the incident interface. After propagating through the plate of thickness ℓ, a phase difference ϵ is developed between the two waves. Let x represent the polarization direction of the ordinary wave and y that of the extraordinary wave. Then the incident (plane) wave can be decomposed into components along the x and y axes.

$$E_x = E_{ox} \cos(\omega t - kz)$$

$$E_y = E_{oy} \cos(\omega t - kz)$$

where $$k = \frac{\omega}{c}$$

assuming that the incident medium is vacuum. It could also be an isotropic medium. In this case, c becomes v, the phase velocity. At the exit of the wave plate,

$$E_x = E_{ox} \cos(\omega t - kz - k_o\ell) \tag{7-28}$$

$$E_y = E_{oy} \cos(\omega t - kz - k_e\ell) \tag{7-29}$$

where $$k_o = \frac{\omega}{c} n_o \tag{7-30}$$

$$k_e = \frac{\omega}{c} n_e \tag{7-31}$$

O: ordinary; e: extraordinary

Thus the phase difference between E_x and E_y is, by eq. (7-5),

$$\epsilon = -(k_e - k_o)\ell$$

or $$\epsilon = \frac{\omega}{c} \ell (n_o - n_e) \tag{7-32}$$

Rewriting eq. (7-28) and (7-29) in the form of eq. (7-3) and (7-4), we have

$$E_x = E_{ox} \cos(\omega t - kz) \tag{7-33}$$

$$E_y = E_{oy} \cos(\omega t - kz + \epsilon) \tag{7-34}$$

which are identical to eq. (7-3) and (7-4). At the exit of the plate, the two waves will exit normally to the surface because the entrance and exit surfaces are parallel. They will recombine vectorially producing different states of polarization depending on the value of ϵ given by eq. (7-32). We look at some common wave plates as follows.

(a) <u>Quarter wave plate</u> ($\lambda/4$ plate)

Conditions: $\epsilon = 2n\pi \pm \pi/2$ (7-35)

These are the conditions for the generation of circular polarizations (see §7.3) if, in addition, $E_{ox} = E_{oy}$ (7-36)

Eq. (7-35) means that (using eq. (7-32))

$$\frac{\omega}{c} \ell \, (n_o - n_e) = 2n\pi \pm \pi/2$$

$$\ell \, (n_o - n_e) = \frac{c}{\omega} (2n\pi \pm \pi/2)$$

or $$\ell \, (n_o - n_e) = \frac{\lambda}{2\pi} (2n\pi \pm \pi/2)$$

$$= \left(n \pm \frac{1}{4}\right)\lambda \qquad n = 0,1,2,\ldots \tag{7-37}$$

i.e. the optical path difference between the O and e waves in traversing the quarter wave plate is $n\lambda$ plus or minus a quarter of a wavelength (λ); "plus" sign for ↻ , "minus" for ↺ . Eq. (7-36) means that the incident polarization should be linear and make an angle of 45° with the x - and y - axes so that $E_{ox} = E_{oy}$.

Because the wave plate is passive, and the o and e waves propagate in the same direction for incident wave normal to the surface, the reverse is also true; i.e. if the incident wave is circularly polarized, the transmitted wave will be linearly polarized. To see this, we decompose the incident wave into

$$E_x = E_o \cos(\omega t - kz) \quad (7\text{-}38)$$

$$E_y = E_o \cos(\omega t - kz + 2n\pi \pm \pi/2) \quad (7\text{-}39)$$

which is the definition of circular polarization. After traversing the plate, an additional phase difference of $(2n\pi \pm \pi/2)$ is added to the phase of E_y.

$$\begin{aligned} E_y &= E_o \cos(\omega t - kz + 4n\pi \pm \pi) \\ &= -E_o \cos(\omega t - kz) \end{aligned} \quad (7\text{-}40)$$

Combining eq. (7-38) and (7-40):

$$\begin{aligned} \vec{E} &= \vec{E}_x + \vec{E}_y \\ &= E_o(\hat{i} - \hat{j})\cos(\omega t - kz) \end{aligned} \quad (7\text{-}41)$$

which is a linear polarization making an angle of 45° with the x and y axes.

Summary: For a quarter wave plate: linear polarization in, circular out and vice versa.

(b) <u>Half-wave plate</u> ($\lambda/2$ plate)

Condition: $\epsilon = (2n + 1)\pi$ (7-42)

This is the condition of linear polarization (§7.2); i.e. the transmitted wave is linearly polarized. This is not a sufficient statement. Let's look more closely. Eq. (7-42) into (7-32) gives

$$(n_o - n_e)\ell = \frac{\lambda}{2\pi} \cdot (2n + 1)\pi = 2n\lambda + \lambda/2 \quad (7\text{-}43)$$

i.e. the optical path difference between the o and e waves in traversing the half wave plate is $2n\lambda$ plus half of a wavelength. If the incident wave is a linearly polarized plane wave with components

$$E_x = E_{ox} \cos(\omega t - kz) \tag{7-44}$$

$$E_y = E_{oy} \cos(\omega t - kz) \tag{7-45}$$

Then the transmitted wave components (relative to each other) become

$$E_x = E_{ox} \cos(\omega t - kz) \tag{7-46}$$

$$E_y = E_{oy} \cos(\omega t - kz + (2n + 1)\pi) \tag{7-47}$$

$$\text{i.e. } E_y = -E_{oy} \cos(\omega t - kz) \tag{7-48}$$

Combining eq. (7-46) and (7-48) we see that the transmitted electric vector becomes

$$\vec{E}_t = (E_{ox}\,\hat{i} - E_{oy}\,\hat{j})\cos(\omega t - kz) \tag{7-49}$$

Compare with the incident vector (eq.(7-44) and (7-45):

$$\vec{E}_i = (E_{ox}\,\hat{i} + E_{oy}\,\hat{j})\cos(\omega t - kz) \tag{7-50}$$

We see that $\vec{E}_i$ has been rotated by an angle 2Θ in traversing the half-wave plate (Fig. 7-10) where Θ is the angle between $\vec{E}_i$ and $\hat{i}$. In practice, Θ is often set at 45°. Then the effect of the half wave plate is to rotate the incident linear polarization by 90°. However, Θ can be any value so that $0 \leq 2\Theta \leq \pi/2$. This means that a $\lambda/2$ plate acts as a polarization rotator.

(c) <u>A general wave plate</u>

For a general value of ϵ, the resultant transmitted wave will be elliptically polarized (§7-4), if the incident wave is linearly polarized.

d) <u>A wave retarder</u> (Babinet compensator)

In contrast to a wave plate which introduces a fixed phase shift, a wave retarder is an anisotropic device whose effect is to introduce a continuously variable relative phase shift between the x and y components of an incident wave. The central design of a wave retarder is to stack two anisotropic wedges on top of each other (Fig. 7-11) with their optic axes perpendicular to each other assuming uniaxial crystals. This is called a <u>Babinet</u>

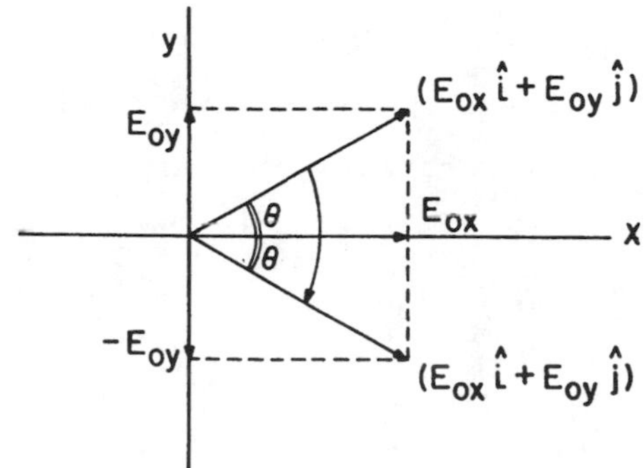

Fig. 7-10. Effect of a half wave plate: polarization rotation.

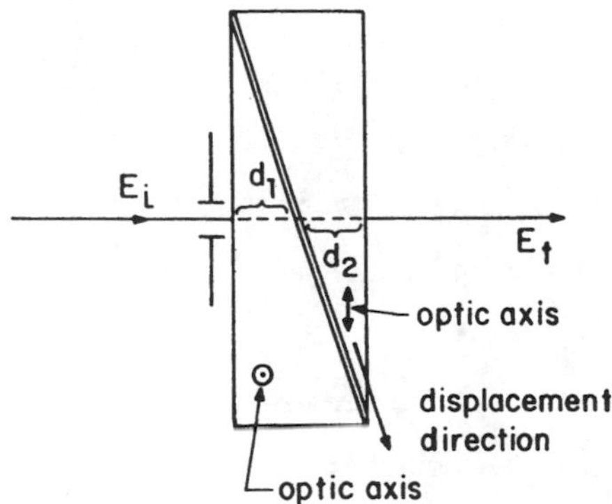

Fig. 7-11. A wave retarder: Babinet compensator.

compensator. A laser beam whose electric field is $\vec{E}_i$ enters the surface at normal incidence into the retarder. Inside the left wedge, there are two waves (O and e) propagating in the same direction. The polarization of the O-wave is perpendicular to the optic axis while the e-wave is parallel. The relative phase shift between the O and e waves after traversing the left wedge through the thickness d_1 is (by eq. 7-32)

$$\epsilon_1 = \frac{\omega}{c} d_1 (n_o - n_e) \tag{7-51}$$

After entering the right wedge, the O-wave becomes an e-wave while the e-wave becomes an O-wave because the optic axis is now parallel to the polarization of the O-wave inside the left wedge. The relative phase difference in traversing the thickness d_2 of the right wedge is (by eq. 7-32)

$$\epsilon_2 = \frac{\omega}{c} d_2 (n_e - n_o) \tag{7-52}$$

after traversing the whole retarder, the total relative phase shift is

$$\epsilon = \frac{\omega}{c} d_1 (n_o - n_e) + \frac{\omega}{c} d_2 (n_e - n_o)$$

$$\epsilon = \frac{\omega}{c} (n_o - n_e) (d_1 - d_2) \tag{7-53}$$

By sliding one wedge with respect to the other along the contact surface, $(d_1 - d_2)$ can be varied continuously, and by eq. (7-53), the total relative phase shift ϵ is also varied continuously.

Different variations of wave retarders have been designed and the reader is referred to an optics book for more detail. (See for example, Hecht.)

§7.8 Jones vectors

The previous section analyzes from the physical point of view the change in the polarization state of an E-M wave that propagates through wave plates and retarders. The procedure is rather time consuming, especially if one needs to analyze the propagation through a series of anisotropic media. Using the Jones vector, a straight forward matrix calculation can be applied to solve the above problem "quickly".

Definition: Jones vector

We first decompose the electric field vector of an E-M wave into the x and y components in a laboratory reference frame. For convenience, we call x horizontal and y vertical. At a fixed position z along the propagation direction,

$$\vec{E}(\vec{r},t) = E_x(\vec{r},t)\hat{i} + E_y(\vec{r},t)\hat{j} \tag{7-54}$$

$$\text{Jones vector} \quad \vec{J} \equiv \begin{pmatrix} \vec{E}_x(\vec{r},t) \\ \vec{E}_y(\vec{r},t) \end{pmatrix} \tag{7-55}$$

As we have seen in the previous section, when the E-M wave propagates through wave plates, the original phase of $\vec{E}$, i.e. $(\omega t - kz)$ was not affected. Only phase shifts were added to the original phase. Hence, we can neglect the phase $(\omega t - kz)$ and write eq. (7-55) as

$$\vec{J} = \begin{pmatrix} E_{ox}e^{i\varphi_x} \\ E_{oy}e^{i\varphi_y} \end{pmatrix} \tag{7-56}$$

Where exponential notation, rather than cosine function, is required here. Therefore, the Jones vector is a complex vector in an abstract space. To obtain the real field, one has to calculate, for example,

$$E_x = \text{Re}\left\{ J_x e^{i(\omega t - kz)} \right\} = \text{Re}\left\{ E_{ox} e^{i(\omega t - kz + \varphi_x)} \right\}$$

$$= E_{ox}\cos(\omega t - kz + \varphi_x) \tag{5-57}$$

Often, one is only interested in the polarization state of the wave after propagation through anisotropic media. Only a relative phase difference is needed (see the previous section). Thus, we can normalize the amplitude of the Jones vector and retain the relative phase difference. Eq. (7-56) thus becomes

$$\vec{J} = e^{i\varphi_x} \begin{pmatrix} E_{ox} \\ E_{oy} e^{i(\varphi_y - \varphi_x)} \end{pmatrix}$$

Omitting $e^{i\varphi_x}$, because it is a common phase factor, we have

$$\vec{J} = \begin{pmatrix} E_{ox} \\ E_{oy} e^{i\epsilon} \end{pmatrix} \tag{7-57}$$

where $$\epsilon \equiv \varphi_y - \varphi_x \tag{7-58}$$

(Here, as in §7.2 to 7.4, we use the x - component as the reference, and consider the phase difference of y with respect to x - components.) To normalize eq. (5-57), let:

$$\vec{J}_N = \begin{pmatrix} E_{Nx} \\ E_{Ny} \end{pmatrix} \tag{7-59}$$

And require $\vec{J}_N \cdot J_N^* = 1$ (7-60)

where the asterisk "*" denotes the complex conjugate. This means

$$\vec{E}_{Nx} \cdot E_{Nx}^* + E_{Ny} \cdot E_{Ny}^* = 1 \tag{7-61}$$

and $$E_{Nx} = \frac{E_{ox}}{(\vec{J} \cdot \vec{J}^*)^{1/2}} \tag{7-62}$$

$$E_{Ny} = \frac{E_{oy} e^{i\epsilon}}{(\vec{J} \cdot \vec{J}^*)^{1/2}} \tag{7-63}$$

Note (1): $\vec{J} \cdot \vec{J}^* = E_{ox} \cdot E_{ox}^* + E_{oy} \cdot E_{oy}^*$

$= E_{ox}^2 + E_{oy}^2$ if E_{ox} , E_{oy} are real (7-64)

Note (2): The definition of the Jones vector (eq. 7-57) is relative. In eq. (7-57), we have used the x - component as our reference similar to the definition of the polarization, eq. (7-3) and (7-4). One can also define the Jones vector using the y - component as the reference. In such a case

$$\vec{J} = \begin{pmatrix} E_{ox}e^{i\epsilon'} \\ E_{oy} \end{pmatrix}, \quad \epsilon' \equiv \varphi_x - \varphi_y .$$

We shall stick to the first definition (7-57).

<u>Example 1</u>: A general <u>linear polarization</u> making an angle ψ with the x-axis.

We first determine the general Jones vector (eq. 7-56). Since the x and y components of such a linear polarization are in phase,

i.e. $\vec{E} = \vec{E}_o \cos(\omega t - kz)$

$$= [(E_o \cos\psi)\hat{i} + (E_o \sin\psi)\hat{j}]\cos(\omega t - kz)$$

The Jones vector is thus

$$\vec{J} = \begin{pmatrix} E_o \cos\psi \\ E_o \sin\psi \end{pmatrix} \tag{7-65}$$

To normalize, we calculate

$$\vec{J} \cdot \vec{J}^* = E_o^2(\cos^2\psi + \sin^2\psi) = E_o^2$$

$$E_{Nx} = \frac{E_o \cos\psi}{E_o} = \cos\psi$$

$$E_{Ny} = \frac{E_o \sin\psi}{E_o} = \sin\psi$$

$$\therefore \quad \vec{J}_N = \begin{pmatrix} \cos\psi \\ \sin\psi \end{pmatrix} \tag{7-66}$$

If $\psi = 0$, we have horizontal polarization, and

$$\vec{J}_{Nh} = \begin{pmatrix} 1 \\ 0 \end{pmatrix} \tag{7-67}$$

If $\psi = \pi/2$, we have vertical polarization and

$$\vec{J}_{NV} = \begin{pmatrix} 0 \\ 1 \end{pmatrix} \tag{7-68}$$

<u>**Example 2:**</u> Circular polarization. From §7.3, the field for a circular polarization is

$$\vec{E} = E_o \cos(\omega t - kz)\hat{i} + E_o \cos(\omega t - kz + 2n\pi \pm \pi/2)\hat{j} \qquad (7\text{-}69)$$

Compare with (7-57), we have

$$\epsilon = 2n\pi \pm \pi/2, \ \varphi_x = 0, \ E_{ox} = E_{oy} \equiv E_o$$

Thus, substituting into eq. (7-57),

$$\vec{J} = \begin{pmatrix} E_o \\ E_o e^{\pm i\pi/2} \end{pmatrix} \qquad (7\text{-}70)$$

To normalize, we calculate

$$\vec{J} \cdot \vec{J}^* = E_o^2 + E_o^2 = 2E_o^2 \qquad (7\text{-}71)$$

$$E_{Nx} = \frac{E_o}{\sqrt{2}\,E_o} = \frac{1}{\sqrt{2}} \qquad (7\text{-}72)$$

$$E_{Ny} = \frac{E_o e^{\pm i\pi/2}}{\sqrt{2}\,E_o} = \frac{1}{\sqrt{2}}(\pm i) \qquad (7\text{-}73)$$

$$\therefore \quad \vec{J}_N = \frac{1}{\sqrt{2}} \begin{pmatrix} 1 \\ \pm i \end{pmatrix} \qquad (7\text{-}74)$$

$$\text{or} \quad \vec{J}_N(\circlearrowright) = \frac{1}{\sqrt{2}} \begin{pmatrix} 1 \\ i \end{pmatrix} \qquad (7\text{-}75)$$

$$\vec{J}_N(\circlearrowleft) = \frac{1}{\sqrt{2}} \begin{pmatrix} 1 \\ -i \end{pmatrix} \qquad (7\text{-}76)$$

The senses of rotation are determined by comparing with the definitions in §7.3;

i.e. $\epsilon = 2n\pi + \pi/2$ for $\circlearrowright$

and $\epsilon = 2n\pi - \pi/2$ for $\circlearrowleft$

<u>**Example 3:**</u> <u>Elliptic polarization</u>

Eq. (7-57) is the general Jones vector; i.e. it represents the most general state of polarization which is elliptical. Normalization of this general vector

$$\vec{J} \equiv \begin{pmatrix} E_{ox} \\ E_{oy} e^{i\epsilon} \end{pmatrix} \qquad (7\text{-}77)$$

gives

$$\vec{J}_N = \frac{1}{\sqrt{1 + \tan^2\Theta}} \begin{pmatrix} 1 \\ e^{i\epsilon} \tan\Theta \end{pmatrix} \qquad (7\text{-}78)$$

$$\text{where } \tan\Theta = \frac{E_{oy}}{E_{ox}} \qquad (7\text{-}79)$$

i.e. Θ is the angle at which the major axis of the ellipse makes with the x - axis. The derivation of (7-78) is left as an exercise to the reader.

§7.9 Propagation through wave plates using Jones matrix formalism

We now consider the propagation through a wave plate. The incident wave can be decomposed into E_x and E_y in the laboratory frame (say, horizontal and vertical, for convenience). The passage of the wave into the wave plate immediately induces the two wave O and e whose polarizations are $\vec{E}_o$ and $\vec{E}_e$ respectively. This change of polarization from $(\vec{E}_x, \vec{E}_y)$ to $(\vec{E}_o, \vec{E}_e)$ is equivalent to a rotation of the coordinates (x, y) into (O, e) through an angle ψ. (Note: $\vec{E}_o \perp \vec{E}_e$, hence (O, e) forms a Cartesian coordinate.) (Fig. 7-12). The O and e polarization directions are often called slow and fast axes in the literature because for negative uniaxial materials, $v_e > v_o$. The O and e waves propagate through the wave plate of thickness ℓ. They suffer phase shifts of $k_o\ell$ and $k_e\ell$ respectively. The relative phase shift or phase difference between the O and e waves is thus

$$\epsilon = (k_o - k_e)\ell \qquad (7\text{-}80)$$

When the waves exit the plate, we measure them in the laboratory (x, y) frame again. This is equivalent to rotating the (O, e) frame back into the (x, y) frame (by an angle $-\psi$).

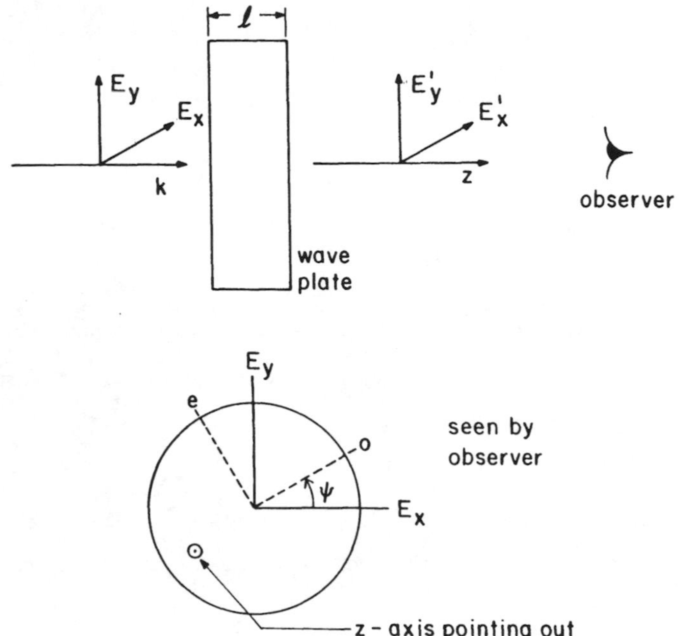

Fig. 7-12. Detailed analysis of the function of a wave plate made of uniaxial material using the Jones vector formalism.

Thus, the propagation of an E-M wave through a wave plate is equivalent to three steps of operation, namely, rotation of the (x, y) frame by an angle ψ (into the (O, e) frame), shifting of phases of the O and e waves and rotation of the (O, e) frame by an angle $-\psi$ (back into the (x, y) frame). Rotation of coordinates can be represented by matrices. Thus, rotation by an angle ψ is given by

$$R(\psi) \equiv \begin{pmatrix} \cos\psi & \sin\psi \\ -\sin\psi & \cos\psi \end{pmatrix} \tag{7-81}$$

while rotation by an angle $(-\psi)$

is
$$R(-\psi) = \begin{pmatrix} \cos\psi & -\sin\psi \\ \sin\psi & \cos\psi \end{pmatrix} \tag{7-82}$$

Let $\begin{pmatrix} E_o \\ E_e \end{pmatrix}$ and $\begin{pmatrix} E_o' \\ E_e' \end{pmatrix}$ be

the Jones vectors representing the o and e waves just inside the incident and exit surfaces of the wave plate. The phase changes of the o and e waves after traversing the thickness of the wave plate are:

$$\begin{pmatrix} E_o' \\ E_e' \end{pmatrix} = \begin{pmatrix} E_o e^{-ik_o \ell} \\ E_e e^{-ik_e \ell} \end{pmatrix} \tag{7-83}$$

where the identical phase factor $e^{i(\omega t - kz)}$ for the two waves at the input side is omitted at the right hand side of eq. (7-83). Eq. (7-83) can be rewritten as

$$\begin{pmatrix} E_o' \\ E_e' \end{pmatrix} = \begin{pmatrix} e^{-ik_o \ell} & 0 \\ 0 & e^{-ik_e \ell} \end{pmatrix} \begin{pmatrix} E_o \\ E_e \end{pmatrix} \tag{7-84}$$

$$\text{Let } M_p \equiv \begin{pmatrix} e^{-ik_o \ell} & 0 \\ 0 & e^{-ik_e \ell} \end{pmatrix} \tag{7-85}$$

It can be seen that M_p represents a phase shift operator (matrix). Summarizing the above, we can write down the total operation W representing the propagation through a wave plate as

$$W = R(-\psi) M_p R(\psi) \tag{7-86}$$

we have

$$\begin{pmatrix} E_x' \\ E_y' \end{pmatrix} = W \begin{pmatrix} E_x \\ E_y \end{pmatrix} \tag{7-87}$$

where the two Jones vectors at the left and right sides of eq. (7-87) are the field components of the transmitted and incident waves in the laboratory frame respectively. The operator W of eq. (7-87) means that the incident vector is first operated by $R(\psi)$, then by M_p and then by $R(-\psi)$. It should be stressed that the order of such operations should start from the right to the left. We can simplify M_p slightly by defining

$$\epsilon \equiv (k_o - k_e)\ell \equiv \text{relative phase shift} \tag{7-88}$$

and
$$\phi = \frac{1}{2}(k_o + k_e)\ell \equiv \text{mean absolute phase change} \tag{7-89}$$

Thus
$$\frac{\epsilon}{2} + \phi = k_o\ell \tag{7-90}$$

and
$$-\frac{\epsilon}{2} + \phi = k_e\ell \tag{7-91}$$

and from eq. (7-85)

$$M_p = e^{-i\phi}\begin{pmatrix} e^{-i\epsilon/2} & 0 \\ 0 & e^{i\epsilon/2} \end{pmatrix} \tag{7-92}$$

Thus,

$$W = e^{-i\phi}\begin{pmatrix} \cos\psi & -\sin\psi \\ \sin\psi & \cos\psi \end{pmatrix}\begin{pmatrix} e^{-i\epsilon/2} & 0 \\ 0 & e^{i\epsilon/2} \end{pmatrix}\begin{pmatrix} \cos\psi & \sin\psi \\ -\sin\psi & \cos\psi \end{pmatrix} \tag{7-93'}$$

and
$$\begin{pmatrix} E'_x \\ E'_y \end{pmatrix} = W\begin{pmatrix} E_x \\ E_y \end{pmatrix} \tag{7-93}$$

It is left an exercise for the reader to show that W is a unitary matrix; i.e. $W^\dagger W = I$. Thus if (E_x, E_y) is normalized, (E'_x, E'_y) is also a normalized vector.

For a dichroic polarizer (by absorption), such as Polaroid sheets, the matrix representation for the propagation of an E-M wave is something else. Let A be the absorbing (non-birefringent) polarizer's operator (matrix). One sees that

$$\begin{pmatrix} E'_{Nx} \\ E'_{Ny} \end{pmatrix} = A\begin{pmatrix} E_{Nx} \\ E_{Ny} \end{pmatrix} \tag{7-94}$$

Since this type of polarizer gives only linear polarization, i.e. either $E'_x = 0$ or $E'_y = 0$, there are only two expressions for A, namely

$$A_x = e^{-i\phi}\begin{pmatrix} 1 & 0 \\ 0 & 0 \end{pmatrix} \tag{7-95}$$

$$A_y = e^{-i\phi}\begin{pmatrix} 0 & 0 \\ 0 & 1 \end{pmatrix} \tag{7-96}$$

where $e^{-i\phi}$ is a general phase factor, and A_x , A_y are the matrices that will transform the incident wave into a wave of linear polarization along the x and y axes respectively. To show this, one simply substitutes eq. (7-95) and (7-96) into (7-94) and obtains

$$\begin{pmatrix} E_{Nx}' \\ \\ E_{Ny}' \end{pmatrix} = \begin{cases} A_x \begin{pmatrix} E_{Nx} \\ E_{Ny} \end{pmatrix} \\ \\ A_y \begin{pmatrix} E_{Nx} \\ E_{Ny} \end{pmatrix} \end{cases} = \begin{cases} e^{-i\phi} E_{Nx} \begin{pmatrix} 1 \\ 0 \end{pmatrix} \\ \\ e^{-i\phi} E_{Ny} \begin{pmatrix} 0 \\ 1 \end{pmatrix} \end{cases}$$

Example 1: Quarter wave plate

Similar to §7.7, the plate introduces a relative phase shift of

$$\epsilon = 2n\pi \pm \pi/2$$

Using eq. (7-93), assuming that the incident wave is linearly polarized in the y - direction, and $\psi = 45°$, we have:

$$\begin{pmatrix} E_x' \\ E_y' \end{pmatrix} = W \begin{pmatrix} 0 \\ 1 \end{pmatrix}$$

$= \ldots.$ (exercise for the reader)

$$= e^{-i\phi} \begin{pmatrix} \cos(\pm\,\pi/4) & -i\sin(\pm\,\pi/4) \\ -i\sin(\pm\,\pi/4) & \cos(\pm\,\pi/4) \end{pmatrix} \begin{pmatrix} 0 \\ 1 \end{pmatrix} \qquad (7\text{-}97)$$

$$= e^{-i\phi} \begin{pmatrix} -i\sin(\pm\,\pi/4) \\ \cos(\pm\,\pi/4) \end{pmatrix}$$

$$= e^{-i\phi} \begin{pmatrix} \mp\, i\,\dfrac{1}{\sqrt{2}} \\ \dfrac{1}{\sqrt{2}} \end{pmatrix}$$

$$= \frac{e^{-i\phi}}{\sqrt{2}} \begin{pmatrix} \mp\, i \\ 1 \end{pmatrix}$$

Using the x-component as the reference, we have

$$\begin{pmatrix} E_x' \\ E_y' \end{pmatrix} = \frac{ie^{-i\phi}}{\sqrt{2}} \begin{pmatrix} 1 \\ \pm\, i \end{pmatrix} \rightarrow \frac{1}{\sqrt{2}} \begin{pmatrix} 1 \\ \pm\, i \end{pmatrix} \qquad (7\text{-}98)$$

where the last step omits the common phase for normalization purposes. In other words, a quarter wave plate transforms a linear polarization to circular polarization if $\psi = 45°$. Since the plate is passive, the reverse should also be true; i.e. it turns circular polarization into a linear one. To prove this, we observe from eq. (7-97) that

$$W = e^{-i\phi} \begin{pmatrix} \cos(\pm \pi/4) & -i\sin(\pm \pi/4) \\ -i\sin(\pm \pi/4) & \cos(\pm \pi/4) \end{pmatrix} \tag{7-99}$$

$$= \frac{e^{-i\phi}}{\sqrt{2}} \begin{pmatrix} 1 & \mp i \\ \mp i & 1 \end{pmatrix} \tag{7-100}$$

$$\begin{pmatrix} E_x' \\ E_y' \end{pmatrix} = W \cdot \frac{1}{\sqrt{2}} \begin{pmatrix} 1 \\ \pm i \end{pmatrix} \qquad \text{(by eq. (7-74))}$$

$$= \frac{e^{-i\phi}}{2} \begin{pmatrix} 1 & \mp i \\ \mp i & 1 \end{pmatrix} \begin{pmatrix} 1 \\ \pm i \end{pmatrix} \tag{7-101}$$

Eq. (7-101) gives:

$$\begin{pmatrix} E_x' \\ E_y' \end{pmatrix} = \left\{ \begin{array}{l} \dfrac{e^{-i\phi}}{2} \begin{pmatrix} 2 \\ 0 \end{pmatrix} = e^{-i\phi} \begin{pmatrix} 1 \\ 0 \end{pmatrix} \\ \dfrac{e^{i\phi}}{2} \begin{pmatrix} 0 \\ \pm 2i \end{pmatrix} = \pm i e^{-i\phi} \begin{pmatrix} 0 \\ 1 \end{pmatrix} \end{array} \right\} \tag{7-102}$$

Omitting the common phases in eq. (7-102) for normalization purposes, we have

$$\begin{pmatrix} E_x' \\ E_y' \end{pmatrix} = \left\{ \begin{array}{l} \begin{pmatrix} 1 \\ 0 \end{pmatrix} \\ \begin{pmatrix} 0 \\ 1 \end{pmatrix} \end{array} \right. = \left\{ \begin{array}{l} \vec{J}_{Nh} \\ \vec{J}_{NV} \end{array} \right. \tag{7-103}$$

which proves the above statement.

Example 2: Half-wave plate

As in §7.7, the plate introduces a relative phase shift of

$$\epsilon = (2n + 1)\pi \ , \ (n = 0,1,2,\ldots) \tag{7-104}$$

For an arbitrary angle ψ between the x and the o - axes, eq. (7-93) becomes, starting with a vertically polarized incident wave,

i.e. $\begin{pmatrix} E_x \\ E_y \end{pmatrix} = \begin{pmatrix} 0 \\ 1 \end{pmatrix}$,

$$\begin{pmatrix} E_x' \\ E_y' \end{pmatrix} = ie^{-i\phi} \begin{pmatrix} \cos(\pi/2 + 2\psi) \\ \sin(\pi/2 + 2\psi) \end{pmatrix} \quad (7\text{-}105)$$

The derivation is left as an exercise to the reader. Omitting the common phase factor for normalization purposes, we obtain

$$\begin{pmatrix} E_x' \\ E_y' \end{pmatrix} = \begin{pmatrix} \cos(\pi/2 + 2\psi) \\ \sin(\pi/2 + 2\psi) \end{pmatrix} \quad (7\text{-}106)$$

This is a linear polarization making an angle $(\pi/2 + 2\psi)$ with the x - axis, (see eq. 7-66). Since the incident wave's polarization makes an angle of $\pi/2$ with the x - axis, the transmitted polarization (being still linear) makes an angle of 2ψ with respect to the incident polarization. That is, a half-wave plate rotates a linear polarization by an angle of $+ 2\psi$. If $\psi = 45°$, the angle of rotation is 90°, which is the case commonly used in experiments. Note that the direction of rotation of the polarization is the same as that of ψ (the rotation of the x - axis into the o - axis, Fig. 7-12).

Some comments on ψ

The angle ψ between the x - axis and the o - axis depends on the orientation of the optic axis. Since, as defined in §7.7, the optic axis of a uniaxial wave plate is perpendicular to z and since the polarization of the o - wave is always perpendicular to the optic axis and the propagation direction (Fig. 7-13a), (See also the section on the index ellipsoid in chapter 6), the polarization of the e - wave is in the plane containing the optic axis. In practice, the wave plate is rotated about the z - axis so that the

O - axis makes an angle of ψ with the laboratory x - axis, (Fig. 7-13b). This angle is usually set at 45° so that a quarter wave plate will transform linear polarization into circular and vice versa and a half-wave plate will rotate a linear polarization by 90°.

The angle ψ can be experimentally determined in the following way. Since ψ is the angle between the O - wave and the x - axis (horizontal for instance), we need to determine the O - axis. The O - axis is the direction

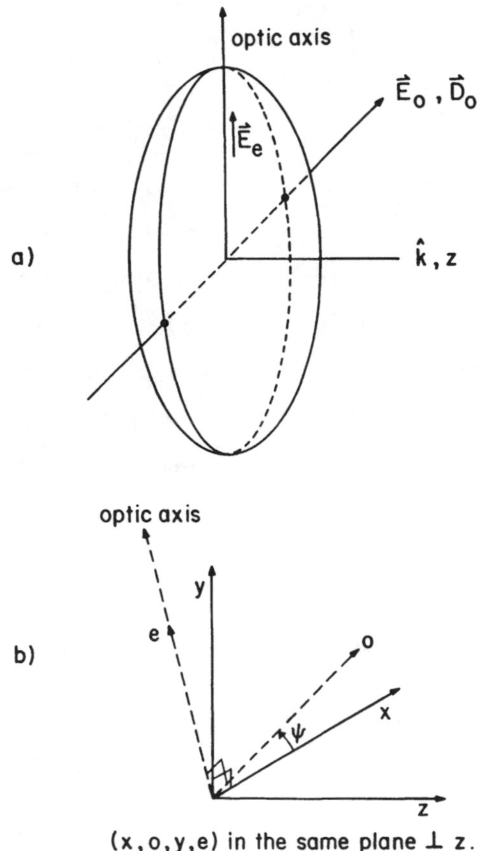

Fig. 7-13. Analysis of the angle ψ between the O-wave and the laboratory x-axis of a uniaxial wave plate.

of polarization of the O - wave. By definition, it is always perpendicular to the optic axis. Hence, we should determine the optic axis. Fig. (7-14) shows one way of determining the optic axis. We start with a pair of crossed polarizers (Fig. 7-14a). A light beam of arbitrary polarization is sent into the crossed polarizers with their polarization transmission axes V and h as indicated. No light will pass through the second polarizer. Now we add our wave plate in between the crossed polarizer. An arbitrarily oriented wave plate will generate two waves (O and e) with orthogonal polarizations so that at the exit of the wave plate W, the polarization state is changed with respect to the input polarization which is perpendicular to the paper. Thus, one can decompose the polarization at the

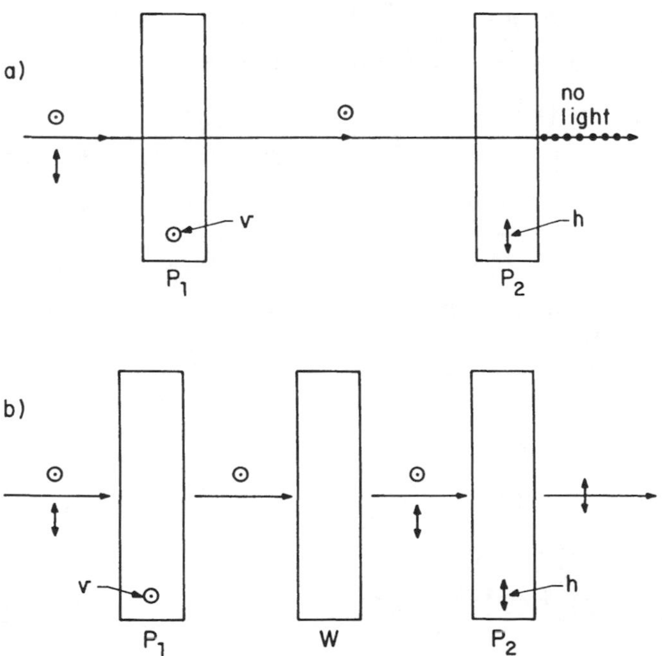

Fig. 7-14. Experimental determination of the angle ψ between the O - wave and the x - axis of the laboratory of a uniaxial wave plate.

output of W into two polarizations preferred by P_1, and P_2, i.e. V and h directions. At the exit of P_2, we detect some light polarized in the h-direction.

We now rotate W so that at the output of P_2, the transmitted intensity becomes zero again. This situation corresponds to two possibilities in W. Either the optic axis is parallel to the propagation direction or it is perpendicular to the propagation direction. When the optic axis is parallel to the direction of propagation, by definition, the polarization state of the light is unchanged in passing through W, as if W is not there. Hence no light will get through P_2. If the optic axis is perpendicular to the propagation direction and, in particular, in the polarization transmission direction h of P_2, the incident V - polarization will pass through W because it is perpendicular to the optic axis and thus represents the O - wave. Since there is no e - component in the incident beam to W. Hence only V - polarization comes out from W and P_2 will absorb or reject it. Similarly, if the optic axis is in the direction V of P_1, the incident V - polarization will pass through W because it is now an e - wave. P_2 again absorbs it.

We now keep W fixed at such a position and rotate P_1 only by an angle $\Theta \neq 90°$ about the z - axis (propagation axis). If the optic axis in W is parallel to the propagation direciton, any polarization will pass through W without any change and P_2 will stop it; i.e. no transmission after P_2. If the optic axis in W is in the h or V - direction, there will be O and e waves generated in W so that P_2 will transmit the h - polarization component. Thus, the optic axis is determined.

Some comments on slow and fast axes

In the beginning of this section, we very briefly mentioned the names fast and slow axes. They can be confusing if one isn't careful. The fast and slow axes represent the polarization directions of the O and e waves inside a uniaxial crystal. If the O - wave propagates faster than the e - wave, (positive uniaxial crystal) the polarization vector of the O - wave is along the fast axis and that of the e - wave is along the slow axis. For a negative uniaxial crystal, the reverse is true. That is, since the e - wave propagates faster than the O - wave, the e - wave's polarization vector is along the fast axis and that of the O - wave along the slow axis.

In all, the fast (slow) axis represents the polarization direction of faster (slower) of the O and e waves.

§7.10 The power of crossed polarisers

A pair of crossed polarizers are simply two polarizers of any type whose transmission axes are perpendicular to each other so that light of some appropriate wavelength and of any polarization will not pass through them. (The light will pass through one but will be blocked by the other.) It becomes essentially a light isolator (Fig. 7-15(a)). If one inserts between the two polarizers an optical element that alters the state of polarization of the light that is inside there, the light whose polarization is altered will be transmitted. Such an idea has been put to good use in many optical systems. The following are some examples.

Fig. 7-15(b) shows a pair of crossed polarizers with an electro-optical element in between. When appropriately biased, the E-O crystal can act as a quarter wave plate, for example, and the output of the ensemble follows the voltage change across the E-O crystal, i.e. modulation. The crossed polarizer can be applied to strain analysis in matter (§9.2, last paragraph), Q-switching, mode locking, pulse slicing, pulse selection, back reflection isolator in laser amplifier chains, etc. Some of the above items will be explained in Chapter VIII.

We discuss the last item, i.e. back reflection isolation. This is shown in Fig. 7-15(c). When the laser pulse (short, less than a few tens of nanoseconds) arrives at the device, the E-O element is quickly biased to V_π so that the laser pulse, polarized vertically, after being transmitted by the first polarizer, is transformed into horizontally polarized light (the

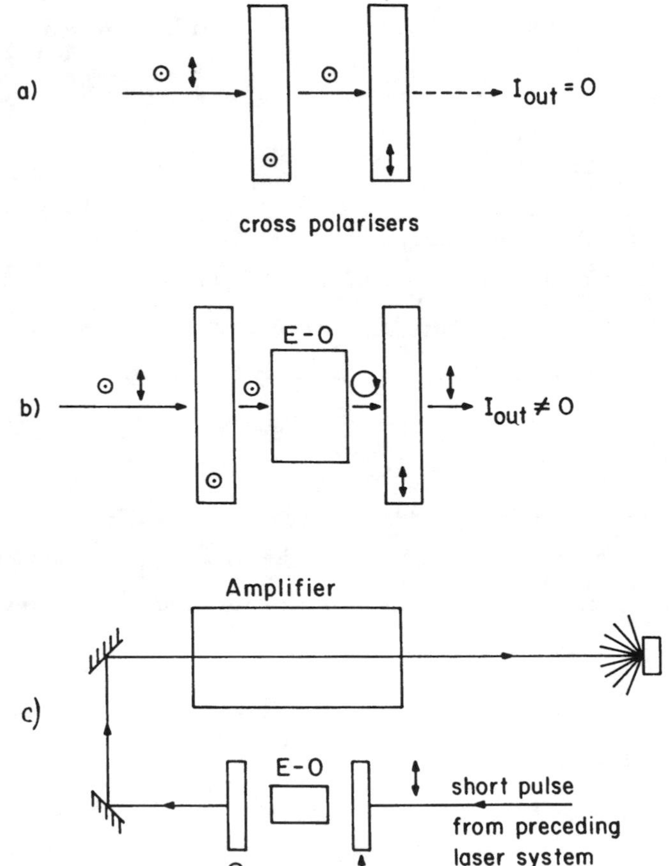

Fig. 7-15. (a), (b) An example of the application of a cross polarizer with an electro-optic element E - O.

(c) A cross polarizer together with an electro-optic element (E - O) acts as a back reflection isolator in a laser oscillator-amplifiers chain.

E - O element is of course properly oriented with its optic axis along the propagation direction. It acts as an electrically induced half wave plate) and is transmitted by the second crossed polarizer. Soon after the passage of the laser pulse, the high voltage across the E-O crystal is turned off. Now, the laser pulse, after passing through the amplifier, hits the target and there is then some reflection that will come back along the original optical path. If the amplifier's gain is still significant at this moment, it will amplify the back scattered pulse significantly. But such an amplified pulse will not go through the isolator because the E-O crystal is now not biased and is isotropic (optic axis along propagation direction). Any light will not go through the crossed polarizer, because the E-O crystal doesn't affect the polarization state of the light passing through it.

Closing comment

This chapter presents a rather detailed discussion on the polarization states of light and how they could be manipulated using wave plates. The Jones matrix formation helps in analyzing and designing systems with both active and passive polarizing components.

CHAPTER VIII

ELECTRIC FIELD INDUCED ANISOTROPY: ELECTRO-OPTICS AND Q-SWITCHING

An external electric field will induce in a dielectric a polarization which in general is anisotropic whether or not the material is naturally anisotropic. Depending on the electric field strength and the material structure, some anisotropy is negligible but some is significant. When light (laser) waves propagate through a transparent dielectric across which an external electric field is applied, the state of polarization of the light waves will be changed because of the natural and induced anisotropy of the material. Because one can vary the external electric field at rather high frequencies, the variation of the state of polarization of the light waves will thus be fast. This means that one can electrically manipulate light waves at a fast rate through the manipulation of the state of polarization or the phase of the light. If one can find a detector to detect such changes, such manipulation or modulation can be applied to high speed signal processing, communication, short laser pulse generation, fast shutter, etc. High speed modulation is possible in principle because the frequency of the optical wave (carrier wave) is very high ($\sim 10^{14}$ Hz). Thus, coupled with laser beams, such so-called electro-optical devices become one of the principal elements in a photonic or optoelectronic system.

§8.1 Electric field induced anisotropy

A D.C. electric field is applied across an anisotropic material. Normally, the external electric field is weak compared to the internal fields of the material (usually crystal). Hence, we can consider the external electric field as a perturbation to the internal field in later calculations.

The external D.C. field induces a polarization in the anisotropic medium. If we now send in an E-M wave, it will "detect" that the original unperturbed dielectric tensor is modified (by the D.C. field induced polarization). This means that the original index ellipsoid describing the response of the unperturbed medium to the E-M wave will be modified. In this chapter, we shall study how the index ellipsoid is changed and how this is used in applications. Essentially, the ellipsoid will be deformed and tilted through a rotation in space with respect to the original ellipsoid. We now analyse carefully what happens.

Originally, the index ellipsoid in the principal coordinate axes is given by (see chapter VI)

$$\frac{x^2}{n_x^2} + \frac{y^2}{n_y^2} + \frac{z^2}{n_z^2} = 1 \tag{8-1}$$

where $$n_i^2 \equiv \frac{\epsilon_i}{\epsilon_o} \qquad (i = x,y,z,) \tag{8-2}$$

$$(x,y,z,) \equiv (D_x \ , \ D_y \ , \ D_z) \frac{1}{\sqrt{\epsilon_o 2W_e}} \tag{8-2}$$

The external D.C. electric field induces a change in the polarization, the dielectric tensor, the indices and the field $\vec{D}$. The new dielectric tensor will no longer be diagonal and the equation of the new index ellipsoid will contain cross terms in the original principal coordinate axes. It's more convenient to use the notion of impermeability to describe the index ellipsoid (eq. (8-1)).

Definition: <u>The impermeability tensor</u>

$$\eta_{ij} \equiv \epsilon_o / \epsilon_{ij} \qquad (8\text{-}4)$$

where $\epsilon_{ii} = \epsilon_x$, ϵ_y , ϵ_z for i = x,y,z respectively. Substituting eq. (8-4) into eq. (8-1), one obtains

$$\eta_{xx}x^2 + \eta_{yy}y^2 + \eta_{zz}z^2 = 1 \qquad (8\text{-}5)$$

With the external electric field applied to the medium, eq. (8-5) now changes into one containing cross terms;

i.e. $$\eta_{ij}x_ix_j = 1 \qquad (8\text{-}6)$$

where η_{ij} is a function of $\vec{E}$, and where repeated indices represent summation over the indices and for simplicity, the coordinates x,y,z are replaced by x_i or x_j where i,j = 1,2,3. In what follows, (x,y,z) and (x_1 , x_2 , x_3) will be interchanged freely, using whichever is more convenient.

η_{ij} is now a function of the external field $\vec{E}$; i.e. $\eta_{ij}(\vec{E})$. Because we have assumed that the external field induces only a perturbation in the permeability, we can express $\eta_{ij}(\vec{E})$ in a Taylor series around the field free value (i.e. at $\vec{E}$ = 0).

$$\eta_{ij}(\vec{E}) = \eta_{ij}(0) + \sum_k \left(\frac{\partial \eta_{ij}}{\partial E_k} \right)_{E_k = 0} (E_k - 0)$$

$$+ \frac{1}{2} \sum_{k,\ell} \left(\frac{\partial^2 \eta_{ij}}{\partial E_k \partial E_\ell} \right)_{E_k = E_\ell = 0} (E_k - 0)(E_\ell - 0)$$

$$+ \ldots . \qquad (8\text{-}7)$$

Rewriting eq. (8-7) as

$$\eta_{ij}(\vec{E}) - \eta_{ij}(0) \equiv \Delta\eta_{ij}(\vec{E})$$
$$= \sum_k (r_{ij})_k E_k$$
$$+ \sum_{k,\ell} (s_{ij})_{k\ell} E_k E_\ell + \ldots. \quad (8-8)$$

where $(r_{ij})_k \equiv \left(\frac{\partial \eta_{ij}}{\partial E_k} \right)_{E_k = 0}$, $(s_{ij})_{k\ell} \equiv \frac{1}{2} \left(\frac{\partial^2 \eta_{ij}}{\partial E_k \, \partial E_\ell} \right)_{E_k = E_\ell = 0}$ (8-9)

Eq. (8-8) represents the electro-optic response of the medium. Using repeated indices to represent summation, eq. (8-8) becomes

$$\Delta\eta_{ij}(\vec{E}) = r_{ijk}E_k + s_{ijk\ell}E_kE_\ell + \ldots \quad (8-10)$$

where $r_{ijk} \equiv (r_{ij})_k \equiv$ <u>linear or Pockels electro-optic coefficient</u>

and $s_{ijk\ell} \equiv (s_{ij})_{k\ell} \equiv$ <u>quadratic or Kerr electro-optic coefficient</u>

We note that when $\vec{E} = 0$, eq. (8-6) becomes

$$\eta_{ij}(0)x_ix_j = 1 \quad (8-11)$$

which should be identical to eq. (8-5).

i.e. $\eta_{ij}(0)x_ix_j = \eta_{xx}x^2 + \eta_{yy}y^2 + \eta_{zz}z^2$

$$\equiv \frac{x^2}{n_x^2} + \frac{y^2}{n_y^2} + \frac{z^2}{n_z^2} \quad (8-13)$$

§8.2 Linear electro-optic effect: Pockels effect

If in the expression for the electro-optic response of the medium (eq. 8-8 or 8-10), the first term at the right hand side is dominant, we can neglect the higher order perturbation terms and obtain

$$\Delta\eta_{ij}(\vec{E}) \cong r_{ijk}E_k \tag{8-14}$$

This is the linear electro-optic response (Pockels effect.) We note that since ϵ_{ij} is a symmetric tensor, η_{ij} (eq. 8-4) must also be symmetric; i.e.

$$\eta_{ij} = \eta_{ji} \qquad (i \neq j) \tag{8-15}$$

Hence

$$r_{ijk} \equiv \left(\frac{\partial\eta_{ij}}{\partial E_k}\right)_{E_k = 0} \qquad \text{(by eq. 8-9)}$$

$$= \left(\frac{\partial\eta_{ji}}{\partial E_k}\right)_{E_k = 0} \qquad \text{(by eq. 8-15)}$$

$$\equiv r_{jik} \tag{8-16}$$

We can now write the equation of the modified index ellipsoid (eq. 8-6) as follows.

$$\eta_{ij}(\vec{E})x_i x_j = 1$$

$$[\eta_{ij}(0) + \Delta\eta_{ij}(\vec{E})]x_i x_j = 1$$

$$\frac{x^2}{n_x^2} + \frac{y^2}{n_y^2} + \frac{z^2}{n_z^2} + r_{ijk}E_k x_i x_j = 1 \tag{8-17}$$

Where eq. (8-13) and (8-14) were used. We expand the fourth term on the left hand side of eq. (8-17).

$$r_{ijk}E_k x_i x_j \equiv \sum_{i,j,k=1}^{3} r_{ijk}E_k x_i x_j$$

$$\equiv \sum_{i,j} x_i x_j \left(\sum_k r_{ijk}E_k\right)$$

$$= \sum_{i,j} x_i x_j \left(r_{ij1}E_1 + r_{ij2}E_2 + r_{ij3}E_3\right)$$

$$= \ldots \quad (\underline{\text{exercise}})$$

$$= (x_1 , x_2 , x_3) \begin{Bmatrix} (x_1 , x_2 , x_3) \begin{bmatrix} (r_{111} , r_{112} , r_{113}) \\ (r_{121} , r_{122} , r_{123}) \\ (r_{131} , r_{132} , r_{133}) \end{bmatrix} \\ (x_1 , x_2 , x_3) \begin{bmatrix} (r_{211} , r_{212} , r_{213}) \\ (r_{221} , r_{222} , r_{223}) \\ (r_{231} , r_{232} , r_{233}) \end{bmatrix} \\ (x_1 , x_2 , x_3) \begin{bmatrix} (r_{311} , r_{312} , r_{313}) \\ (r_{321} , r_{322} , r_{323}) \\ (r_{331} , r_{332} , r_{333}) \end{bmatrix} \end{Bmatrix} \begin{bmatrix} E_1 \\ E_2 \\ E_3 \end{bmatrix} \qquad (8\text{-}18)$$

The matrix multiplication of eq. (8-18) should be done from right to left. We use the following short hand notation to simplify the triple subscripts of r.

(ij)	SHORT HAND NOTATION
(11)	1
(22)	2
(33)	3
(23) ≡ (32)	4
(13) ≡ (31)	5
(12) ≡ (21)	6

The last three rows are permitted because of the symmetry of the permeability tensor (eq. 8-15, 16). With these, eq. (8-18) can be simplified. (This is an <u>exercise</u> for the reader.)

$$M \equiv r_{ijk}E_k x_i x_j$$

$$\equiv (x_1^2\ ,\ x_2^2\ ,\ x_3^2\ ,\ 2x_2x_3\ ,\ 2x_1x_3\ ,\ 2x_1x_2) \begin{pmatrix} r_{11} & r_{12} & r_{13} \\ r_{21} & r_{22} & r_{23} \\ r_{31} & r_{32} & r_{33} \\ r_{41} & r_{42} & r_{43} \\ r_{51} & r_{52} & r_{53} \\ r_{61} & r_{62} & r_{63} \end{pmatrix} \begin{pmatrix} E_1 \\ E_2 \\ E_3 \end{pmatrix} \qquad (8\text{-}19)$$

(The matrix r_{ij} is called the electro-optic tensor or coefficient.) Hence the general index ellipsoid under the linear electro-optic effect is, from eq. (8-17), using $(x,y,z) \equiv (x_1\ ,\ x_2\ ,\ x_3)$, for the sake of uniformity:

$$\frac{x_1^2}{n_{x_1}^2} + \frac{x_2^2}{n_{x_2}^2} + \frac{x_3^2}{n_{x_3}^2} + M = 1 \qquad (8\text{-}20)$$

where the 4th term, M, on the left hand side is given by eq. (8-19). Normally, the matrix r_{ij} in eq. (8-19) is known for an electro-optical material.

Example: a KDP crystal

KDP ≡ Potassium dihydrogen phosphate or KH_2PO_4.

This is a very popular crystal and its r_{ij} matrix is known:

$$r_{ij} = \begin{pmatrix} 0 & 0 & 0 \\ 0 & 0 & 0 \\ 0 & 0 & 0 \\ r_{41} & 0 & 0 \\ 0 & r_{41} & 0 \\ 0 & 0 & r_{63} \end{pmatrix} \qquad (8\text{-}21)$$

Values of the elements of r_{ij} for various practical crystals have been tabulated. (See Yariv and Yeh).

Substituting into eq. (8-19) gives

$$r_{ijk}E_k x_i x_j = 2r_{41}E_1x_2x_3 + 2r_{41}E_2x_1x_3 + 2r_{63}E_3x_1x_2 \tag{8-22}$$

Substituting into eq. (8-20), and changing (x_1, x_2, x_3) back to (x,y,z) for the sake of convenience, we have

$$\frac{x^2}{n_x^2} + \frac{y^2}{n_y^2} + \frac{z^2}{n_z^2} + 2r_{41}E_1yz + 2r_{41}E_2xz + 2r_{63}E_3xy = 1 \tag{8-23}$$

KDP is a uniaxial crystal;

$$\therefore \quad n_x = n_y \equiv n_o \quad , \quad n_z \equiv n_e \tag{8-24}$$

Eq. (8-23) becomes

$$\frac{x^2}{n_o^2} + \frac{y^2}{n_o^2} + \frac{z^2}{n_e^2} + 2r_{41}E_1yz + 2r_{41}E_2xz + 2r_{63}E_3xy = 1 \tag{8-25}$$

Case 1: Electric field in the z - direction

i.e. $\quad E_1 = E_2 = 0 \quad , \quad E_3 \equiv E_z \neq 0 \qquad (8\text{-}26)$

(See Fig. 8-1, a) Eq. (8-25) becomes

$$\frac{x^2}{n_o^2} + \frac{y^2}{n_o^2} + \frac{z^2}{n_e^2} + 2r_{63}E_zxy = 0 \tag{8-27}$$

This is the equation of an ellipsoid. Rotation of the axes (see below) will bring eq. (8-27) into the following familiar form:

$$\frac{x'^2}{n_{x'}^2} + \frac{y'^2}{n_{y'}^2} + \frac{z'^2}{n_{z'}^2} = 1 \tag{8-28}$$

where (x', y', z') is the new principal coordinate axes with $n_{x'}$, $n_{y'}$, $n_{z'}$ the corresponding indices. To derive eq. (8-28), we need to find the analytic expressions for $n_{x'}$ etc. We make the following observations.

(a) In both the equations (8-27) and (8-28), the xz and yz terms are missing. The rotation does not involve any rotation of the z - axis; hence $z = z'$. Thus $n_e^2 = n_{z'}^2$. (Fig. 8-1, b, c)

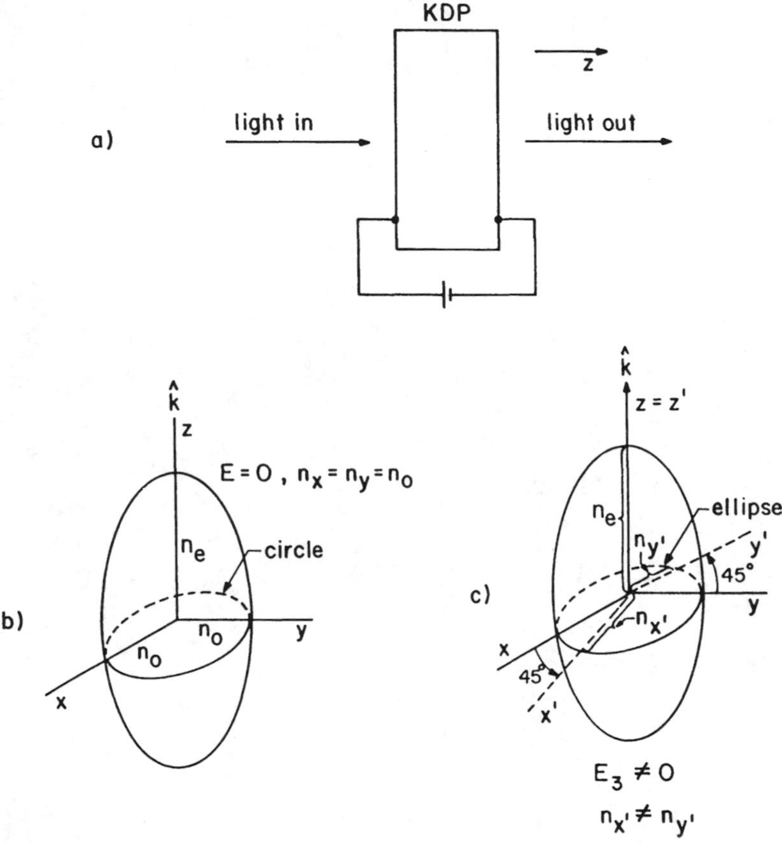

Fig. 8-1. Analysis of a KDP crystal which acts as an electro-optic wave plate (Pockels effect).

(b) The rotation is thus a rotation of the x,y axes about the z - axis (from observation (a)). Let the angle of rotation be α. From analytic geometry, every quadratic expression

$$Ax^2 + Bxy + Cy^2 \qquad (B \neq 0) \tag{8-29}$$

can be reduced to the form

$$A'x'^2 + C'y'^2$$

by rotating the axes through an angle α ($0 < \alpha < 90°$) where

$$\cot 2\alpha = \frac{A - C}{B} \tag{8-30}$$

From eq. (8-27) and (8-29), we see

that $\quad A = C$ in (8-30)

Hence $\cot 2\alpha = 0$

$$\alpha = 45° \tag{8-31}$$

and
$$x = x'\cos\alpha - y'\sin\alpha \tag{8-32}$$

$$y = x'\sin\alpha + y'\cos\alpha \tag{8-33}$$

Substituting eq. (8-32), (8-33) with (8-31) into eq. (8-27) gives (exercise)

$$\left(\frac{1}{n_o^2} + r_{63}E_z\right) x'^2 + \left(\frac{1}{n_o^2} - r_{63}E_z\right) y'^2 + \frac{z'^2}{n_e^2} = 1 \tag{8-34}$$

Alternative way of deriving eq. (8-34):

This is a general way of diagonalizing the symmetric matrix

$$S = \begin{pmatrix} a & d & e \\ d & b & f \\ e & f & c \end{pmatrix} \tag{8-35}$$

of the general quadratic equation

$$ax^2 + by^2 + cz^2 + 2dxy + 2exz + 2fyz = g \tag{8-36}$$

The diagonalization requires that

$$| S - \lambda I | = 0 \tag{8-37}$$

where I is a unitary matrix and λ is a root of the diagonal matrix of S;

i.e.
$$S_D \equiv \begin{pmatrix} \lambda_1 & 0 & 0 \\ 0 & \lambda_2 & 0 \\ 0 & 0 & \lambda_3 \end{pmatrix} \tag{8-38}$$

The reader can solve eq. (8-37) as an exercise using the definitions in (8-35), (8-36) and comparing with (8-27).

Comparing eq. (8-34) and (8-28), we have

$$\frac{1}{n_{x'}^2} = \frac{1}{n_o^2} + r_{63}E_z \qquad (8\text{-}39)$$

$$\frac{1}{n_{y'}^2} = \frac{1}{n_o^2} - r_{63}E_z \qquad (8\text{-}40)$$

$$n_{z'} = n_e \qquad (8\text{-}41)$$

Since the change in the index by the electric field is assumed to be a perturbation (i.e. small), we can write

$$\frac{1}{n_i^2} - \frac{1}{n_o^2} = \Delta\left(\frac{1}{n^2}\right)\Bigg|_{n = n_o} \quad (\ \Delta : \text{differential}, \quad i = x' \text{ or } y') \qquad (8\text{-}42)$$

$$= -\ 2n^{-3}\Big|_{n = n_o} \ \Delta n$$

$$= -\ 2n_o^{-3}(n_i - n_o)$$

$$n_i - n_o = -\frac{1}{2}\, n_o^3 \left(\frac{1}{n_i^2} - \frac{1}{n_o^2}\right) \qquad (8\text{-}43)$$

Substituting eq. (8-39), (8-40) into (8-43) separately, one obtains

$$n_{x'} = n_o - \frac{1}{2}\, n_o^3\, r_{63}\, E_z \qquad (8\text{-}44)$$

$$n_{y'} = n_o + \frac{1}{2}\, n_o^3\, r_{63}\, E_z \qquad (8\text{-}45)$$

$$n_{z'} = n_e \qquad (8\text{-}46)$$

Case 2: Electric field in the x - direction

i.e. $\qquad E_1 \equiv E_x \neq 0 \quad , \qquad E_2 = E_3 = 0 \qquad$ (8-47')

Substituting into eq. (8-25), we have

$$\frac{x^2}{n_o^2} + \frac{y^2}{n_o^2} + \frac{z^2}{n_e^2} + 2r_{41}E_x yz = 1 \qquad (8\text{-}47)$$

Using a similar analysis as in case 1, one can rotate the y,z axes about the x - axis and transform eq. (8-47) into the "normal" form of an ellipsoid.

The detail is left as an exercise. Let α be the angle through which the y,z axes are rotated, one obtains

$$\tan 2\alpha = \frac{2r_{41}E_x}{\frac{1}{n_o^2} - \frac{1}{n_e^2}} \tag{8-48}$$

and the ellipsoid is

$$\frac{x'^2}{n_o^2} + y'^2 \left\{ \frac{1}{n_o^2} + \frac{\tan^2\alpha}{n_e^2} + 2r_{41}E_x\tan\alpha \right\} \cos^2\alpha$$

$$+ z'^2 \left\{ \frac{1}{n_e^2} + \frac{\tan^2\alpha}{n_o^2} - 2r_{41}E_x\tan\alpha \right\} \cos^2\alpha = 1 \tag{8-49}$$

From tabulated values, (Yariv and Yeh).

$$r_{41} \approx 9 \times 10^{-12} \quad m/V \qquad \text{at } \lambda \doteq 0.55\ \mu m$$

Hence, even at a very high D.C. voltage,

e.g. $\quad E_x = 10^6 \quad V/m$

$r_{41}E_x \approx 10^{-5}$, a very small value,

Hence $\quad \tan 2\alpha \approx 0$ (8-50)

$$\Rightarrow \begin{cases} \cos^2\alpha \approx 1 & (8\text{-}51) \\ \tan^2\alpha \approx 0 & (8\text{-}52) \end{cases}$$

Using eq. (8-51) and, (8-52) in (8-49) yields

$$\frac{x'^2}{n_o^2} + y'^2 \left\{ \frac{1}{n_o^2} + 2r_{41}E_x\tan\alpha \right\}$$

$$+ z'^2 \left\{ \frac{1}{n_e^2} - 2r_{41}E_x\tan\alpha \right\} = 1 \tag{8-53}$$

Again, using the same procedure as in eq. (8-42) and (8-43), the new indices are obtained (exercise).

$$n_{x'} = n_o \tag{8-54}$$

$$n_{y'} = n_o - \frac{1}{2} n_o^3 r_{41} E_x \tan\alpha \quad (8-55)$$

$$n_{z'} = n_e + \frac{1}{2} n_e^3 r_{41} E_x \tan\alpha \quad (8-56)$$

§8.3 Application to electrical modulation of light waves: electro-optic modulator

Consider a z - cut[1] KDP crystal plate (the propagation direction of the E-M wave is also along the z - axis.) We apply an a.c. electric field across it.

$$\vec{E}(t) = E_z(t)\ \hat{z} \quad (8-57)$$

The index ellipsoid of the crystal will thus change in time as the electric field varies. We study in this book only cases in which the frequency of the electric field is not too high so that the index change "follows" faithfully the variation of the external electric field; i.e. the index change and the electric field are in phase.

Case 1 D.C. field

Assume for the moment that the external field is still D.C. From §8.2, the modified index ellipsoid of the KDP crystal under a D.C. electric field directed along the z - direction is given by eq. (8-34), and eq. (8-54 to 56) give the modified indices (see also eq. 8-28). Since the propagation direction $\hat{k}$ is along the principal z - axis and z = z', we obtain the intersection ellipse between the x' , y' plane perpendicular to $\hat{k}$ and the modified index ellipsoid shown in Fig. 8-1b. Note that the modified ellipsoid does not possess a rotational symmetry about the z' - axis because

---

1 z - cut: the principal z - axis of the index ellipsoid of the crystal is perpendicular to the crystal (plate) surface. In this case of KDP which is a uniaxial crystal, the z - axis is also the optic axis (also called the c - axis) around which the index ellipsoid has a rotational symmetry. (See chap. VI).

$n_{x'} \neq n_{y'}$. Thus the E-M wave propagating in the z - direction sees two indices $n_{x'}$, and $n_{y'}$; i.e. there are two waves propagating at two different velocities $c/n_{x'}$ and $c/n_{y'}$ in the crystal. Since the incident direction is perpendicular to the crystal plate surface, the two waves in the crystal propagate in the same direction (z - axis). The relative phase retardation (shift) between the two waves is (see §7.7, eq. 7-32)

$$\epsilon = (k_{y'} - k_{x'})d \tag{8-58}$$

where d is the thickness of the plate.

$$\text{i.e.} \qquad \epsilon = \frac{\omega}{c}(n_{y'} - n_{x'})d \tag{8-59}$$

$$= \frac{\omega}{c} n_o^3 r_{63} E_z d \qquad \text{using eq. (8-44, 45)} \tag{8-60}$$

$$= \frac{\omega}{c} n_o^3 r_{63} V \tag{8-61}$$

where $V = E_z d$ = voltage applied across the plate surfaces. We now calculate the polarization of the output wave. Assume that the input (incident) wave is linearly polarized and whose electric field vector in the laboratory (x,y) coordinate is horizontal; i.e. in Jones vector representation,

$$\vec{E}_i = \begin{pmatrix} 1 \\ 0 \end{pmatrix} \tag{8-62}$$

We assume also that the modified (x' , y') coordinates of the crystal make an angle of 45° with the (x,y) coordinates. Assuming that $\epsilon = \pi$, we can calculate the output polarization using eq. (7-93) and eq. (8-62). This is equivalent to a half-wave plate (§7.9, example 2) except that in this case, the relative phase shift ϵ is induced by the external field. From eq. (7-106), the output polarization is rotated by 90° ($\because \psi = 45°$) (Fig. (8-2)). The voltage at which $\epsilon = \pi$ is thus called half-wave voltage, V_π , and by eq. (8-61),

$$V_\pi = \frac{c\pi}{\omega n_o^3 r_{63}} = \frac{\lambda}{2n_o^3 r_{63}} \tag{8-63}$$

In terms of V_π , eq. (8-61) can be rewritten as

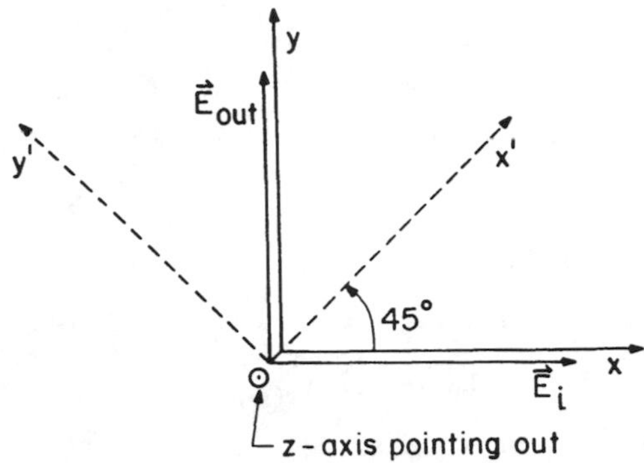

Fig. 8-2. Polarization rotation by a half-wave voltage.

$$\epsilon = \pi V/V_{\pi} \tag{8-64}$$

Thus, one can apply this technique to turn a laser beam's linear polarization by 90°. Of course, if the angle is ψ, this electrically induced half-wave plate will turn the polarization by an angle of 2ψ (§7.7 and 7.9).

Another voltage of interest is the <u>quarter wave voltage</u> ($V_{1/4} \equiv V_{\pi/2}$) at which the crystal is equivalent to a quarter wave plate (i.e. $\epsilon = \pi/2$). One sees that by eq. (8-61) $V_{1/4} = 1/2\ V_{\pi}$. Circular polarization will be obtained after one pass through the E - O crystal.

<u>Case 2. A C field: amplitude modulation</u>

We now vary the external field sinusoidally at the modulation frequency ω_m. Assume ω_m is much smaller than the optical frequency ω; i.e. $\omega_m \ll \omega$. We use a linear polarizer to analyse the output (Fig. 8-3). In general, the input light beam can be at any polarization state. We have chosen elliptical polarization in the figure. After passing through the A C field biased z - cut KDP crystal, the polarization of the output is in a different general state, represented in the figure by another elliptical polarization;

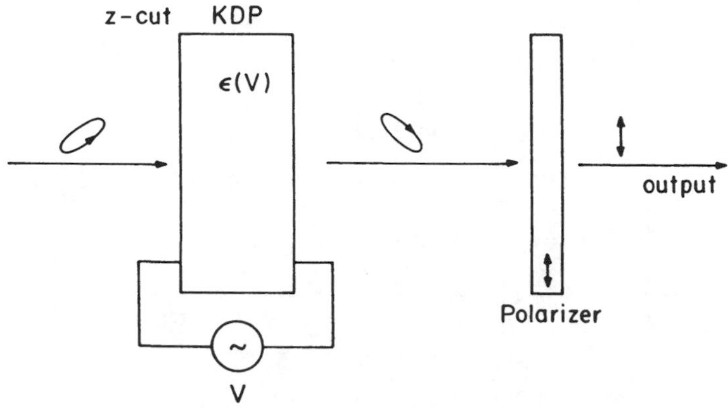

Fig. 8-3. A.C. field modulation across a KDP electro-optic crystal.

the polarizer transmits only the vertical component of the polarization. As the applied voltage changes with time, the output's polarization state changes so that the amplitude of its vertical component also changes with time. This leads to amplitude modulation.

In practice, one adds a quarter wave plate between the output polarizer (or analyser) and the KDP crystal, (Fig. 8-4). The KDP is biased as in the discussion of eq. (8-26 to 46.) We note from eq. (8-44) and (8-45), for positive E_z , $n_{y'} > n_{x'}$; i.e. the velocity of the wave in the crystal whose polarization is in the y' - direction is smaller than that in the x' - direction ($\because$ $v = \frac{c}{n}$, hence greater n means smaller v). Thus, y' and x' are the induced slow and fast axes respectively. The angle between x' and x is 45° (eq. 8-31). In Fig. 8-4(b) and (c), the following is chosen. The slow (s) axis of the quarter wave plate is parallel to the slow (y') axis of the KDP and the fast (f) parallel to the fast (x'). Thus, the quarter wave plate acts as if it is part of the KDP crystal with a fixed relative phase shift of $\pi/2$. The total relative phase shift is $\epsilon' = \epsilon + \pi/2$.

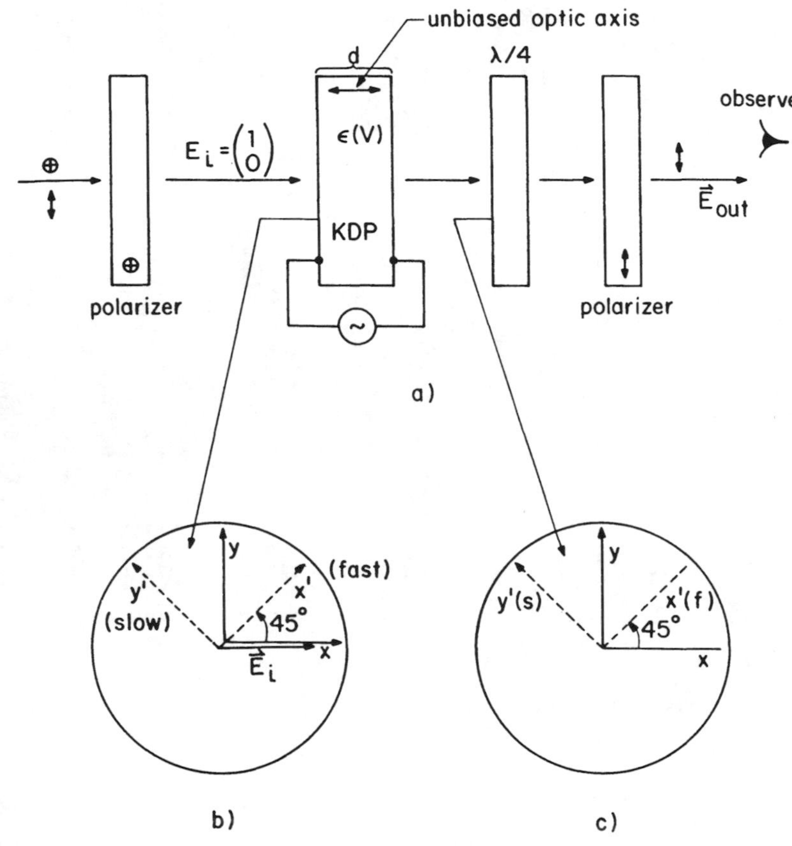

Fig. 8-4. Electro-optic amplitude modulation of a light wave.

The polarizer transmits polarization in the vertical direction perpendicular to the z - direction as indicated. Its Jones matrix is from eq. 7-96,

$$A_y = e^{-i\phi} \begin{pmatrix} 0 & 0 \\ 0 & 1 \end{pmatrix} \tag{8-65}$$

We can calculate $\vec{E}_{out}$ using eq. (7-93), (7-93'), (8-65), (8-62).

$$W = R\,(-\,45°)\; M_p\; R(45°)\; A \tag{8-66}$$

where
$$R\,(45°) = \frac{1}{\sqrt{2}} \begin{pmatrix} 1 & 1 \\ -1 & 1 \end{pmatrix} \tag{8-67}$$

$$R\,(-45°) = \frac{1}{\sqrt{2}}\begin{pmatrix} 1 & -1 \\ 1 & 1 \end{pmatrix} \tag{8-68}$$

$$M_p = e^{-i\phi}\begin{pmatrix} e^{-i\epsilon'/2} & 0 \\ 0 & e^{i\epsilon'/2} \end{pmatrix} \tag{8-69}$$

$$A = \begin{pmatrix} 0 & 0 \\ 0 & 1 \end{pmatrix} \qquad \text{(omitting the phase)} \tag{8-70}$$

$$\vec{E}_{out} = W\vec{E}_i = W\begin{pmatrix} 1 \\ 0 \end{pmatrix}$$

$$= i\,\sin(\epsilon'/2)e^{-i\phi}\begin{pmatrix} 0 \\ 1 \end{pmatrix} \tag{8-71}$$

The power transmission is then

$$T = \frac{\vec{E}_{out}^{\,*}\cdot\vec{E}_{out}}{\vec{E}_i^{*}\cdot\vec{E}_i} = \frac{\sin^2(\epsilon'/2)}{1}$$

or $$T = \sin^2(\epsilon'/2) \tag{8-72}$$

Now $$\epsilon' = \epsilon + \pi/2 \tag{8-73}$$

and if $$\epsilon \equiv \epsilon_m \sin\omega_m t \tag{8-74}$$

we obtain

$$T = \sin^2\left(\pi/4 + \frac{\epsilon_m}{2}\sin\omega_m t\right)$$

$$= \left\{\frac{1}{2} - \cos\left(\frac{\pi}{2} + \epsilon_m \sin\omega_m t\right)\right\}$$

$$= \frac{1}{2}\{1 + \sin(\epsilon_m \sin\omega_m t)\} \tag{8-75}$$

For $\epsilon \ll 1$,

$$T = \frac{1}{2}\left\{1 + \epsilon_m \sin\omega_m t - \ldots.\right\}$$

$$\simeq \frac{1}{2}(1 + \epsilon_m \sin\omega_m t) \tag{8-76}$$

The transmission follows ω_m or the transmitted power is in phase with the modulation frequency. The amplitude ϵ_m is given by eq. (8-61):

$$\epsilon_m = \frac{\omega}{c} n_o^3 r_{63} V \tag{8-77}$$

In terms of the half-wave voltage V_π (eq. 8-63),

$$\epsilon_m = \frac{\pi V}{V_\pi} \tag{8-78}$$

Eq. (8-76) becomes

$$T = \frac{1}{2}\left[1 + \frac{\pi V}{V_\pi} \sin \omega_m t \right] \tag{8-79}$$

Figure (8-5) shows such a modulated transmission with respect to the D.C. transmission

$$T_{D.C.} = \sin^2(\epsilon/2) \qquad \text{(using eq. 8-72)}$$

$$= \sin^2 \frac{\pi}{2}\left[\frac{V}{V_\pi} \right] \qquad \text{(using eq. 8-64)} \tag{8-80}$$

Case 3: A C field: phase modulation (FM)

The same z - cut KDP with the same bias voltage can be used except that the input wave's polarization is parallel to the x' - axis of the crystal

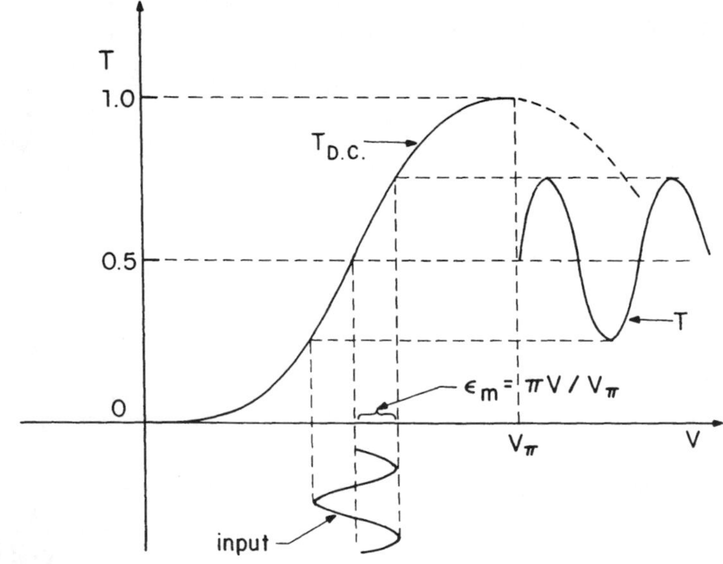

Fig. 8-5. Result of electro-optic amplitude modulation.

(Fig. 8-6). Since the input polarization is already parallel to one of the principal axes of the crystal (Fig. 8-6b), the state of polarization of the wave will remain the same in propagating through the crystal. However, its phase will be retarded. From eq. (8-44), the expression for $n_{x'}$ which is the index of the wave whose polarization is in the x' - direction, we see that if $E_z = 0$, the incident wave sees the index n_o. Hence, the phase of the transmitted wave will change when E_z varies from zero to a finite value.

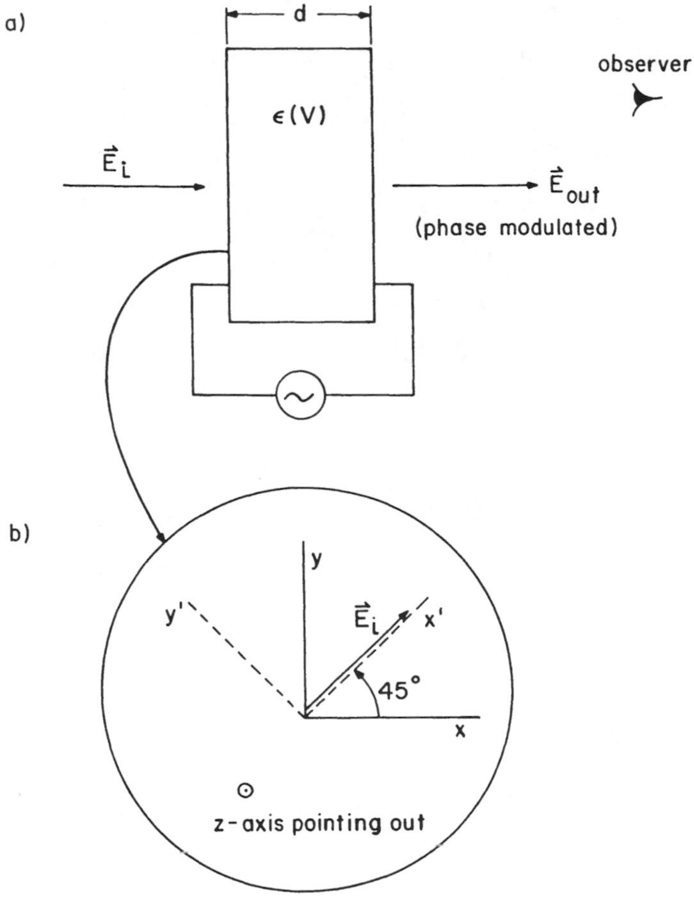

Fig. 8-6. Electro-optic phase modulation.

If $\vec{E}_i = A\cos(\omega t - kz)$, assuming no loss,

$$\vec{E}_{out} = A\cos(\omega t - kz - k'd)$$

where

$$k'd = \frac{\omega}{c} n_{x'} d$$

$$= \frac{\omega}{c} d(n_o - \frac{1}{2} n_o^3 r_{63} E_z) \qquad \text{by eq. (8-44)}$$

$$= \frac{\omega n_o d}{c} - \frac{\omega n_o^3 r_{63}}{2c} V \qquad (V = E_z d)$$

If $V = V_m \sin\omega_m t$,

$$\vec{E}_{out} = A\cos\left(\omega t - kz - \frac{\omega n_o d}{c} + \delta\sin\omega_m t \right) \tag{8-81}$$

where

$$\delta \equiv \frac{\omega n_o^3 r_{63}}{2c} V_m \tag{8-82}$$

Eq. (8-81) is an equation of phase modulation.

§8.4 Quadratic electro-optic effect

We start with eq. (8-10), the Taylor expansion of the change of impermeability due to an external field:

$$\Delta\eta_{ij}(\vec{E}) = r_{ijk}E_k + s_{ijk\ell}E_kE_\ell + \ldots. \tag{8-83}$$

If, because of some reason, such as inversion symmetry in the crystal, $r_{ijk} = 0$, we consider then the contribution of the quadratic term in eq. (8-83).

i.e.

$$\Delta\eta_{ij}(\vec{E}) \equiv \eta_{ij}(\vec{E}) - \eta_{ij}(0)$$

$$= s_{ijk\ell}E_kE_\ell \tag{8-84}$$

where from eq. (8-9),

$$s_{ijk\ell} \equiv \frac{1}{2}\left(\frac{\partial^2\eta_{ij}}{\partial E_k \partial E_\ell} \right)_{E_k = E_\ell = 0} \tag{8-85}$$

Since the order of the partial differentiation is not important, we can interchange k and ℓ.

Thus

$$s_{ijk\ell} = s_{ij\ell k} \qquad (8\text{-}86)$$

Also, because of symmetry of the index tensor $\epsilon_{ij} = \epsilon_{ji}$

and thus $\eta_{ij} = \eta_{ji}$ (eq. 8-15)

Eq. (8-85) is unchanged with i and j interchanged.

i.e. $$s_{ijk\ell} = s_{jik\ell} \qquad (8\text{-}87)$$

Because of eq. (8-86 and 87), we are allowed to use the following short hand notation.

(ij) or (kℓ)	SHORT HAND NOTATION
(11)	1
(22)	2
(33)	3
(23) = (32)	4
(13) = (31)	5
(12) = (21)	6

The index ellipsoid (eq. 8-6)

$$\eta_{ij}(\vec{E})x_i x_j = 1$$

now becomes (using eq. (8-84))

$$(\eta_{ij}(0) + s_{ijk\ell}E_k E_\ell)x_i x_j = 1 \qquad (8\text{-}88)$$

From now on, (x,y,z,) and (x_1 , x_2 , x_3) are freely interchanged whenever convenient. Eq. (8-88) becomes

$$\frac{x^2}{n_x^2} + \frac{y^2}{n_y^2} + \frac{z^2}{n_z^2} + s_{ijk\ell}E_k E_\ell x_i x_j = 1 \qquad (8\text{-}89)$$

Using the short hand notation and expanding the summation, the fourth term in eq. (8-89) becomes, after some tedious but straight forward algebra, (exercise)

$$s_{ijk\ell}E_kE_\ell x_ix_j = (x_1^2\ ,\ x_2^2\ ,\ x_3^3\ ,\ 2x_2x_3\ ,\ 2x_1x_3\ ,\ 2x_1x_2)\ \cdot$$

$$\begin{pmatrix} s_{11} & s_{12} & s_{13} & s_{14} & s_{15} & s_{16} \\ s_{21} & s_{22} & s_{23} & s_{24} & s_{25} & s_{26} \\ s_{31} & s_{32} & s_{33} & s_{34} & s_{35} & s_{36} \\ s_{41} & s_{42} & s_{43} & s_{44} & s_{45} & s_{46} \\ s_{51} & s_{52} & s_{53} & s_{54} & s_{55} & s_{56} \\ s_{61} & s_{62} & s_{63} & s_{64} & s_{65} & s_{66} \end{pmatrix} \begin{pmatrix} E_1^2 \\ E_2^2 \\ E_3^2 \\ 2E_3E_2 \\ 2E_1E_3 \\ 2E_1E_2 \end{pmatrix}$$

$$\equiv\ M \tag{8-90}$$

The ellipsoid for the second order (quadratic) electro-optic effect is (from eq. 8-89)

$$\frac{x^2}{n_x^2} + \frac{y^2}{n_y^2} + \frac{z^2}{n_z^2} + M = 1 \tag{8-91}$$

where M is given by eq. (8-90)

<u>**Example**</u>: <u>An isotropic medium: Kerr effect</u>

We consider an isotropic medium containing polarizable molecules such as liquid CS_2. The CS_2 polar molecules can be aligned in an electric field. Of course, not all the CS_2 molecules are aligned along the field direction. Only a tiny fraction of the molecules are either lined up or partially lined up, the rest being "destroyed" randomly by collisions. The result is the creation of a net field induced polarization in the liquid in the direction of the applied field. This makes the liquid anisotropic. The quadratic E - O coefficient or matrix s_{ij} (eq. 8-90) of an isotropic medium can be found from tabulated sources (eg. Yariv and Yeh). Thus, we have

$$M = \left(x_1^2 , x_2^2 , x_3^2 , 2x_2x_3 , 2x_1x_3 , 2x_1x_2 \right) \cdot \begin{pmatrix} s_{11} & s_{12} & s_{12} & 0 & 0 & 0 \\ s_{12} & s_{11} & s_{12} & 0 & 0 & 0 \\ s_{12} & s_{12} & s_{11} & 0 & 0 & 0 \\ 0 & 0 & 0 & \frac{1}{2}(s_{11}-s_{12}) & 0 & 0 \\ 0 & 0 & 0 & 0 & \frac{1}{2}(s_{11}-s_{12}) & 0 \\ 0 & 0 & 0 & 0 & 0 & \frac{1}{2}(s_{11}-s_{12}) \end{pmatrix} \cdot \begin{pmatrix} E_1^2 \\ E_2^2 \\ E_3^2 \\ 2E_3E_2 \\ 2E_1E_3 \\ 2E_2E_1 \end{pmatrix} \tag{8-92}$$

and the index ellipsoid is

$$\frac{x^2}{n^2} + \frac{y^2}{n^2} + \frac{z^2}{n^2} + M = 1 \tag{8-93}$$

Note the isotropic index n = constant.

<u>Special case</u>: $\vec{E} = E_3 \hat{z}$

i.e. $E_1 = E_2 = 0$

On simplifying eq. (8-92), eq. (8-93) becomes

$$x_1^2 \left\{ \frac{1}{n^2} + s_{12}E_3^2 \right\} + x_2^2 \left\{ \frac{1}{n^2} + s_{12}E_3^2 \right\} + x_3^2 \left\{ \frac{1}{n^2} + s_{11}E_3^2 \right\} = 1 \tag{8-94}$$

Let
$$n_o^2 \equiv \left[\frac{1}{n^2} + s_{12}E_3^2 \right]^{-1} \tag{8-95}$$

$$n_e^2 \equiv \left[\frac{1}{n^2} + s_{11}E_3^2 \right]^{-1} \tag{8-96}$$

eq. (8-94) becomes

$$\frac{x^2}{n_o^2} + \frac{y^2}{n_o^2} + \frac{z^2}{n_e^2} = 1 \qquad (8\text{-}97)$$

This is an equation for a uniaxial medium. Eq. (8-95) and (8-96) can be simplified to yield expressions of n_o and n_e under the assumption of small index change. The derivation is similar to that of eq. (8-42, 43) and is left as an exercise. The result is

$$n_o = n - \frac{1}{2} n^3 \, s_{12} E_3^2 \qquad (8\text{-}98)$$

$$n_e = n - \frac{1}{2} n^3 \, s_{11} E_3^2 \qquad (8\text{-}99)$$

From the form of eq. (8-97) and the knowledge gained in chapter VI when considering optical anisotropy, we conclude that the induced 0 and e waves in the normally isotropic liquid (of CS_2) have orthogonal polarizations. Such liquid in a cell with plane windows acts as a wave plate. The relative phase shift of the two waves in traversing the liquid is given by eq. (7-32) in which the index difference, or birefringence is (using eq. 8-98, 99)

$$n_e - n_o = \frac{1}{2} \, n^3 \, (s_{12} - s_{11}) \, E_3^2 \qquad (8\text{-}100)$$

This is often expressed as

$$n_e - n_o = K \, \lambda \, E^2 \qquad (8\text{-}101)$$

where $\lambda \equiv$ vacuum wavelength of the light

$K \equiv$ Kerr constant

The values of K of some practical media can be found in tabulated sources (e.g. Yariv and Yeh).

§8.5 Electro-optical shutter: short laser pulse slicer and Q-switching lasers

The basic idea of using a shutter is simply that one would like to open and close the shutter at precise moments so that only during the open phase of the shutter will light pass through it. For high speed events, one would like to be able to open and close the shutter in as short a time as possible so as to transmit only a very short light pulse. A mechanical shutter is too slow.

The reasons why one needs to have very short pulses, especially laser pulses, are manifold. From the point of view of generating very intense instantaneous power of laser radiation, it is easy to arrive at a very high peak power using short pulses, because the power is defined as energy/pulse duration. For a constant energy content, the shorter the laser pulse is, the higher the peak power will be. From the point of view of data processing and communication, short pulses permit us to increase the data sending rate. From the point of view of observing phenomena that last for only a very short time, we need short pulses to illuminate and observe the phenomena, similar to using an extremely fast stroboscope. From the point of view of "instantaneous" excitation of matter, short pulses are required. Finally, just for the sake of pushing back the frontier of ultrashort pulse technology, one would like to produce the shortest pulses ever. The shortest laser pulse at around 0.62 μm is about a few femto-seconds (1 femtosecond = 10^{-15} sec). A new question is now posed, "How does an extremely short (femtosecond) and extremely intense laser pulse interact with matter? Could there be new physics discovered? Such are some research problems at present concerning intense short pulses.

In this section, we make use of the fact that we can change the state of polarization of a laser or light beam in an E - O crystal and try to turn it into a fast shutter ($<$ 1 nanosecond or 10^{-9} sec at the limit). We again use the same KDP as an example (§8.3). Two cases will be discussed qualita-tively. The KDP shutter is either inside or outside the laser cavity.

Case 1: Extra-cavity Electro-optic shutter: slicing short laser pulses

Fig. 8-7 shows the schematic setup. The KDP crystal is placed between two crossed polarizers. The one at the left (incident side) is not necessary if the long laser pulse is already polarized in the direction allowed by the polarizer. The uniaxial KDP's optic axis is along the propagation axis, as in the case of §8.3. When a long laser pulse polarized perpendicular to the paper arrives at the crystal, no high voltage is applied to the KDP yet. Thus, the laser light passes without any change in the state of polarization because the laser propagates along the optic axis. This light, when reaching the Glan prism, (or Glan Thompson prism or a stack of plates of isotropic material set at the Brewster angle (Chapter III etc.) will be reflected entirely and the transmission is zero. When the pulse's maximum passes through the KDP, a high voltage pulse of peak voltage equal to V_π (§8.3), the half-wave voltage, and of very short duration (say ~ 1 ns) is applied across the crystal. During this brief period of time the KDP acts as a halfwave plate and the polarization of the transmitted laser pulse is turned 90° with respect to the incident polarization so that the Glan prism

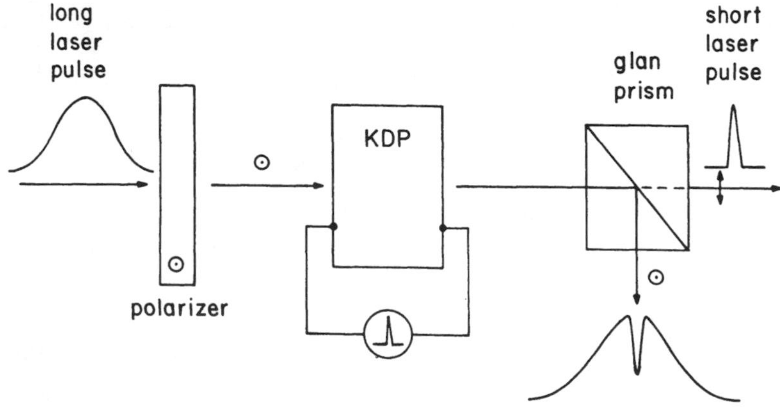

Fig. 8-7. Electro-optic laser pulse slicer (principle).

now transmits the pulse. After the short voltage pulse is over, the laser is again unperturbed and is completely reflected by the Glan-prism. In Fig. 8-7, the form of the reflected and transmitted pulses are shown.

The circuit for the creation of the short high voltage pulse is not trivial to understand although the circuit design is simple. Fig. 8-8 shows a circuit using the technique of a laser triggered spark gap. Fig. 8-8(a) shows a schematic drawing of a real laboratory setup while Fig. 8-8(b) is the abbreviated circuit diagram of (a). The reader should have a knowledge of electric (voltage) wave propagation in transmission lines in order to appreciate fully what is described here. (c.f. any textbook on applied electromagnetism for engineers.)

Referring to Fig. 8-8(b), the high voltage source (H.V.) is used to charge up the pulse forming 50Ω transmission line (PFTL) to a voltage of $2V_\pi$ (V_π is the half wave voltage for the crystal KDP.) After this, the contact C is open isolating the PFTL from the H.V. A laser pulse arrives some time later and is focussed into the laser triggered spark gap (LTSG) creating a plasma inside the cell (Spark gap). The rise time of the plasma formation is very fast ($<$ 1 ns). Now the PFTL "discharges" across the remaining part of the circuit. A high voltage "square" wave of peak voltage V_π and duration $2L/v$ is created, where L is the length of the PFTL and v is the wave propagation velocity in the cable. The pulse travels across the KDP and terminates at the 50 Ω cables, of course. In principle, any cable can be used so long as the terminating resistance is equal to the impedance of the cable. The terminating cable after the KDP crystal is very long because one wants to avoid residual pulse reflection arriving at the KDP before the short pulse is over. When this pulse travels across the KDP crystal the latter behaves like a half-wave plate. The laser polarization passing through it during

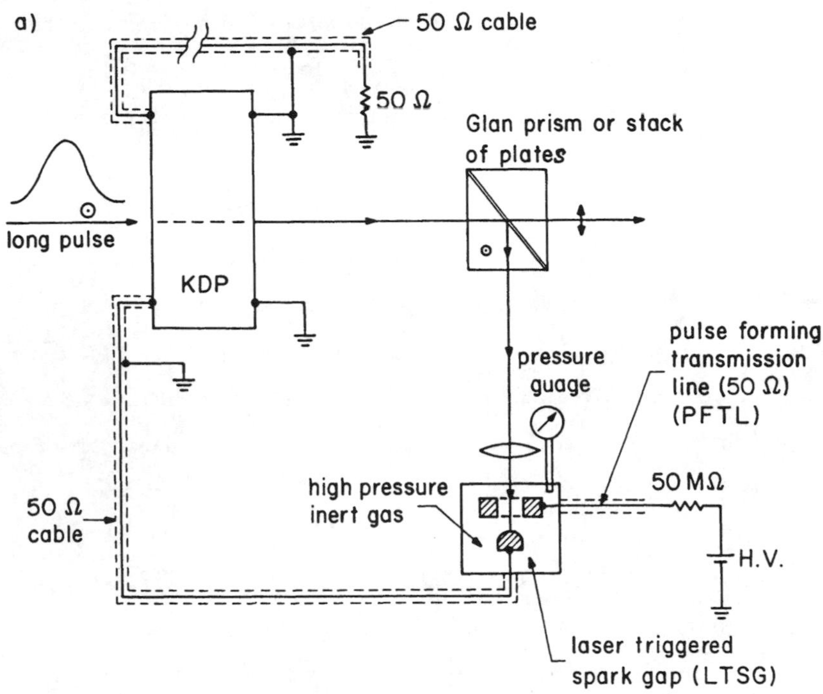

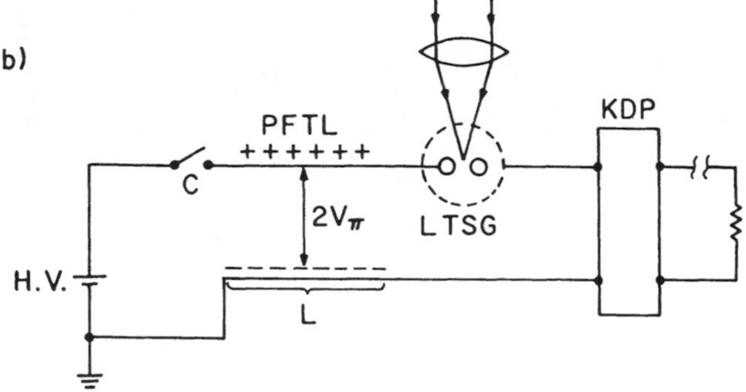

Fig. 8-8. (a) Schematic optic lay-out and electrical circuit for an electro-optic laser pulse slicing system.

(b) Transmission line analysis of an E-O laser pulse slicing system.

this time will be turned by 90° and is transmitted by the Glan prism (Fig. 8-8 (a). Before and after, all laser light that passes through the KDP crystal along its optic axis will be rejected by the Glan prism.

The duration of the electrical pulse, being 2L/v, can be varied by changing the length of the pulse forming cable (PFTL). Although in principle one can shorten the cable indefinitely, at least two difficulties limit the pulse duration to go shorter than $\lesssim$ 1 ns. The first is mechanical, i.e. it is not so easy to connect a very short cable in a practical system. But this is secondary. The more important difficulty is the rise time of the pulse which is controlled by the rise time of the plasma formation in the spark gap. The latter is at best ~ 0.05 – 0.1 ns so that the voltage pulse cannot be shorter than about 0.1 ns. But in reality, stray capacitance including that of the crystal limits the rise time to ~ 1 ns. The pulse forming cable for such a duration is still long enough (order of a few cm) for easy connection. The resulting laser pulse transmitted by the Glan prism is of course of the same duration as the voltage pulse and thus is limited to ~ 1 ns.

Case 2: Intra-cavity Electro-optic shutter and Q-switching a laser

We show here two common intra-cavity techniques of switching out intense short (~ 10 ns) laser pulses. One uses a quarter wave voltage across the E – O crystal, the other, a half-wave voltage. Such techniques are called Q-switching.

Before going into them, let's define Q-switching and its principle.

Definition: Q-switching

Because Q or the Q of the cavity, (eq. 2-4") represents the loss in the cavity, Q-switching means switching the cavity from a high loss state (low Q) that prevents laser oscillation to a low loss state that favors oscillation. The word Q-spoiling was also used in older literature.

Definition: Giant pulse

The laser pulse coming out of a Q-switched laser, being very intense, (a lot of energy in a very short time), is sometimes called a giant pulse. A few hundred mili-joules in about 10 ns from a Q-switched Nd-YAG laser oscillator of beam diameter ~ 5mm is about the maximum before laser damage of optical components in the cavity takes place. (Hence, one needs amplification by first expanding the beam diameter using telescopes and spatial filters.)

Principle of Q-switching process

The basic idea is the following. Let's insert an optical element or a group of optical elements inside the cavity. The property of such an element (or group of elements) is to prevent the laser cavity from oscillating at low photon flux inside the cavity. That is, it represents a high loss (low Q). Note: in Chapter II, the Q or quality factor is defined as the inverse of the loss (eq. 2-4"). Now we start pumping the active medium inside the cavity. In Fig. (8-9), light propagating into B is generated from the active medium A through spontaneous emission plus stimulated emission in A after at most one round trip as shown by the U-arrowed path. But B has a high loss by definition and rejects this light either through absorption or scattering or reflection. Thus, there is no feedback and the tendency of oscillation is blocked.

Meanwhile, the pump source keeps on pumping A, increasing its population inversion above the normal threshold lasing level. At the same time, light

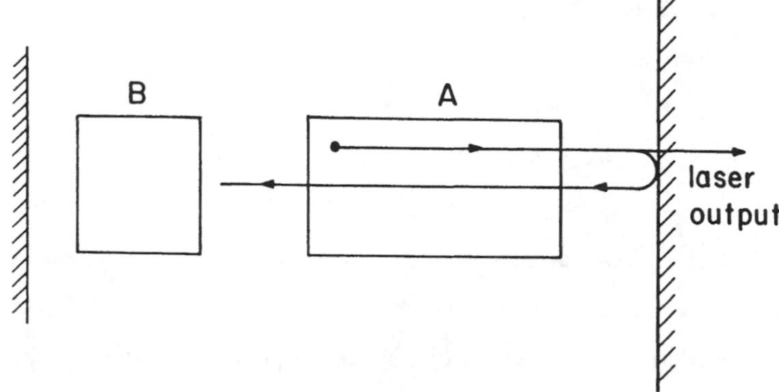

Fig. 8-9. Q-switching schematic with a saturable absorber B.

waves, one after another (or wave after wave) emanating from A hit B with stronger and stronger intensities because the population inversion keeps increasing so that each successive wave will see a higher gain. Yet they are all absorbed, until the inversion becomes very much above threshold. Then, "suddenly", B lets go an incoming strong wave of light because it cannot "stand" the intensity anymore, and the wave starts bouncing back and forth between the two mirrors. Because of the very high population inversion, the gain is very high so that the radiation density builds up extremely fast. In the time for just a few round trips, the population inversion is completely depleted and the output is a pulse containing a lot of energy in a very short time (~ 10 ns). This is a giant pulse. If by this time, the pumping is also terminated or nearly so, there is only one pulse in the output.

(A) Passive Q-switching

In passive Q-switching, the element B in Fig. 8-9 is a saturable absorber which could be a solid or liquid or gas depending on the wavelength of the laser and the availability of appropriate materials. The principle of operation is the same as before. B absorbs all weak radiation coming from A while the latter builds up its population inversion. When the inversion reaches a certain high level, the next wave of light going toward B from within A is sufficiently amplified (in at most two passes) reaching an intensity called the <u>saturation intensity</u>, I_S of the absorber. This forces B to let go the wave as if B is punched through by the strong wave. Fig. (8-10) illustrates the transmission characteristics of a saturable absorber. Oscillation is started and a giant pulse is created.

We now consider the E - O techniques.

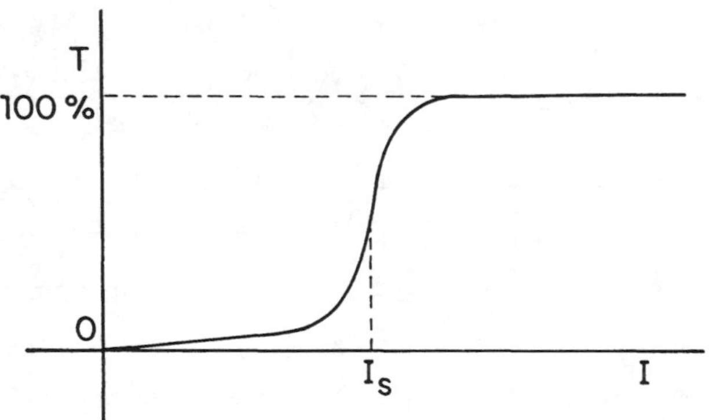

Fig. 8-10. Transmission curve of a saturable absorber.

<u>(B) Quarter Wave Voltage Q-switching</u>

Fig. 8-11 shows the technique of using a quarter wave voltage. (a) shows the schematic setup. The same KDP crystal is set between the totally reflecting end mirror of the laser cavity and a Glan prism. The high voltage $V_{1/4}$ across the KDP is turned on at t = 0 when the laser fires. (b) shows the path of a light beam coming from the active medium after the voltage across the KDP is turned on. Assuming the light beam is randomly polarized. Let's

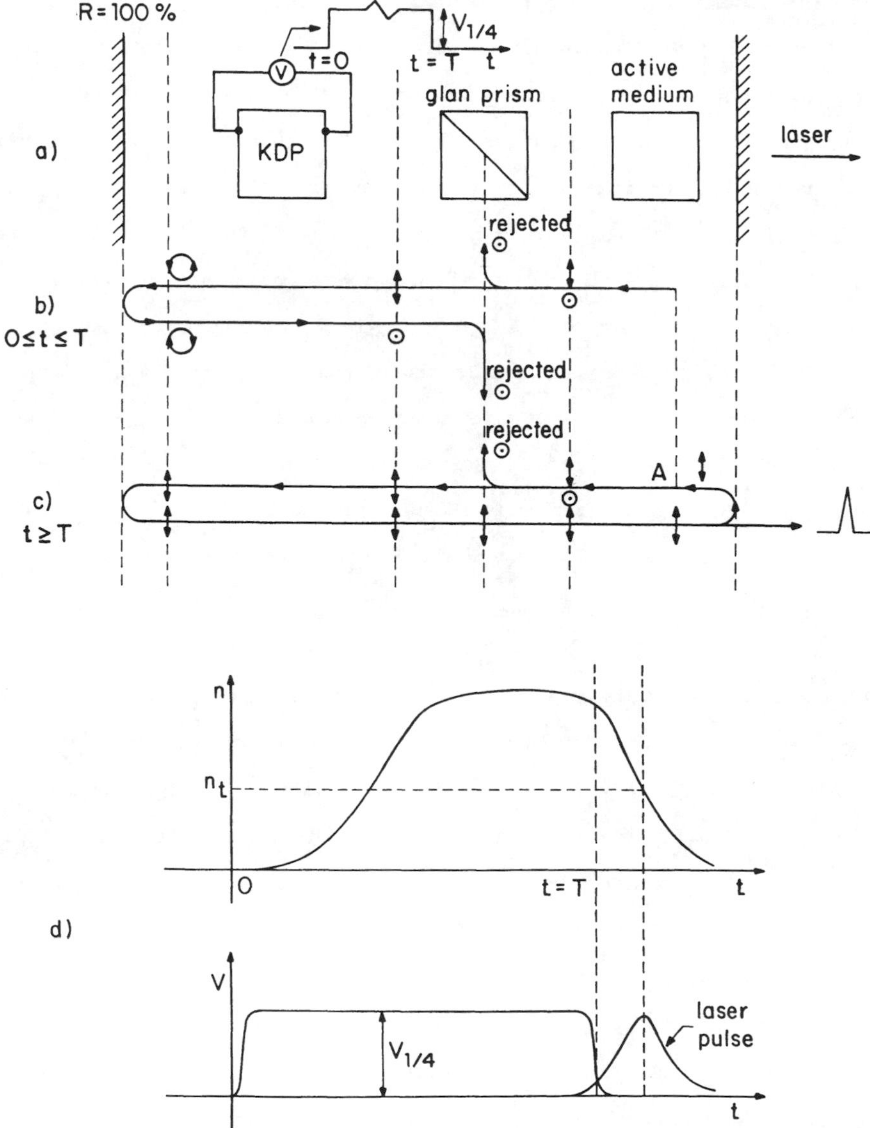

Fig. 8-11. Quarter wave voltage Q-switching of a laser.

call the linear polarization perpendicular to the page the P-polarization and the other (↕), V-(vertical) polarization. The P-polarization part of the light is rejected by the Glan prism while the V-polarization is transmitted. After passing once through the KDP towards the end mirror, the V-polarization is transformed into circular polarization because at the bias voltage of $V_{1/4}$, the KDP acts as a quarter wave plate. The beam is then reflected, still circularly polarized, and passes again through the KDP with the voltage $V_{1/4}$ still on. The circular polarization is now turned into a P-polarization. The reader should use either the Jones matrix or the analytical method to show that such double passage through a quarter wave plate produces indeed a rotation of the linear polarization by 90° (i.e. exercise). This P-polarization is again reflected by the Glan prism. (The above means that the Glan prism and the KDP together constitute the element B in Fig. 8-9.)

The above situation discourages any laser oscillation to occur. Meanwhile the pumping of the active medium keeps on going and the population inversion builds up significantly above the threshold value n_t. (Fig. 8-11d, top part) The voltage pulse evolution V(t) is shown in the bottom part of (d). Soon after the population inversion reaches the maximum, the H.V. pulse is terminated at t = T and the laser starts oscillating as shown in (c). Now, the KDP is inactive so that the V-polarization that passes through it is unaffected since the propagation is along the KDP's optic axis. The Glan prism transmits the V-polarization which then propagates through the active medium and is amplified. The reflected beam from the front mirror after being amplified again by the active medium will pass through the Glan prism thus closing the "circle" of oscillation. During this time, the pumping is significantly reduced or near zero.

Since the active medium is highly inverted, the amplification of light is very significant in every pass thus depleting the inversion quickly. After a few round-trip passes, the inversion is depleted to below the threshold value. (No replenishment of the inversion is possible now because the pumping is nearly zero by design.) The laser pulse transmitted through the output coupling mirror is at its maximum at the time when the rate of change of the inversion $\frac{dn}{dt}$ is still very large. It can be shown that this occurs at $n = n_t$, the threshold inversion. The bottom part in (d) shows the laser pulse. It shows also that even after n_t is reached, the strong radiation pulse that has been quickly released is still "sucking" energy out of the remaining inversion in the active medium. This is because the radiation in the cavity is now so strong that it takes very little time to "clean up" the remaining inversion through stimulated emission. The time is so short that the passive loss in the cavity, being a statistical quantity (i.e. average over a long time), didn't fully affect the gain. (Note that the threshold inversion is in reality defined as a balance between absolute inversion and passive statistical loss. If the depletion of the inversion is very fast, the laser radiation doesn't see the full passive loss; i.e. the loss is effectively reduced. Hence, amplification continues beyond $n = n_t$ for a short while.)

(c) Half-wave Voltage Q-switching

Fig. (8-12) shows a similar analysis as in Fig.8-11. It is left as an exercise for the reader to follow the figure and analyse it accordingly. We just point out that the KDP and the two cross polarizers constitute the Q-switch element B in Fig. 8-9. The high voltage V_π is turned on at time $t = T$ soon after the inversion reaches the maximum and the pumping is significantly reduced.

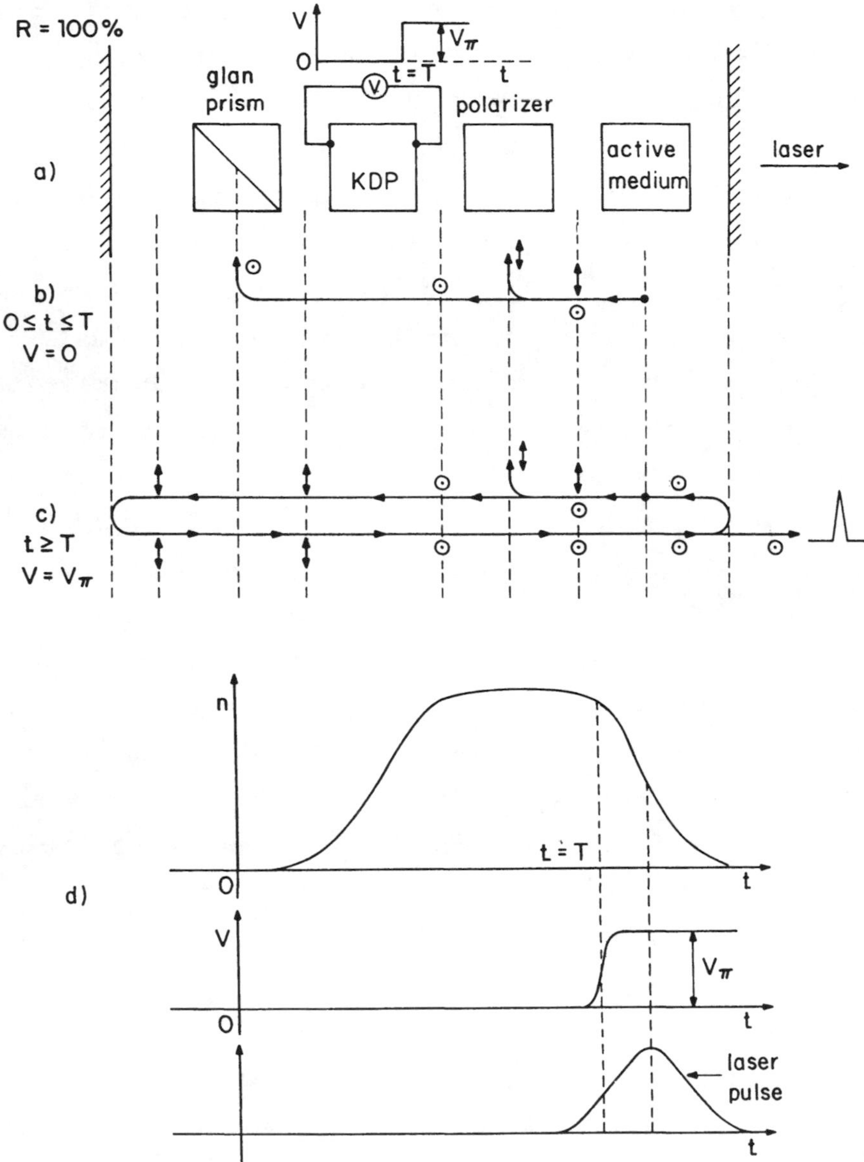

Fig. 8-12. Half-wave voltage Q-switching of a laser.

(D) Other Q-switching techniques

These are discussed in Chapter XII.

§8.6 Transverse biasing of E - O crystal

So far, in §8.5, for pedagogical reasons, we have always used the longitudinally biased KDP (i.e. $\vec{E} = E_3\hat{z}$, eq. (8-26)) to illustrate modulation, the rapid shutter (pulse slicer) and Q-switching. In general, the crystal doesn't have to be longitudinally biased in order to achieve the above applications. It could be transversely biased as mentioned briefly in the example following eq. (8-47'), in which the KDP crystal is biased in the x - direction ($\vec{E} = E_1\hat{x}$). (We always keep in mind that the z - direction is the propagation direction by definition.) Under such a condition, the laser beam will still see the electrically induced anisotropy so that its phase and/or polarization state will be changed.

To see this, we consider eq. (8-47). When $\vec{E} = 0$, eq. (8-47) gives the ellipsoid shown in Fig. (8-13(a)). With the propagation direction $\hat{k}$ in the z - axis, the laser beam passing through it will not experience any change in phase nor polarization. When $\vec{E} = E_1\hat{x} \neq 0$, the ellipsoid becomes tilted (rotated about the x - axis) and deformed as given by eq. (8-53) and illustrated in Fig. 8-13(b). Propagation in the z - axis will cause the laser beam to "see" a uniaxial crystal; i.e. $n_{x'} = n_o$ and $n_{y'}$ given by eq. (8-55). The phase difference of the two components in the crystal is

$$\epsilon = \frac{\omega}{c}\left(n_{x'} - n_{y'} \right)\ell$$

where ℓ is the length of the crystal. Using eq. (8-54), (8-55),

$$\epsilon = \frac{\omega\ell}{2c} n_o^3 r_{41} E_3 \tan\alpha$$

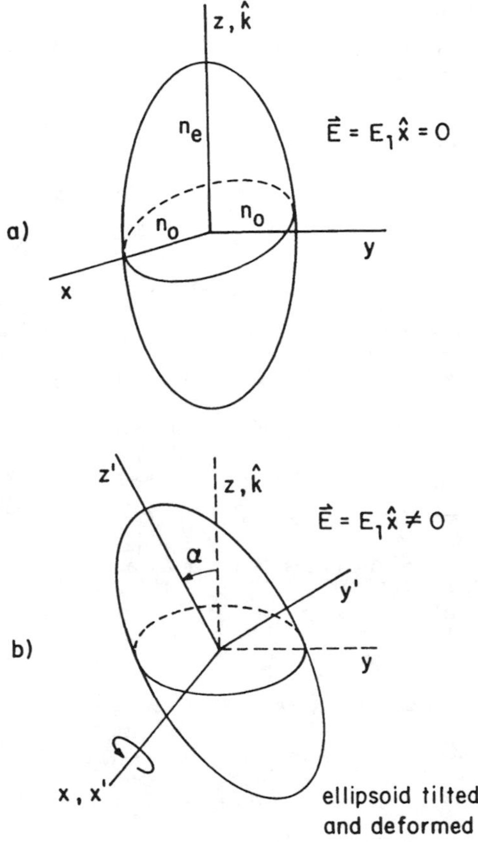

Fig. 8-13. Transverse biasing of an E-O crystal.

Thus controlling the length ℓ and the field E_3 will give us any phase difference we want and hence any phase and polarization change of the laser beam.

Many other E - O crystals use transverse biasing. Examples: CdTe and GaAs are both transversely biased for use with TEA-CO_2 laser beams, mostly for slicing pulses. We won't go into any detail here.

§8.7 Closing remark

The electro-optic effect is an electric field induced anisotropy in materials. When we use this effect in different applications, the E - O material functions essentially as an active wave plate; active in the sense that the electric field induced anisotropy can be turned on and off at will, thus modifying the phase and polarization state of the transmitted light (laser) wave. When the E - O material is used as a light modulator or fast shutter, one would certainly want to have it operate at very high frequencies or very high speed and there is a limit to the speed (or frequency response) of course.

From the basic design, when we apply a voltage across a pair of electrodes sticking on two surfaces of an E - O medium, the ensemble becomes a capacitor. Thus the capacitance of the E - O cell together with the impedance of the other electrical elements constitute a resultant impedance that limits the speed of the system. Since this becomes an electrical design problem, we will not go into futher detail here.

CHAPTER IX

MECHANICAL FORCE INDUCED ANISOTROPY AND ACOUSTO-OPTICS

In nature, there is no ideally rigid body; i.e. all existing materials are deformable bodies. When we apply a mechanical force onto any medium in nature, it will in general induce a change in the body. This general change can be decomposed into three types of fundamental motions, namely translation, rotation and strain. Translation and rotation represent the change of an ideally rigid body whereas the strain represents the body's deformation with respect to its original state. Such a deformation (sometimes called "elastic change") can be put into use in modern opto-electronic applications. Qualitatively, imagine that a strain is created in an anisotropic medium. Its internal molecular and atomic structure is thus modified, even very slightly. Such a slight change is sufficient to induce change in the material density and thus index. It means that the original index ellipsoid of the medium is slightly modified similar to the case of electric field induced change of the index ellipsoid in Chapter VIII. A laser (light) beam being very sensitive, will feel the change in passing through the medium thus changing its polarization state. One can take advantage of such a change in the polarization state of the laser (light) beam to generate modulation to the laser. Similar to the effect of electric field, a strain induced in an isotropic medium will generate in it an anisotropy that can be used advantageously. There are many ways of creating strains in a medium. In modern opto-electronics applications, the most popular way is to create acoustic (elastic) waves; hence acousto-optics (or opto-elastic effect).

§9.1 The strain matrix

It is instructive to review briefly the derivation of the strain from first principles. Consider, for the ease of understanding, a solid body. (It could be liquid or gas.) We focus our attention on only two neighboring points O and P inside the body (Fig. 9-1). Let O initially coincides with the origin of a Cartesian coordinate system which is fixed in space. The position of P is at $\vec{r} = (x,y,z)$. A mechanical force is applied to the body so that it has undergone a slight but general change. This means that there is a displacement of both P and O with respect to the origin. We denote the displacement of P by (ξ, η, ζ) and that of O by (ξ_o, η_o, ζ_o). The most general infinitesimal relative motion of P with respect to O is

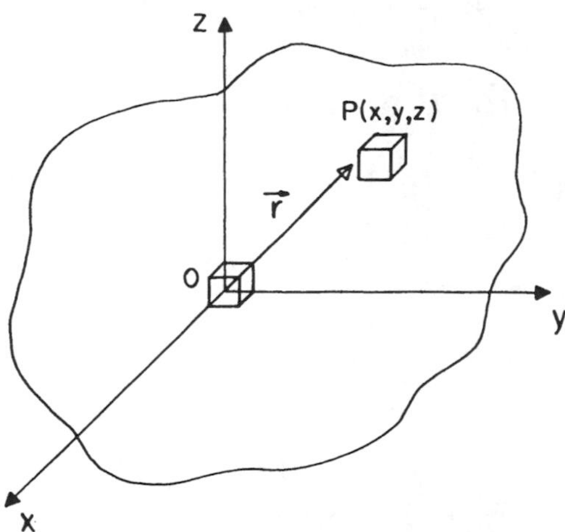

Fig. 9-1. Geometrical relationship between two elementary volumes at two neighboring points O and P inside a deformable body. The spatial Cartesian coordinate axis are fixed in space and the point O coincides with the origin before deformation.

given by the Taylor's expansion. (Note again that P is a neighboring point of O by definition so that the coordinates (x,y,z) of P are small quantities.)

$$\xi = \xi_o + \frac{\partial \xi}{\partial x} x + \frac{\partial \xi}{\partial y} y + \frac{\partial \xi}{\partial z} z + \text{higher order terms} \tag{9-1}$$

$$\eta = \eta_o + \frac{\partial \eta}{\partial x} x + \frac{\partial \eta}{\partial y} y + \frac{\partial \eta}{\partial z} z + \text{higher order terms} \tag{9-2}$$

$$\zeta = \zeta_o + \frac{\partial \zeta}{\partial x} x + \frac{\partial \zeta}{\partial y} y + \frac{\partial \zeta}{\partial z} z + \text{higher order terms} \tag{9-3}$$

Because we have assumed only small changes, we can neglect the higher order terms and rewrite eq. (9-1) to (9-3) in the following form.

$$\begin{pmatrix} \xi \\ \eta \\ \zeta \end{pmatrix} = \begin{pmatrix} \xi_o \\ \eta_o \\ \zeta_o \end{pmatrix} + \begin{pmatrix} 0 & \varphi_{12} & \varphi_{13} \\ \varphi_{21} & 0 & \varphi_{23} \\ \varphi_{31} & \varphi_{32} & 0 \end{pmatrix} \begin{pmatrix} x \\ y \\ z \end{pmatrix} + \begin{pmatrix} \epsilon_{11} & \epsilon_{12} & \epsilon_{13} \\ \epsilon_{21} & \epsilon_{22} & \epsilon_{23} \\ \epsilon_{31} & \epsilon_{32} & \epsilon_{33} \end{pmatrix} \begin{pmatrix} x \\ y \\ z \end{pmatrix} \tag{9-4}$$

$$\equiv \vec{s}_o + \vec{s}_1 + \vec{s}_2 \tag{9-5}$$

where
$$\varphi_{ik} \equiv \frac{a_{ik} - a_{ki}}{2} = -\varphi_{ki} \; ; \qquad i,k \equiv 1,2,3 \tag{9-5}$$

$$\epsilon_{ik} \equiv \frac{a_{ik} + a_{ki}}{2} = \epsilon_{ki} \; ; \qquad i,k \equiv 1,2,3 \tag{9-6}$$

(Thus,
$$\left(a_{ik} = \varphi_{ik} + \epsilon_{ik} \right) \tag{9-7}$$

$$a_{1k} \equiv \frac{\partial \xi}{\partial x_k} \qquad k = 1,2,3 \; ; \; \left(x_1 , x_2 , x_3 \right) \equiv (x,y,z) \tag{9-8}$$

$$a_{2k} \equiv \frac{\partial \eta}{\partial x_k} \qquad k = 1,2,3 \; ; \; \left(x_1 , x_2 , x_3 \right) \equiv (x,y,z) \tag{9-9}$$

$$a_{3k} \equiv \frac{\partial \zeta}{\partial x_k} \qquad k = 1,2,3 \; ; \; \left(x_1 , x_2 , x_3 \right) \equiv (x,y,z) \tag{9-10}$$

Note: (x,y,z,) and (X_1 , x_2 , x_3) will be interchanged freely whenever convenient in what follows. (The reader should do the exercise of substituting the above definitions into eq. (9-4) and obtain (9-1) to 9-3)). The first term on the right hand side of eq. (9-4) represents a translation of

P with respect to the origin fixed in space; i.e. $\vec{s}_o$ in eq. (9-5). The second term is the displacement $\vec{s}_1$ of the tip of the vector $\vec{r}$ by a rotation through a small angle. This represents the rotation of the body. Let $\vec{\phi}$ be the axial vector representing a rotation by an angle ϕ around an axis (Fig. 9-2), where

$$\phi \equiv \begin{pmatrix} \varphi_x \\ \varphi_y \\ \varphi_z \end{pmatrix} \qquad (9\text{-}11)$$

and where
$$\varphi_x \equiv \frac{a_{32} - a_{23}}{2} \qquad (9\text{-}12)$$

$$\varphi_y \equiv \frac{a_{13} - a_{31}}{2} \qquad (9\text{-}13)$$

$$\varphi_z \equiv \frac{a_{21} - a_{12}}{2} \qquad (9\text{-}14)$$

such that
$$\vec{s}_1 = \vec{\phi} \times \vec{r} \qquad (9\text{-}15)$$

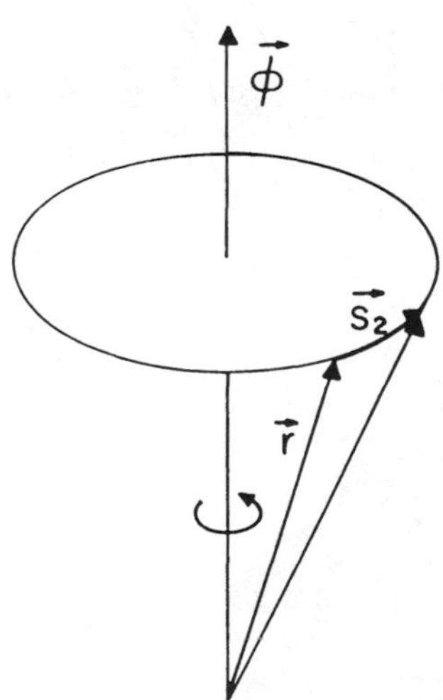

Fig. 9-2. Relation between the axial vector and a rotation.

As an <u>exercise</u>, the reader is asked again to substitute eq. (9-11) to (9-14) into (9-15) and obtain the second term on the right hand side of eq. (9-4).)

The third term on the right hand side of eq. (9-4) is the displacement $\vec{s}_2$ caused by the <u>strain</u> (or deformation or elastic change.) The matrix (ϵ_{ik}) is the <u>strain matrix</u> (tensor). A slightly modified definition follows.

Consider the diagonal elements of the strain matrix.

$$\epsilon_{11} = a_{11} \quad \text{by eq. (9-6)}$$

$$= \frac{\partial \xi}{\partial x} \quad \text{by eq. (9-8)} \qquad (9\text{-}16)$$

Similarly,

$$\epsilon_{22} = \frac{\partial \eta}{\partial y} \qquad (9\text{-}17)$$

$$\epsilon_{33} = \frac{\partial \zeta}{\partial z} \qquad (9\text{-}18)$$

They correspond to the increments in <u>length</u> per unit <u>original length</u> (or <u>extensions</u>) in the directions of x,y and z respectively. The off diagonal terms can be similarly interpreted, but they represent the increment in one direction per unit original length in the other two directions. For example,

$$\epsilon_{12} \equiv \frac{a_{12} + a_{21}}{2} \qquad \text{from eq. (9-6)}$$

$$= \frac{1}{2}\left(\frac{\partial \xi}{\partial y} + \frac{\partial \eta}{\partial x}\right) \qquad \text{from eq. (9-8)}$$

$$2\epsilon_{12} = \frac{\partial \xi}{\partial y} + \frac{\partial \eta}{\partial x} \qquad (9\text{-}19)$$

This type of combined "cross increments" (eq. 9-19) is defined as <u>shearing</u> and we define all "cross increments" of the type (9-19) as the <u>shearing strains s_{ij}</u>: or the <u>shear</u>. They were given by

$$S_{ij} \equiv 2\epsilon_{ij} \qquad (i \neq j) \tag{9-20}$$

$$S_{ii} \equiv \epsilon_{ii} \qquad (i = j) \tag{9-21}$$

Thus, a modified strain tensor (matrix) is given by

$$S \equiv \begin{pmatrix} S_{11} & S_{12} & S_{13} \\ S_{21} & S_{22} & S_{23} \\ S_{31} & S_{32} & S_{33} \end{pmatrix} \tag{9-22}$$

where S_{ij} $(i \neq j)$ is given by eq. (9-20) and S_{ii} $(i = j)$ are given by eq. (9-16) to (9-18). The difference between (ϵ_{ij}) and eq. (9-22) is simply eq. (9-20)). Since $\epsilon_{ij} = \epsilon_{ji}$ (from eq. (9-6)),

$$S_{ij} = S_{ji} \tag{9-23}$$

i.e. S is symmetric.

<u>Example 1</u>: If there is no translation nor rotation, but only strain, and

$$\vec{s}_2 = \hat{z} \cdot A\cos(\Omega t - kz) \tag{9-24}$$

calculate the strain tensor.

<u>Solution</u>: Since there is no translation nor rotation, the first two terms on the right hand side of eq. (9-4) are zero. We have

$$\begin{pmatrix} \xi \\ \eta \\ \zeta \end{pmatrix} = \vec{s}_2 \tag{9-25}$$

Comparing eq. (9-24) and (9-25), we see that

$$\zeta = A\cos(\Omega t - kz) \tag{9-26}$$

$$\xi = \eta = 0$$

i.e. ζ is a function of z. Thus, all the matrix elements of S are zero except the one containing $(\partial\zeta/\partial z)$. And this element is

$$S_{33} \equiv \frac{\partial \zeta}{\partial z} \quad \text{(from eq. (9-21) and (9-18))}$$

$$= AK\sin(\Omega t - Kz) \tag{9-27}$$

and

$$S = \begin{pmatrix} 0 & 0 & 0 \\ 0 & 0 & 0 \\ 0 & 0 & AK\sin(\Omega t - Kz) \end{pmatrix} \quad (9\text{-}28)$$

Example 2: Again assuming no translation and rotation of a body, but the strain displacement $\vec{s}_2$ is

$$\vec{s}_2 = \hat{y}\, A\cos(\Omega t - Kz) \quad (9\text{-}29)$$

Find the strain matrix.

Solution: Again, eq. (9-24) is valid.

We have

$$\xi = \zeta = 0 \quad (9\text{-}30)$$

$$\eta = A\cos(\Omega t - Kz) \quad (9\text{-}31)$$

i.e. η is a function of z and only elements of S containing $(\partial\eta/\partial z)$ is non-zero. This term is

$$S_{23} \equiv \frac{\partial\eta}{\partial z} + \frac{\partial\zeta}{\partial y}$$

$$= AK\sin(\Omega t - Kz) + 0 \quad (9\text{-}32)$$

$$S = \begin{pmatrix} 0 & 0 & 0 \\ 0 & 0 & AK\sin(\Omega t - Kz) \\ 0 & 0 & 0 \end{pmatrix} \quad (9\text{-}33)$$

Note: When an acoustic wave is applied to a medium, it doesn't in general induce any rigid body displacement and rotation. Thus $\vec{s}_o = \vec{s}_1 = 0$, unless the wave is extremely strong.

Before ending this section, we remind ourselves that 0 and P in Fig. (9-1) are two neighboring points. Thus, $\vec{s}_2$ is a local resultant displacement of an infinitesimal volume due to the strain. An example is the propagation of an acoustic wave in a medium. Depending on the material structure, an acoustic wave propagation induces at some local point a resultant displacement of a very small volume of the matter not necessarily in the direction of propagation of the acoustic wave. This displacement is $\vec{s}_2$. If we define the propagation direction as the z-direction and the local "particle" (very small volume) displacement is $\vec{U}(z,t)$,

then

$$\vec{s}_2 = \vec{U}(z,t) = U(z,t)\hat{z}$$

If $U(z,t) = A\cos(\Omega t - Kz)$ which is a propagating wave, we have

$$\vec{s}_2 = \hat{z}A\cos(\Omega t - Kz) \tag{9-34}$$

§9.2 Mechanically induced anisotropy

We assume that some mechanical force has induced only a strain in a medium. Such a strain leads to a change in local densities and thus a change in the refractive index of the medium. Such index change is described by the change in the optical impermeability η_{ij} defined by eq. (8-4) in chapter VIII: i.e. $\eta_{ij} \equiv \epsilon_o/\epsilon_{ij}$. The change in η_{ij} can be expressed in terms of the strain tensor S because it is the strain that has induced the change in η_{ij}. Thus

$$\Delta\eta_{ij} = p_{ijk\ell}\, S_{k\ell} + \text{higher order terms } (i,j,k,\ell = 1,2,3), \tag{9-35}$$

where repeated indices means summation. $S_{k\ell}$ are the matrix elements of S, $p_{ijk\ell}$ are the matrix elements of a proportionality matrix. In eq. (9-35), it is assumed that $\Delta\eta_{ij}$ is proportional to the strain to the first order. Since in the definition of strain in §9.1, there were higher order terms that were neglected, we write down, for being general, "higher order terms" in eq. (9-35). We will neglect them immediately because the first order term is generally small.

The physical meaning of eq. (9-35) is that the change in each of the impermeability matrix (tensor) element (i.e. $\Delta\eta_{ij}$, for the $(ij)^{th}$ element) is the combined consequences of each of the mechanically induced shear strain element $s_{k\ell}$; hence the summation over k and ℓ. Let $\eta_{ij}(S)$ be the $(ij)^{th}$ element of the perturbed impermeability and $\eta_{ij}(0)$, the unperturbed one. Thus

$$\Delta\eta_{ij} = \eta_{ij}(S) - \eta_{ij}(0) \tag{9-35}$$

The perturbed index ellipsoid (c.f. eq. (8-6)) is

$$\eta_{ij}(S)x_i x_j = 1 \tag{9-36}$$

(Note again, repeated indices means summation.) Using eq. (9-35), this becomes

$$\eta_{ij}(0)x_i x_j + \Delta\eta_{ij}x_i x_j = 1 \tag{9-37}$$

The first term is again the unperturbed expression given by the left hand side of eq. (8-5). The second term is

$$\Delta\eta_{ij}x_i x_j = p_{ijk\ell}\, S_{k\ell}\, x_i x_j \tag{9-38}$$

$$\equiv M \tag{9-39}$$

We again use short hand notations. Because η_{ij} and S_{ij} are both symmetric tensors,

i.e. $\eta_{ij} = \eta_{ji}$ from eq. (8-15)

and $S_{ij} = S_{ji}$ from eq. (9-23)

we can make the following simplification.

(ij) or ($k\ell$)	SHORT HAND NOTATION
(11)	1
(22)	2
(33)	3
(23) or (32)	4
(13) or (31)	5
(12) or (21)	6

From eq. (9-38),

$$\Delta\eta_{ij} \Rightarrow \Delta\eta_m = p_{mn}\, S_n \qquad (m,n = 1,2,\ldots 6) \tag{9-40}$$

Note that now in eq. (9-40) (m,n) run from 1 to 6. Rewriting eq. (9-37) using eq. (8-5) and eq. (9-39),

$$\frac{x_1^2}{n_1^2} + \frac{x_2^2}{n_2^2} + \frac{x_3^2}{n_3^2} + M = 1 \tag{9-41}$$

where M can be calculated from expanding eq. (9-40). This is left as an <u>exercise</u> to the reader. (This is similar to the case of Kerr E - O effect; i.e. $p_{ijk\ell} \leftrightarrow S_{ijk\ell}$ and $S_{i\ell} \leftrightarrow E_k E_\ell$)

$$M = \left(x_1^2 \ , \ x_2^2 \ , \ x_3^2 \ , \ 2x_2x_3 \ , \ 2x_1x_3 \ , \ 2x_1x_2 \right) \cdot \begin{pmatrix} p_{11} & p_{12} & p_{13} & p_{14} & p_{15} & p_{16} \\ p_{21} & p_{22} & p_{23} & p_{24} & p_{25} & p_{26} \\ p_{31} & p_{32} & p_{33} & p_{34} & p_{35} & p_{36} \\ p_{41} & p_{42} & p_{43} & p_{44} & p_{45} & p_{46} \\ p_{51} & p_{52} & p_{53} & p_{54} & p_{55} & p_{56} \\ p_{61} & p_{62} & p_{63} & p_{64} & p_{65} & p_{66} \end{pmatrix} \begin{pmatrix} S_1 \\ S_2 \\ S_3 \\ S_4 \\ S_5 \\ S_6 \end{pmatrix} \tag{9-42}$$

The tensor elements of (p_{mn}) are known as <u>elasto-optic coefficients</u>, and the matrices for many useful materials are tabulated according to their symmetry. (See for example, Yariv and Yeh.) We should recall that the index ellipsoid (as well as the Fresnel wave normals or ray surfaces) describes <u>local response</u> of a medium and every point in the medium has the same response (see chapter VI). So does the strain (see end of previous section.)

Example 1: Acoustic wave propagating in water

An acoustic wave propagates in water in the z - direction. The local particle (very small volume) displacement induced by the acoustic wave is

$$\vec{U}(z,t) = \hat{z}A\cos(\Omega t - Kz) \tag{9-43}$$

Find the index change.

Solution: Since $\vec{s}_2$ is the local displacement (cf. last paragraph of §9.1)

$$\vec{s}_2 = \vec{U}(z,t) = \hat{z}A\cos(\Omega t - Kz) \tag{9-44}$$

This is indentical to eq. (9-24). Thus, from eq. (9-28), the only non-zero matrix element of S is

$$\begin{aligned} S_{33} &\equiv S_3 \qquad \text{(by the short hand notation)} \\ &= AK\sin(\Omega t - Kz) \end{aligned} \tag{9-45}$$

Water is isotropic. The elasto-optic tensor (p_{mn}) is given by

$$\left(p_{mn} \right) \equiv \begin{pmatrix} p_{11} & p_{12} & p_{12} & 0 & 0 & 0 \\ p_{12} & p_{11} & p_{12} & 0 & 0 & 0 \\ p_{12} & p_{12} & p_{11} & 0 & 0 & 0 \\ 0 & 0 & 0 & \frac{1}{2}(p_{11}- p_{12}) & 0 & 0 \\ 0 & 0 & 0 & 0 & \frac{1}{2}(p_{11}- p_{12}) & 0 \\ 0 & 0 & 0 & 0 & 0 & \frac{1}{2}(p_{11}- p_{12}) \end{pmatrix} \tag{9-46}$$

$$S = \begin{pmatrix} 0 \\ 0 \\ S_3 \\ 0 \\ 0 \\ 0 \end{pmatrix} \tag{9-47}$$

From eq. (9-42), using (9-46) and (9-47),

$$M = \left(x_1^2 \, , \, x_2^2 \, , \, x_3^2 \, , 2x_2x_3 \, , 2x_1x_3 \, , \, 2x_1x_2 \right) \begin{pmatrix} p_{12} \, S_3 \\ p_{12} \, S_3 \\ p_{11} \, S_3 \\ 0 \\ 0 \\ 0 \end{pmatrix}$$

$$= p_{12}S_3x_1^2 + p_{12} \, S_3x_2^2 + p_{11}S_3x_3^2 \tag{9-48}$$

Substituting eq. (9-48) into eq. (9-41), using

$n_1 = n_2 = n_3 \equiv n_o$ (water is isotropic)

$$x_1^2 \left(\frac{1}{n_o^2} + p_{12}S_3 \right) + x_2^2 \left(\frac{1}{n_o^2} + p_{12}S_3 \right) + x_3^2 \left(\frac{1}{n_o^2} + p_{11}S_3 \right) = 1 \tag{9-49}$$

Eq. (9-49) is a uniaxial index ellipsoid if $p_{11} \neq p_{12}$; i.e. under the propagation of an acoustic wave, an isotropic medium becomes a uniaxial material in general. From eq. (9-49),

$$\left(n_x^2 \right)^{-1} = \left(n_y^2 \right)^{-1} = \frac{1}{n_o^2} + p_{12} \, S_3 = \frac{1}{n_o^2} + p_{12}AK\sin(\Omega t - Kz) \tag{9-50}$$

$$\left(n_z^2 \right)^{-1} = \frac{1}{n_o^2} + p_{11}S_3 = \frac{1}{n_o^2} + p_{11}AK\sin(\Omega t - Kz) \tag{9-51}$$

Under the assumption of small index change, approximation can be made to eq. (9-50) and (9-51) in a similar way as that used in eq. (8-42, 43). This is left as an exercise. The result is

$$n_x = n_y = n_o - \frac{1}{2} n_o^3 \, p_{12} \, AK\sin(\Omega t - Kz) \tag{9-52}$$

$$n_z = n_o - \frac{1}{2} n_o^3 \, p_{11} \, AK\sin(\Omega t - Kz) \tag{9-53}$$

Fig. (9-3) shows the resultant index ellipsoid. From this figure, we see that if we send an E - M wave of arbitrary polarization along the z - direction, the water will still behave isotropically because $n_x = n_y$. But n_x and n_y are both sinusoidal functions. Freezing everything at a

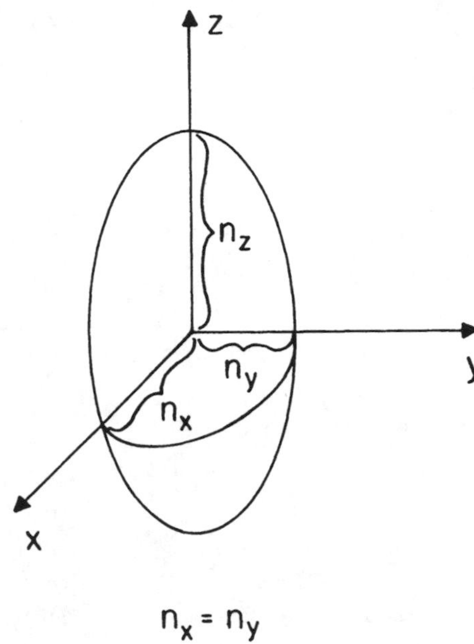

Fig. 9-3. An isotropic material becomes uniaxial under the propagation of an acoustic wave in the material. The figure shows the resultant index ellipsoid.

certain time t, both the polarization components $\vec{E}_x$ and $\vec{E}_y$ of the E - M wave will see a medium of periodically varying index in the z - direction; i.e. each polarization (wave) sees an index grating. If the E - M wave is polarized in the x - direction (i.e. $\vec{E}_y = o$), it will see only the index grating (variation) of (eq. 9-52). Of course, if the E - M wave propagates in an arbitrary direction, we should treat the whole affair as if it propagates in an anisotropic medium. There will be o and e waves created. Each wave will see an index grating of the same Ω and K but of different amplitudes, because of eq. (9-52) and (9-53).

<u>Example 2</u>: <u>Acoustic wave propagating in germanium</u>

Germanium is also isotropic. However, its crystal symmetry dictates that if we propagate an acoustic wave in the direction of one of its crystal axes

(< 001 > direction, defined as z - axis), the shear strain induces a local particle displacement $\vec{s}_2$ in the y - direction (the < 010 > direction of the crystal).

$$\vec{s}_2 = \hat{y}A\cos(\Omega t - kz) \tag{9-54}$$

This is identical to eq. (9-29). From eq. (9-33) the only non-vanishing element of the strain matrix is

$$S_{23} \equiv S_4 \qquad \text{(using short hand notation)}$$
$$= AK\sin(\Omega t - Kz) \tag{9-55}$$

The photo-elastic coefficients are, for Germanium,

$$(p_{mn}) = \begin{pmatrix} p_{11} & p_{12} & p_{12} & 0 & 0 & 0 \\ p_{12} & p_{11} & p_{12} & 0 & 0 & 0 \\ p_{12} & p_{12} & p_{11} & 0 & 0 & 0 \\ 0 & 0 & 0 & p_{44} & 0 & 0 \\ 0 & 0 & 0 & 0 & p_{44} & 0 \\ 0 & 0 & 0 & 0 & 0 & p_{44} \end{pmatrix} \tag{9-56}$$

$$S = \begin{pmatrix} 0 \\ 0 \\ 0 \\ S_4 \\ 0 \\ 0 \end{pmatrix} \tag{9-57}$$

Eq. (9-42) becomes

$$M = \left(x_1^2 \, , \, x_2^2 \, , \, x_3^2 \, , \, 2x_2x_3 \, , \, 2x_1x_3 \, , \, 2x_1x_2 \right) \begin{pmatrix} 0 \\ 0 \\ p_{44}\, S_4 \\ 0 \\ 0 \end{pmatrix}$$

$$= 2p_{44}S_4x_2x_3 \tag{9-58}$$

and eq. (9-41) becomes, ($\because n_1 = n_2 = n_3 \equiv n_o$, Ge being isotropic)

$$\frac{x_1^2}{n_o^2} + \frac{x_2^2}{n_o^2} + \frac{x_3^2}{n_o^2} + 2p_{44}S_4x_2x_3 = 1 \tag{9-59}$$

This equation is similar to eq. (8-27). We need to make a rotation of the (x_1x_2) axes around the x_3 - axis similar to what was done to eq. (8-27). The result is: after a rotation of 45° about the x_3 - axis, we obtain an ellipsoid along the new principal (x_1', x_2', x_3') axes:

$$\frac{x_1'^2}{n_{x_1'}^2} + \frac{x_2'^2}{n_{x_2'}^2} + \frac{x_3'^2}{n_{x_3'}^2} = 1 \tag{9-60}$$

After an approximation similar to the first example (<u>exercise</u> again)

$$n_{x_1'} = n_o \tag{9-61}$$

$$n_{x_2'} = n_o - \frac{1}{2} n_o^3 p_{44} AK\sin(\Omega t - Kz) \tag{9-62}$$

$$n_{x_3'} = n_o + \frac{1}{2} n_o^3 p_{44} AK\sin(\Omega t - Kz) \tag{9-63}$$

The medium becomes biaxial because $n_{x_1'} \neq n_{x_2'} \neq n_{x_3'}$. At the same time, because the induced indices $\left(n_{x_2'}, n_{x_3'} \right)$ vary periodically, an E - M wave propagating through the medium will see a "grating" or periodic index variation. The polarization of the E - M wave and the propagation direction determine the output characteristics.

Before ending this section, it should be stressed that any mechanical force acting on any matter will in principle generate some anisotropy in the matter so long as there is a strain in the material apart from rigid body translation and rotation. For instance, if one compresses a piece of glass (isotropic), it becomes uniaxial. If such kind of strained material is place between two crossed polarisers, and a white light beam is sent through them, beautiful patterns representing the strain topography can be seen at the output. This is because the entrance polarizer transmits only light linearly polarized in its preferred direction. The polarization state of this transmitted light, after passing through the strained material, will be altered locally so that the exit polarizer (analyser) transmits partially only the light that has passed through a strained region of the material.

§9.3 Fundamentals of acousto-optic interaction

(a) Fundamentals of grating's diffraction.

As mentioned in the previous section, when there is an acoustic wave propagating in a medium, an E - M wave passing through the medium will see a 3D index grating. The E - M wave will thus be diffracted by the grating. Although it is presumed that the readers are familiar with the formula of a diffraction grating, it is always instructive to review quickly the fundamentals of the subject.

The fundamental analysis of diffraction from a grating is the superposition of two E - M waves scattered by two adjacent lines of the grating. In other words, we need only to know the consequence of the superposition of two scattered or diffracted waves (plane waves for simplicity) with the same propagation vector $\vec{k}$ and the same frequency ω having undergone different optical paths. Their polarizations, assumed linear, are parallel. The two waves are

$$\vec{E}_1 = \vec{E}_{o1} \cos(\omega t - \vec{k} \cdot \vec{r}_1) \tag{9-64}$$

$$\vec{E}_2 = \vec{E}_{o2} \cos(\omega t - \vec{k} \cdot \vec{r}_2) \tag{9-65}$$

The total field is

$$\vec{E} = \vec{E}_1 + \vec{E}_2 \tag{9-66}$$

and the intensity is

$$I \propto < \vec{E} \cdot \vec{E} > \tag{9-67}$$

We neglect the constant of proportionality

$$I = < \vec{E} \cdot \vec{E} > \equiv \frac{1}{T} \int_o^T \vec{E} \cdot \vec{E} \, dt \tag{9-68}$$

where $$T \gg 2\pi/\omega \tag{9-69}$$

Now, $$\vec{E}\cdot\vec{E} = (\vec{E}_1+\vec{E}_2)\cdot(\vec{E}_1+\vec{E}_2)$$

$$= E_1^2 + E_2^2 + 2E_1E_2 \quad (\because \vec{E}_1 \parallel \vec{E}_2) \qquad (9\text{-}70)$$

The cross term is the interference term and will be zero if $\vec{E}_1$ is perpendicular to $\vec{E}_2$, i.e. waves of perpendicular polarizations will not interfere.)

From (9-68) and (9-70),

$$I = < E_1^2 > + < E_2^2 > + 2< E_1E_2 >$$

$$= I_1 + I_2 + 2< E_1E_2 > \qquad (9\text{-}71)$$

$$E_1E_2 = E_{o1}E_{o2}\cos(\omega t - \vec{k}\cdot\vec{r}_1)\cos(\omega t - \vec{k}\cdot\vec{r}_2)$$

$= \ldots.$ (<u>exercise</u>)

$$= E_{o1}E_{o2}(\cos^2\omega t \cos\vec{k}\cdot\vec{r}_1 \cos\vec{k}\cdot\vec{r}_2 + \sin^2\omega t \sin\vec{k}\cdot\vec{r}_1 \sin\vec{k}\cdot\vec{r}_2 - \sin\omega t \cos\omega t \sin\vec{k}\cdot\vec{r}_2 \cos\vec{k}\cdot\vec{r}_1 - \sin\omega t \cos\omega t \sin\vec{k}\cdot\vec{r}_1 \cos\vec{k}\cdot\vec{r}_2)$$

$\left\langle \begin{cases} \cos^2\omega t \\ \sin^2\omega t \end{cases} \right\rangle = \ldots.$ (<u>exercise</u>)

$$= \frac{1}{2}$$

$\langle \sin\omega t \cos\omega t \rangle = \ldots.$ (<u>exercise</u>)

$$= 0$$

$\therefore < E_1E_2 > = \ldots.$ (<u>exercise</u>)

$$= \frac{E_{o1}\,E_{o2}}{2}\cos[\vec{k}\cdot(\vec{r}_1 - \vec{r}_2)] \qquad (9\text{-}72)$$

Let $$\epsilon \equiv \vec{k}\cdot(\vec{r}_1 - \vec{r}_2) \qquad (9\text{-}73)$$

$\equiv$ phase difference

$$\therefore < E_1E_2 > \equiv \frac{E_{o1}\,E_{o2}}{2}\cos\epsilon \qquad (9\text{-}74)$$

$\therefore\ I = I_1 + I_2 + E_{o1}E_{o2}\cos\epsilon$ (from (9-71) and (9-74))

$$= \begin{cases} \text{maximum when } \cos\epsilon = +1 & (\underline{\text{constructive interference}}) \\ \text{minimum when } \cos\epsilon = -1 & (\underline{\text{destructive interference}}) \end{cases}$$

$$= \left\{ \begin{array}{l} \text{maximum when } \epsilon = 2m\pi \\ \text{minimum when } \epsilon = (2m+1)\pi \end{array} \right\} \qquad (9\text{-}75)$$

where $m = 0,1,2,\ldots.$

If $\vec{r}_1 \parallel \vec{r}_2$ and $\vec{k}$ are all in the z - direction, we have

$$\epsilon = k\,(z_1 - z_2)$$

$$\equiv k\,\Delta L \qquad (\Delta L \equiv \underline{\text{optical path difference}})$$

$$= \frac{2\pi}{\lambda}\,\Delta L$$

Hence
$$I = \begin{cases} \text{maximum when } \frac{2\pi}{\lambda}\Delta L = 2m\pi \\ \text{minimum when } \frac{2\pi}{\lambda}\Delta L = (2m+1)\pi \end{cases}$$

$$= \left\{ \begin{array}{l} \text{maximum when } \Delta L = m\lambda \\ \text{minimum when } \Delta L = \frac{2m+1}{2}\lambda \end{array} \right\} \qquad (9\text{-}76)$$

This result will be used in the following sub-section.

(b) <u>Diffraction by a stationary acoustic grating in an isotropic medium</u>

Assume that an acoustic wave is launched in an <u>isotropic</u> medium (e.g. water) and that the local particle displacement is

$$\vec{U}_z \sim \hat{z}\cos(\Omega t - Kz)$$

(see §9.2, example 1)

A sinusoidal index grating is thus generated in the medium. Let the incident plane E - M wave be linearly polarized in the x - direction and the wave vector $\vec{k}$ in the y - z plane as shown in Fig. 9-4(a). The polarization $\vec{E}_x$ thus sees a sinusoidal index grating planes propagating in the z - direction (see §9.2, example 1). Since acoustic wave propagation velocity is much smaller than the speed of light ($v_{acoustic}/c \sim 10^{-5}$) we can consider the acoustic index grating as stationary in a first approximation. Such a grating is shown as horizontal lines in Fig. (9-4)(b). We assume that the scattered light wave is also plane which implies that the acoustic grating surfaces are plane and the interaction region is large. As is well known, when a light beam crosses an interface separating two materials of indices n_1 and n_2 ($n_1 \neq n_2$), Fresnel reflection takes place together with refraction (Chapter III). The phenomenon is due to the difference in indices across

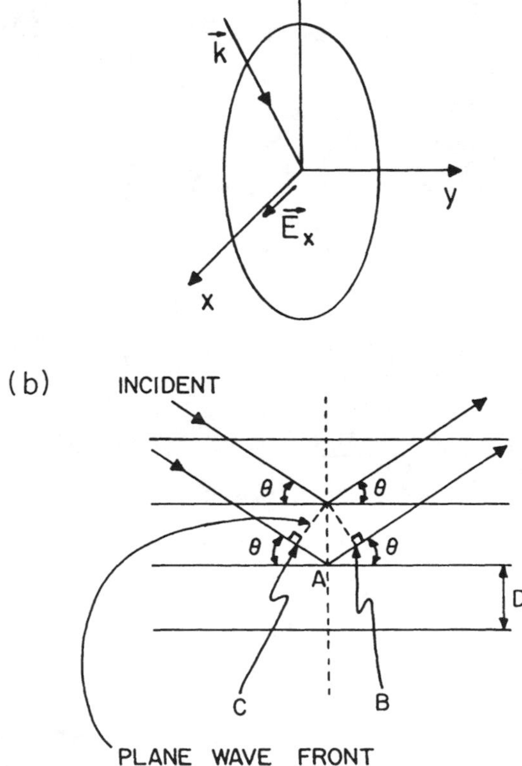

Fig. 9-4. Diffraction by a stationary acoustic grating in an isotropic medium.
(a) Relationship between $\vec{k}$, $\vec{E}_x$ of an incident E-M wave and resultant index ellipsoid of the material.
(b) Acoustic diffraction.

the interface. Thus, in the case of acoustic plane index grating, there are periodic changes in index. Higher index regions normally correspond to higher density regions of the same material. Light wave crossing these higher index regions will be reflected also. We idealize the periodic higher index regions as very thin planes (Fig. 9-4b). A plane wave incident onto two adjacent index planes is shown as two rays reflected by these index planes. (Refraction is ignored in our discussion.) The two reflected beams will be superimposed, giving rise to interference. Constructive interference takes place if the optical path difference between the two beams satisfies the first of eq. (9-76).

i.e. $$\Delta L = m\lambda \ , \ (m = 1,2,\ldots) \qquad (9\text{-}77)$$

or $$AC + AB = m\lambda$$

$$\because \ AC = AB = 2D\sin\Theta$$

$$\therefore \ 2D\sin\Theta = m\lambda \qquad (9\text{-}78)$$

or $$2\pi \cdot \frac{2\sin\Theta}{\lambda} = (m/D) \cdot 2\pi$$

i.e. $$2k\sin\Theta = m K \qquad (9\text{-}79)$$

where $k = \frac{2\pi}{\lambda}$ is the E - M wave vector and $K = \frac{2\pi}{D}$ is the acoustic wave vector. Two interpretations can be considered.

(i) We interpret eq. (9-79) as the condition under which there is constructive interference between the two light beams reflected by two adjacent plane acoustic index grating planes whose wave vector is
$mK = \frac{2\pi}{(D/m)}$ $(m = 1,2,3,\ldots)$.

If $m \geq 2$, it means that higher harmonics (2nd, 3rd etc.) of the acoustic wave of wavelength D/m have contributed to the constructive interference. In the case of a pure sinusoidal acoustic wave, such harmonics are absent and m = 1.

Hence, $$2k\sin\Theta = K \qquad (9\text{-}80)$$

we note that the wavelength λ in eq. (9-78) is the wavelength of light in the medium, <u>not</u> in vacuum. $\lambda = v/\nu = (c/n)/\nu = \frac{1}{n}\lambda_o$

i.e. $$\lambda_o = n\lambda \tag{9-81}$$

where n = index of the material

λ_o = wavelength of light in vacuum

(ii) Another more classical interpretation of eq. (9-78) or (9-79) is that m is the order of diffraction, even if λ and D are the monochromatic wavelengths of the light and acoustic wave respectively. But there is a constraint. Eq. (9-78) can be re-written as

$$\sin\Theta = \frac{m}{2}\frac{\lambda}{D}$$

Therefore, either D cannot be too small or m cannot be too large. Otherwise, $\sin\Theta$ will become greater than one which is impossible. For high frequency acoustic wave, D is small, and is not much larger than λ. Thus, $m = 1$ or at best 2. In this section, we assume $m = 1$.

In summary, constructive interference occurs between light waves reflected by two adjacent acoustic index planes if eq. (9-78), (9-79) or (9-80) is satisfied. Since the distance D is the same between any pair of adjacent planes, all pairs of light waves reflected by different pairs of adjacent acoustic index planes will interfere constructively and the combination of all these waves forms the diffracted wave with constructive interference. If D is kept constant and λ is changed, then Θ has to be changed also in order to have constructive interference (i.e. satisfy eq. 9-78, 79 or 80). Or if $\vec{k}$ (and thus Θ and λ) is kept constant, one has to vary D (the acoustic wavelength) in order to obtain constructive interference in the reflection. Under the situation of constructive interference of the reflected light waves, the transmission of light through the medium is reduced significantly because of the conservation of energy. The transmission is the complement of reflection. Constructive interference in the reflection means destructive interference in the transmission. Such a change in the transmission of light can be put to good use in photonic devices (e.g. Q-switching, mode-locking etc.).

The diffraction discussed above is usually called Bragg diffraction and eq. (9-78, 79, 80) are the conditions of Bragg diffraction.

(c) Diffraction by a travelling acoustic grating - Doppler shift in an isotropic medium

If we consider the real situation in which a travelling acoustic wave is launched in the medium and if we do not make any approximation, then the diffraction of the light wave will suffer a Doppler shift. That is to say, the observer in the laboratory reference frame will see that the diffracted wave's frequency will be shifted. Again, it is instructive to first review the elementary idea of Doppler shift which is then applied to the diffraction.

(i) Classical picture of Doppler shift

As shown in Fig. (9-5(a)), an observer moves at a velocity v towards a light source emitting light waves that travel at a velocity c towards the observer. The situation can be "decomposed" into two situations, one in which the observer is at rest and another in which the light wave is at rest (Fig. 9-5(b) and (c) respectively). In (b), the observer at rest "sees" c/λ sections of waves (each of length λ) passing him in one second; i.e. he "sees" a light frequency $\nu = c/\lambda$ when he is at rest. In (c), the moving observer "sees" a stationary light wave and he passes v/λ sections of wave (each of length λ) in one second. Combining (b) and (c), (adding), the observer

(a)

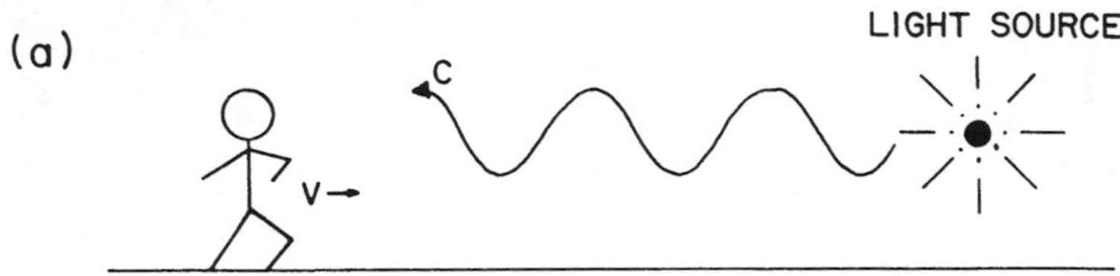

(b)

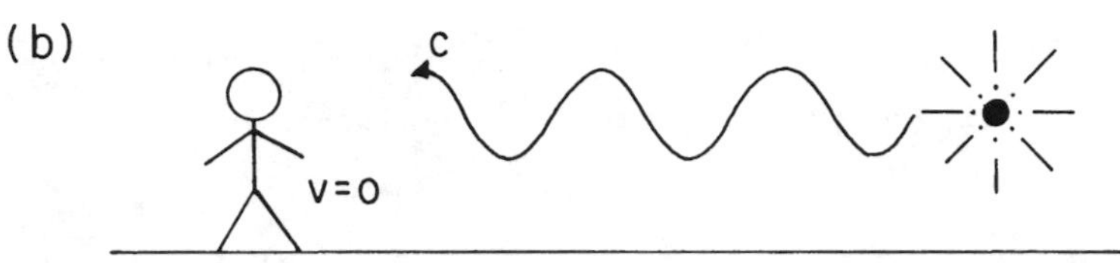

(c)

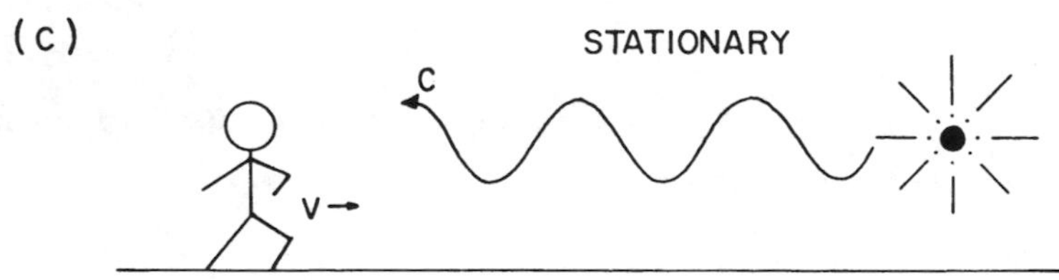

Fig. 9-5. Schematic illustration of the Doppler effect.

"sees" $\left(\frac{c}{\lambda} + \frac{v}{\lambda}\right)$ sections of waves passing him in one second (case of (a)). i.e. he "sees" a light frequency

$$\nu' = \frac{c}{\lambda} + \frac{v}{\lambda}$$

$$= \nu + \frac{v}{\lambda} \tag{9-82}$$

i.e. there is a frequency shift of

$$\Delta\nu \equiv \nu' - \nu = \frac{v}{\lambda} \tag{9-83}$$

Eq. (9-83) is the <u>Doppler up-shift</u> of the light's frequency measured by an observer moving <u>towards</u> the light source. If he moves away from the light source, v become negative and the frequency is <u>Doppler down-shifted</u>. In general, we can write the Doppler shift as either positive or negative:

$$\Delta\nu = \pm \frac{v}{\lambda} \tag{9-84}$$

Now, in the case of diffraction by the acoustic wave, the observer "sees" the light waves reflected from the travelling acoustic index planes. We thus need to determine the frequency shift of one of the reflected "beams" from one of the acoustic index planes. This index plane can be considered as a "mirror" as shown in Fig. (9-6(a)). The observer "sees" that the "mirror" moves in the vertical direction at a velocity v and the light wave moves from the image source towards the observer at a velocity c. Again, the situation can be "decomposed" into two, one with the "mirror" stationary, the other with the light wave stationary and the mirror moving, as shown in Fig. 9-6(b) and (c), respectively. In Fig. 9-6(b), the observer counts $\frac{c}{\lambda}$ sections of waves (each of length λ) in one second, i.e. he measures a frequency $\nu = c/\lambda$. In Fig. 9-6(c), the image moves at twice the speed of the mirror (in the non-relativistic regime). This can be seen from Fig. 9-6(c):

$$h + 2\ell = 2d$$

$$h = 2(d - \ell) \tag{9-85}$$

Assume h is the displacement of the image in one second. Then v is the displacement of the "mirror" in one second. From the figure, $d = \ell + v$ or $d - \ell = v$. Substituting in eq. (9-85), we have

$$h = 2v \tag{9-86}$$

i.e. the image speed is twice the "mirror" speed. To the observer, the image moves towards him at a speed of

$$v' = h\cos(90° - \Theta)$$

$$= 2v\sin\Theta \tag{9-87}$$

This is equivalent to the observer moving towards a stationary light source at the speed v'. Thus, the situation in Fig. 9-6(c) is such that the observer passes $\frac{v'}{\lambda}$ sections of waves. Combining (b) and (c) gives the total number of sections of waves measured by the observer:

$$\nu' = \frac{c}{\lambda} + \frac{v'}{\lambda}$$

$$= \nu + \frac{2v\sin\Theta}{\lambda} \qquad (9\text{-}88)$$

If the acoustic wave is propagating in the opposite direction (towards the bottom), v is negative. Thus the general shift of the frequency of the diffracted light is

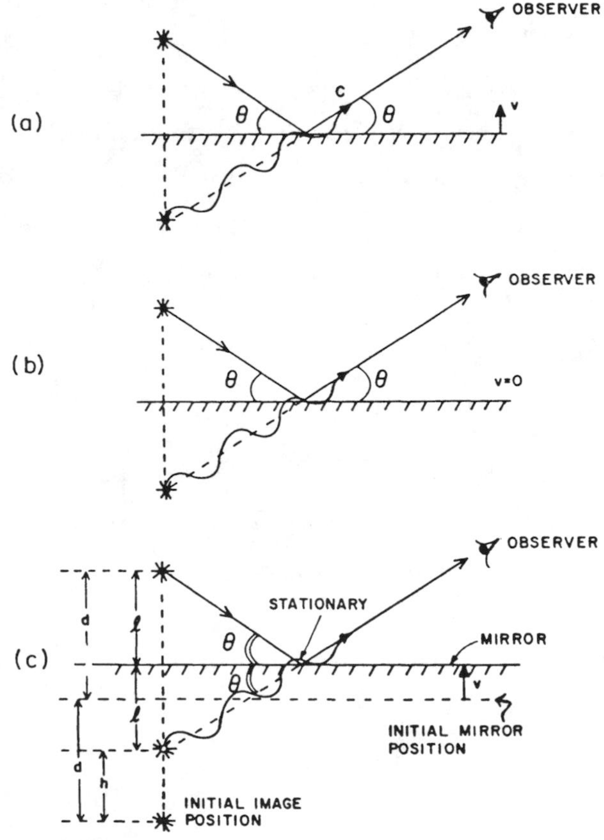

Fig. 9-6. Doppler effect due to the motion of a mirror.

$$\Delta\nu = \nu' - \nu = \pm \frac{2v\sin\theta}{\lambda}$$

$$= \pm \frac{m}{D} v \qquad \text{from eq. (9-78)}$$

or

$$2\pi \, \Delta \nu = \pm \, m \frac{2\pi}{D} v$$

i.e.

$$\Delta\omega = \pm \, mKv \tag{9-89}$$

The acoustic wave velocity is related to K and Ω of the acoustic wave by

$$v = \frac{\Omega}{K} \qquad \text{(see chapter I)} \tag{9-90}$$

Eq. (9-89) thus becomes

$$\Delta\omega = \pm \, m\Omega \; (m = 1,2,3,\ldots) \tag{9-91}$$

(ii) <u>Quantum Picture of Doppler shift</u>

We can also describe the Doppler shift by using the quantum picture. Here, we assume that both the light wave and the acoustic wave are quantized so that we can describe them as photons and phonons respectively Such quantizations can be found in advanced books on quantum optics and solid state physics. No detail will be given here. The reader is thus asked to believe that the E - M wave consists of a stream of photons each of energy $\hbar\omega$ and momentum $\hbar\vec{k}$ while the acoustic wave consists of a stream of phonons each of energy $\hbar\Omega$ and momentum $\hbar\vec{K}$. ($\vec{K}$ is in the direction of $\vec{v}$, the acoustic wave velocity.) The diffracted light wave consists of a stream of photons each of energy $\hbar\omega'$ and momentum $\hbar\vec{k}'$. Conservation of energy requires that

$$\hbar\omega' = \hbar\omega + \hbar\Omega \tag{9-92}$$

$$\hbar\vec{k}' = \hbar\vec{k} + \hbar\vec{K} \tag{9-93}$$

Fig. (9-7) shows the relation (9-93) which becomes

$$\vec{k}' = \vec{k} + \vec{K} \tag{9-94}$$

while (9-92) becomes

$$\omega' - \omega = \Omega$$

or

$$\Delta\omega = \Omega \tag{9-95}$$

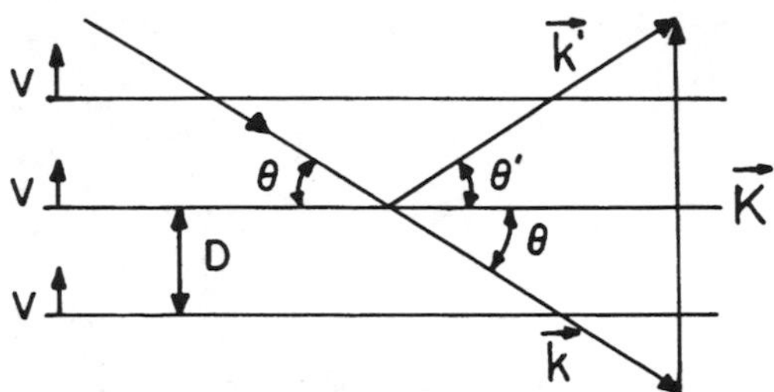

Fig. 9-7. Quantum picture illustrating Doppler effect in the diffraction of an E-M wave by a travelling acoustic plane wave.

If the acoustic wave travels in the $-\vec{v}$ direction, $\vec{K}$ in Fig. 9-7 will reverse direction also and we have (from Fig. (9-7), with $\vec{K}$ reversed in direction)

$$\vec{k} = \vec{K} + \vec{k}' \tag{9-96}$$

and we have to write the conservation of energy as

$$\hbar\omega = \hbar\omega' + \hbar\Omega$$

or

$$\omega' - \omega = -\Omega \tag{9-97}$$

Combining eqs. (9-95) and (9-97), we have

$$\Delta\omega = \pm\Omega \tag{9-98}$$

If the acoustic wave vector $|\vec{K}|$ is replaced by mK, it means we are dealing with the acoustic harmonics. Hence the phonon energy is $m\hbar\Omega$ and eq. (9-98) becomes

$$\Delta\omega = \pm m\Omega \tag{9-99}$$

this equation is identical to eq. (9-91). Eq. (9-93) and Fig. (9-7) gives

$$k'\sin\Theta' + k\sin\Theta = K \tag{9-100}$$

In general, $k' \neq k$, $\Theta' \neq \Theta$. But for very small change, we can approximate $k' = k$, $\Theta' = \Theta$, eq. (9-100) becomes

$$2k\sin\Theta = K \quad \text{(No Doppler shift approximation)} \tag{9-101}$$

which is identical to eq. (9-80).

§9.4 <u>Diffraction by an acoustic wave in an anisotropic medium</u>

If the medium is anisotropic, the index of refraction of the medium "seen" by the light beam is a function of the light propagation direction (Chapter VI). Since $\vec{k}$ and $\vec{k}'$ are not in the same direction, each will see a different index. It means that the condition for Bragg diffraction (eq. 9-78, 79, 80) is not $\Theta = \Theta'$ but $\Theta \neq \Theta'$ (Fig. 9-7).

We first analyse it qualitatively. Assuming that the polarization of the incident wave is linear and is oriented parallel to the polarization of the extraordinary wave inside the anisotropic medium. The diffracted wave, travelling in another direction $\hat{k}'$, will "see" two indices or two values of k' , as shown in Fig. 9-8(a) in which we show the surface of wave normals of a uniaxial crystal. The lengths of $\vec{k}$ and the two possible values of k' are shown. However, for a fixed acoustic wave vector $\vec{K}$, only one value of k' will satisfy the Bragg diffraction condition. This is shown in Fig. 9-8(b) and (c). Either of these two conditions can be fulfilled separately but not simultaneously. This can be achieved by changing $\vec{K}$, the acoustic wave vector; i.e. changing its direction and frequency. Note: the law of reflection (chapter III) requires that $\vec{k}$ and $\vec{k}'$ lie in the same plane.

Quantitatively, let us assume that we have set the acoustic $\vec{K}$ such that Fig. (9-8(b)) is valid. It is re-drawn in Fig. (9-9). From trigonometric relations, we see from Fig. (9-9) that

$$\begin{cases} k'^2 = k^2 + K^2 - 2kK\cos\alpha \\ k^2 = k'^2 + K^2 - 2k'K\cos\beta \end{cases}$$

since

$$\alpha = \pi/2 - \Theta$$

$$\beta = \pi/2 - \Theta'$$

the two relations become, after dividing them by K,

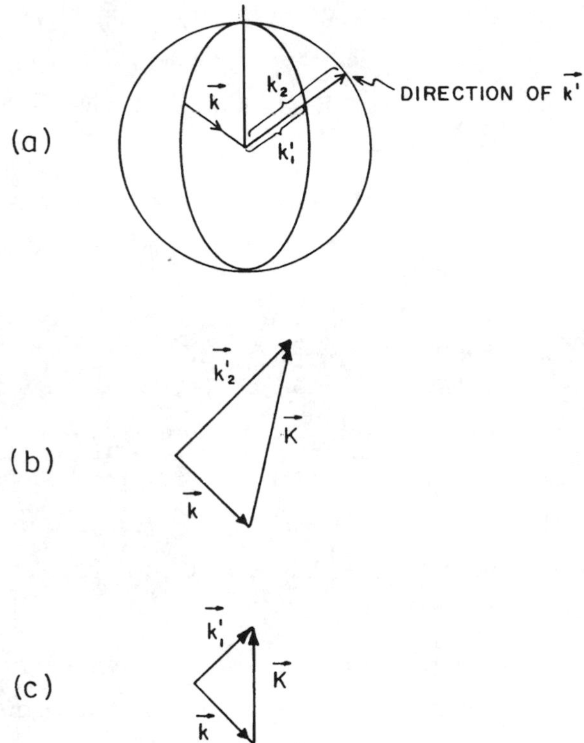

Fig. 9-8. Diffraction of light by an acoustic wave in an anisotropic medium (uniaxial in the figure).

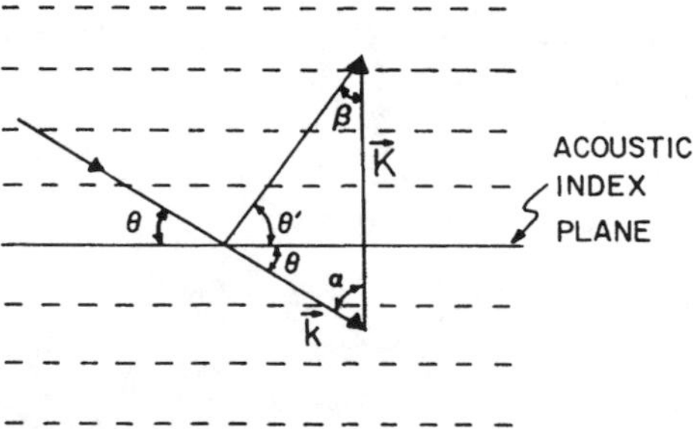

Fig. 9-9. Relationship between the incident and the acoustically diffracted angles θ and θ' in an anisotropic medium.

$$\begin{cases} 2k\sin\Theta = \dfrac{k^2 - k'^2}{K} + K & (9\text{-}102) \\ 2k'\sin\Theta' = \dfrac{k'^2 - k^2}{K} + K & (9\text{-}103) \end{cases}$$

Using

$$K = \frac{2\pi}{D} \tag{9-104}$$

$$k' = \frac{n'\omega'}{c} \tag{9-105}$$

$$k = \frac{n\omega}{c} \tag{9-106}$$

and assuming $\omega' \approx \omega$, because the Doppler shift is very small $(\omega' - \omega = \Omega \ll \omega)$ which gives

$$k' \approx \frac{n'\omega}{c} \tag{9-107}$$

Eq. (9-102) and (9-103) become

$$\begin{cases} 2\ D\sin\Theta = \dfrac{\lambda_o}{n} - \dfrac{D}{n\lambda_o}\left(n'^2 - n^2\right) & (9\text{-}108) \\ 2\ D\sin\Theta' = \dfrac{\lambda_o}{n'} - \dfrac{D}{n'\lambda_o}\left(n'^2 - n^2\right) & (9\text{-}109) \end{cases}$$

where now λ_o is the wavelength of light in vacuum, D is the acoustic wavelength, and n and n' are the refractive indices "seen" by the incident and diffracted waves, respectively. Eq. (9-108) and (9-109) are the general conditions for Bragg diffraction of light by an acoustic index wave in an anisotropic medium. The reason why they are general can be seen by referring back to Fig. 9-8(a). Although we have assumed only one value of the incident $\vec{k}$, the general case of two values of incident $\vec{k}$ is still the same for only <u>one</u> of the two values of k' to satisfy the momentum conservation condition (Fig. 9-8(b) <u>or</u> (c)) thus leading to eq. (9-108) and (9-109). The crystal can be biaxial instead of uniaxial and all the above arguments still apply, meaning that eq. (9-108) and (9-109) are indeed general conditions for Bragg diffraction in an anisotropic medium. If there are acoustic harmonics present, we can write eq. (9-104) as

$$K \to K' \equiv Km \qquad m = 1,2,3,\ldots$$

$$\equiv \frac{2\pi}{D} m$$

$$= \frac{2\pi}{(D/m)}$$

$$= \frac{2\pi}{D_m} \qquad (9\text{-}120)$$

where $$D_m \equiv D/m \qquad (9\text{-}121)$$

Substituting into (9-108), (9-109), we have

$$\begin{cases} 2D_m \sin\Theta = \dfrac{\lambda_o}{n} - \dfrac{D_m}{n\lambda_o}\left(n'^2 - n^2\right) & (9\text{-}122) \\[2ex] 2D_m \sin\Theta' = \dfrac{\lambda_o}{n'} - \dfrac{D_m}{n'\lambda_o}\left(n'^2 - n^2\right) & (9\text{-}123) \end{cases}$$

where D_m is given by (9-121). Both eq. (9-122) and (9-123) reduce to eq. (9-78) in the special case of an isotropic medium, i.e. $n' = n$; $\Theta \simeq \Theta'$. Under this condition, they become

$$2D_m \sin\Theta = \frac{\lambda_o}{n}$$

or $$2\,\frac{D}{m}\sin\Theta = \lambda \qquad \text{(using eq. (9-81) and (9-121))}$$

or $$2D\sin\Theta = m\lambda\ .$$

Eq. (9-108), (109) or (9-122), (123) determine the incident and diffracted angles Θ and Θ' defined in Fig. (9-9). Not all values of λ and D will satisfy the real situation. In other words, there is a range of validity of λ and D as illustrated by the following example.

<u>Example</u>: Let the medium be uniaxial. The incident wave sees an index n_e (i.e. its polarization is parallel to that of the extraordinary wave in the

uniaxial medium). The diffracted wave sees the ordinary index n_o. Also, $n_o > n_e$ (negative uniaxial). We like to find a range of validity for the Bragg diffraction. Consider only the case m = 1; i.e. eq. (9-108), (109), which can be re-written as

$$\left\{\begin{array}{ll} \sin\Theta = \frac{1}{2n_e}\left\{\frac{\lambda_o}{D} - \frac{D}{\lambda_o}(n_o^2 - n_e^2)\right\} & \quad (9\text{-}124) \\ \sin\Theta' = \frac{1}{2n_o}\left\{\frac{\lambda_o}{D} + \frac{D}{\lambda_o}(n_o^2 - n_e^2)\right\} & \quad (9\text{-}125) \end{array}\right.$$

These conditions are valid only if Θ and Θ' are real, or

$$-1 \le \left\{\begin{array}{c} \sin\Theta \\ \sin\Theta \end{array}\right\} \le 1 \tag{9-126}$$

Using eq. (9-124), (125), eq. (9-126) becomes

$$\left\{\begin{array}{ll} -1 \le \frac{D}{\lambda_o}\cdot\frac{1}{2n_e}\left\{\left(\frac{\lambda_o}{D}\right)^2 - (n_o^2 - n_e^2)\right\} \le 1 & \quad (9\text{-}127) \\ -1 \le \frac{D}{\lambda_o}\cdot\frac{1}{2n_o}\left\{\left(\frac{\lambda_o}{D}\right)^2 + (n_o^2 - n_e^2)\right\} \le 1 & \quad (9\text{-}128) \end{array}\right.$$

Let
$$q \equiv \lambda_o/D \tag{9-129}$$

Eq. (9-127) becomes two inequalities:

$$q^2 - 2n_e q - (n_o^2 - n_e^2) \le 0 \tag{9-130}$$

and
$$q^2 + 2n_e q - (n_o^2 - n_e^2) \ge 0 \tag{9-131}$$

while (9-128) becomes

$$q^2 - 2n_o q + (n_o^2 - n_e^2) \le 0 \tag{9-132}$$

for $(n_o > n_e)$.

It is now left as an <u>exercise</u> for the reader to obtain the following condition of validity using eq. (9-130) to (132).

$$|n_o - n_e| \le \frac{\lambda_o}{D} \le |n_o + n_e| \tag{9-133}$$

This is shown in Fig. (9-10) (shaded region).

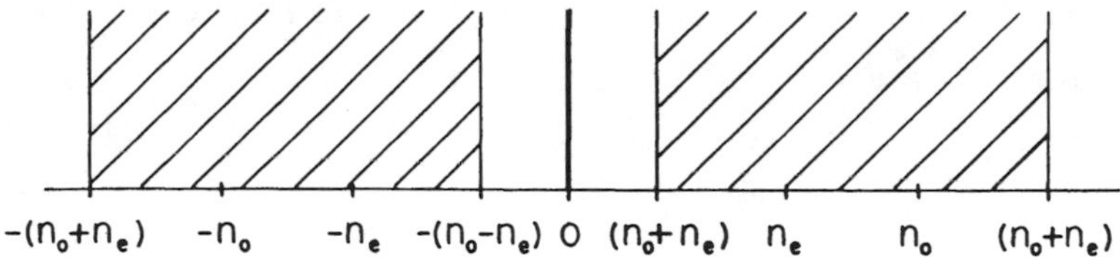

Fig. 9-10. Range of validity (shaded region) during the diffraction of light by an acoustic wave in a uniaxial medium.

§9.5 Higher order diffraction by an acoustic wave

So far, in the previous two sections, we consider only the lowest order diffraction of infinitely large plane light wave by infinitely large plane acoustic waves; namely, m = 1 in eq. (9-78) or (9-79). Higher order diffraction $m \geq 2$ will take place under any one of the following conditions. We shall analyse them only qualitatively.

(i) The interaction between the light wave and the acoustic wave is limited in a very small region. This implies that even if the total acoustic wave is still a plane wave across a large region of the medium, the scattered light wave cannot be plane any more because the small interaction region cannot re-radiate plane waves. By Huygen's principle, each tiny element in the interaction region radiate spherical waves which will re-combine with one another resulting in higher order constructive interference at a distance away. Think of this as similar to the diffraction of a plane light wave by a small aperture. The phase of the optical wave is modified giving rise to higher order patterns (fringes) in the far field. Thus the acoustic limiting the transmission of the light wave into the interaction region by an aperture very close to the diffraction region.

(ii) If the frequency of the acoustic wave is very low so that the acoustic wavelength D is very large, then, even if both the optical and acoustic waves are plane waves, eq. (9-78) and (9-79) are also valid at values of m larger or equal to 2.

(iii) If the acoustic wave is not a plane wave, even if the light wave is plane and the interaction region large, higher order diffraction occurs in the following sense. Referring to Fig. 9-11(a), we assume that the acoustic index wave fronts are curved surfaces. These surfaces are equivalent to the combination of a continuous set of plane waves which are tangent to the curved surfaces. A plane light wave incident onto these curved acoustic index waves will suffer multiple diffraction (scattering) by successive plane acoustic waves (Fig. 9-11(b)). This is equivalent to the wave vector picture of multiple photon-phonon scattering (Fig. 9-11(c)) in which every wave vector ($\vec{k}$ or $\vec{K}$) represents a plane wave. The successive change in the directions of $\vec{K}_1$, $\vec{K}_2$, $\vec{K}_3$ etc. represents the curvature of the acoustic wavefront. Comparing Fig. 9-11(c) and Fig. (9-7), using eq. (9-92), we see that (assuming that the frequency of all $\vec{K}$'s is Ω_o),

$$\omega_1 = \omega_o + \Omega_o$$

$$\omega_2 = \omega_1 + \Omega_o = \omega_o + 2\Omega_o$$

$$\omega_{m+1} = \omega_o + m\Omega_o$$

That is to say, the multiple diffraction gives rise to many high order constructive interference. In general, one speaks of orders $\pm 1, \pm 2, \pm 3, \ldots \pm m \ldots$ as shown in Fig. 9-11(d).

All the above three cases that generate high order diffractions are sometimes called "Raman-Nath" diffraction. One physical way to distinguish Raman-Nath (or high order) diffraction from the lowest order Bragg diffraction is the following. Consider Fig. 9-12. The limit at which multiple

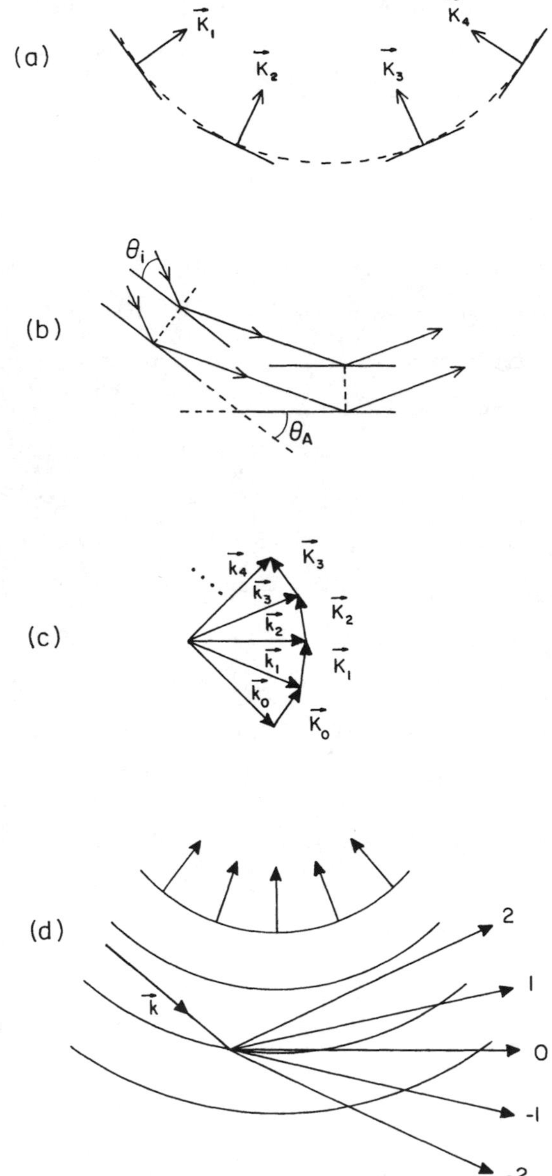

Fig. 9-11. Higher order acoustic diffraction.

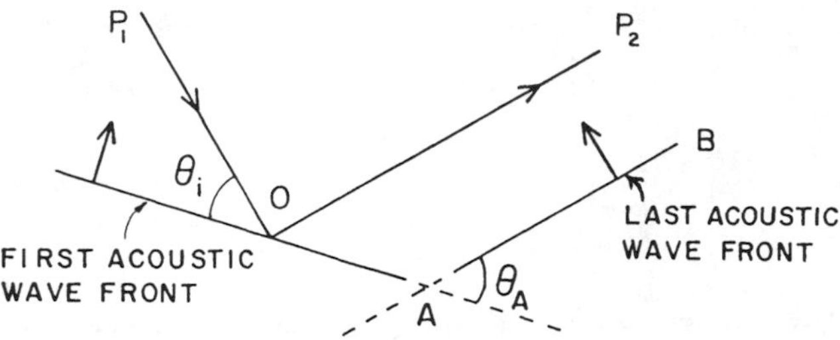

Fig. 9-12. Illustrating the limiting condition for Raman-Nath diffraction.

scattering is impossible is that the reflected beam OP_2 is parallel to the last plane acoustic wave surface AB. This means that

$$\Theta_i = \Theta_A$$

where Θ_i is the angle between the incident beam and the first acoustic index plane it encounters and Θ_A is the angle between the first and the last acoustic index plane (assuming a concave acoustic surface, see Fig. 9-12). Θ_A is essentially the divergence (convergence) angle of the acoustic wavefront. Thus, when the incident angle Θ_i (as defined in Fig. 9-12) is equal or greater than the divergence angle of the acoustic wavefront, no multiple diffraction is possible. Hence, the following criteria are physically sound:

$$\Theta_i \geq \Theta_A \qquad \text{Bragg diffraction}$$

$$\Theta_i < \Theta_A \qquad \text{Raman-Nath diffraction}$$

However, as can be judged from Fig. (9-12), if it happens that Θ_i is only <u>slightly smaller</u> than Θ_A, the beam OP_2 will intersect the last acoustic plane AB at a very far away point. This in practice is impossible. Thus, for practical purposes, $\Theta_i < \Theta_A$ is not sufficient.

We can analyse this problem more closely by looking at a specific example. Assuming that the acoustic wave has a Gaussian spherical converging wave front similar to an optical Gaussian spherical wave. Using the same concept for defining the angle of divergence of an optical spherical wave in the far field (Chapter V, eq. (5-37)), the full angle of divergence of the acoustic wave in the far field is

$$\Theta_A \doteq \frac{2D}{\pi w_A} \tag{9-134}$$

where D is the acoustic wavelength, $2w_A$ is the acoustic beam waist (diameter). As shown in Fig. (9-13), the spherical acoustic index wave front is represented by the curve PQR. That is to say, the acoustic wave has a limited extent. Assume the extreme case in which a light beam is incident at P. From the figure, the reflected beam PQ hits the acoustic wave front again at the point Q. This is the condition for Raman-Nath multiple scattering. The limit for multiple scattering is when the reflected beam hits the other extremity of the acoustic wave, namely the point R. We can thus say that the condition for Raman-Nath multiple diffraction is (See Fig. 9-13).

$$\frac{\pi}{2} - \Theta_r \geq \frac{\pi}{2} - \Theta_A/2 \qquad \text{(Raman-Nath diffraction)} \tag{9-135}$$

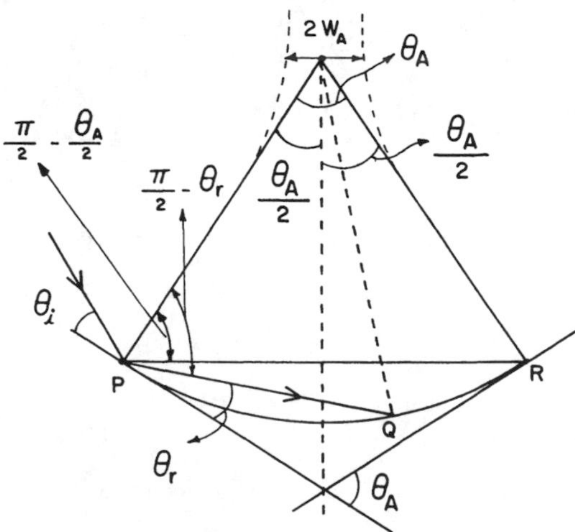

Fig. 9-13. Geometrical analysis of acoustic diffraction of various orders.

The equal sign corresponds to when the reflected beam hits the point R. Since $\Theta_r = \Theta_i$ (law of reflection), eq. (9-135) becomes

$$\frac{\Theta_A}{2} \geq \Theta_i \qquad \text{(Raman-Nath diffraction)} \qquad (9\text{-}136)$$

The incident angle Θ_i has to satisfy the lowest order Bragg diffraction condition (eq. 9-78, m = 1)

$$2D\sin\Theta_i = \lambda \qquad (9\text{-}137)$$

Substituting eq. (9-137) and (9-134) into (9-136), we obtain

$$\frac{D}{\pi W_A} \geq \sin^{-1}\left(\frac{\lambda}{2D}\right) \qquad \text{(Raman-Nath diffraction)} \qquad (9\text{-}138)$$

If Θ_i is very small, $\Theta_i \sim \sin\Theta_i = \frac{\lambda}{2D}$, (from (9-137))

Eq. (9-138) becomes

$$\frac{D}{\pi W_A} \geq \frac{\lambda}{2D}$$

$$\frac{2D^2}{\pi\lambda W_A} \geq 1 \qquad \text{(Raman-Nath diffraction)} \qquad (9\text{-}139)$$

In summary, the conditions for Bragg and Raman-Nath diffractions by Gaussian spherical acoustic waves of beam waist W_A and acoustic wavelength D are:

from (9-136):

$$\Theta_i \begin{cases} > \Theta_A/2 & \text{Bragg} \\ \leq \Theta_A/2 & \text{Raman-Nath} \end{cases}$$

or from (9-138):

$$\sin^{-1}\left(\frac{\lambda}{2D}\right) \begin{cases} > \dfrac{D}{\pi W_A} & \text{Bragg} \\ \leq \dfrac{D}{\pi W_A} & \text{Raman-Nath} \end{cases}$$

For small Θ_i , from (9-139)

$$\frac{2D^2}{\pi\lambda W_A} \begin{cases} < 1 & \text{Bragg} \\ \geq 1 & \text{Raman-Nath} \end{cases}$$

§9.6 Closing remarks

While the electro-optic effect is the consequence of electric field induced anisotropic polarization in a medium, the acousto-optic effect is the consequence of mechanical (acoustic) force induced elastic anisotropy. In both cases, one should pay attention to the polarization state of the light (laser) wave that passes through the medium because of the induced anisotropy. The electro-optic medium is essentially an electric field induced wave plate while the acousto-optic medium can be considered as a mechanical (acoustic) force induced grating. In the latter case, light wave diffracted by the acoustic index grating can be used for laser beam scanning and modulation, mode-locking, Q-switching, spectrum analysis etc. For example, in the case of Q-switching, the diffraction of the laser radiation in an acousto-optic medium inside the laser cavity acts as a loss mechanism thus prohibiting laser oscillation. This allows the Q of the cavity to build up (see Chapter VIII). When the acoustic field is turned off after a while, oscillation starts and a giant laser pulse is generated.

No attempt is made in this chapter to go into the detail of any application of the acousto-optic effect. The reader is referred to sources such as Yariv and Yeh.

CHAPTER X

MAGNETIC FIELD INDUCED ANISOTROPY

When mechanical force and electric field can both induce optical anisotropy in a medium, it is natural to ask if a magnetic field will also induce anisotropy. The answer is yes. The magnetic field can be a static one or the magnetic component of the incident E - M field. Both give rise to a change in some special media resulting in a rotation of the electric field vector (polarization state) of the E - M wave that passes through them. This short chapter explains qualitatively the physical phenomena underlying the magnetic field induced anisotropy. There are two types of magnetic field induced anisotropy, one is natural and one is "artificial". The natural one is traditionally called optical activity and the "artificial" one, Faraday rotation.

§10.1 Optical activity

The phenomenon occurs in some natural material that doesn't have an inversion symmetry, while the molecular structure in the material have some kind of spiral nature. Thus, apart from the normal dielectric response of the material to the electric component of an E - M wave propagating in it (Chapter VI) the spiral structure of the molecules will also respond to the magnetic component of the E - M wave. This is because some non-bonding electrons on such spiral molecules will move around the spiral when there is a time varying magnetic field passing through it. This is similar to a solenoid (or coil) put in a time varying magnetic field (Fig. 10-1). A current will flow through the solenoid according to the Faraday's law:

$$- \frac{\partial \vec{B}}{\partial t} = \nabla \times \vec{E} \qquad (10\text{-}1)$$

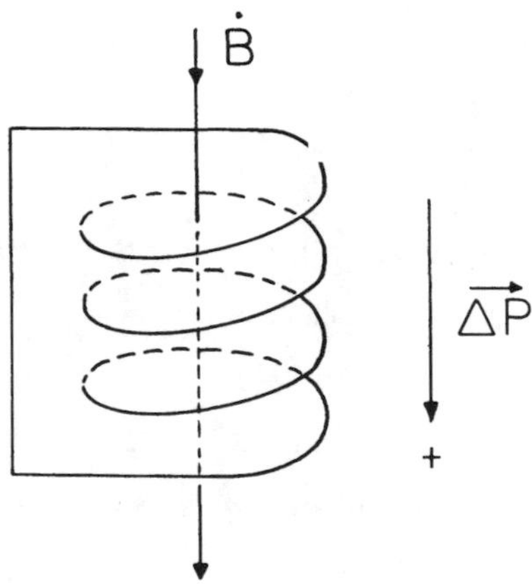

Fig. 10-1. Similarity of response of a spiral molecule to a solenoid when placed in a time varying magnetic field.

Because the magnetic field in an E - M wave varies rapidly, the accumulation of charges (+ and -) at the top and bottom of the solenoid (Fig. 10-1) changes sign also rapidly. Thus, there is a time varying induced dipole moment. If we consider that the medium consists of a distribution of many molecular solenoids, then for every dielectric response of the medium (Chapter VI) in a local small volume, there is also a magnetic response which gives rise to a change in polarization, $\Delta\vec{P}$. From eq. (10-1) and Fig. (10-1), we see that

$$\nabla \vec{P} \propto \left(- \frac{\partial \vec{B}}{\partial t} \right) = \nabla \times \vec{E} \qquad (10\text{-}2)$$

We assume a plane E - M wave propagating through the medium (c.f. eq. (6-26) and (6-27)), so that

$$\nabla = - i\vec{k}$$

Then eq. (10-2) becomes

$$\Delta\vec{P} \propto -\, i\vec{k} \times \vec{E}$$

$$\Delta\vec{P} = -\, i\alpha\vec{k} \times \vec{E} \tag{10-3}$$

where α is the proportionality constant. If we define

$$\vec{G} \equiv -\frac{\alpha\vec{k}}{\epsilon_o} \tag{10-4}$$

Eq. (10-3) becomes

$$\Delta\vec{P} = +\, i\epsilon_o\vec{G} \times \vec{E} \tag{10-5}$$

The total field $\vec{D}$ in the medium is now

$$\begin{aligned}\vec{D} &= \vec{P} + \epsilon_o\,\vec{E} + \Delta\vec{P} \\ &= \epsilon\vec{E} + \Delta\vec{P} \\ &= \epsilon\vec{E} + i\epsilon_o\vec{G} \times \vec{E}\end{aligned} \tag{10-6}$$

Here, ϵ is the dielectric tensor of the medium.

§10.2 Faraday rotation

Consider the propagation of a plane E - M wave in a medium. The electric field of the wave sets the electrons in the path of the E - M wave in motion. Let the instantaneous velocity of the electron be v. Thus

$$m\dot{v} = e\vec{E} \propto e^{i\omega t}$$

$$v \propto \int e^{i\omega t} dt = \frac{1}{i\omega}\, e^{i\omega t} \propto -\, i\vec{E} \tag{10-7}$$

If we now apply an external constant magnetic field $\vec{B}$ along the propagation direction $\hat{k}$ of the E - M wave. The moving electron will experience a magnetic force due to the external field $\vec{B}$ and is given by the Lorentz force

$$\vec{F} = e\vec{v} \times \vec{B} \tag{10-8}$$

The electronic charge e is negative. Such a force will have the effect of generating an additional separation of positive and negative charge as shown in Fig. (10-2). There is then an additional polarization in a local volume,

$$\Delta\vec{P} \propto -\,\vec{F} = -\, e\vec{v} \times \vec{B} \tag{10-9}$$

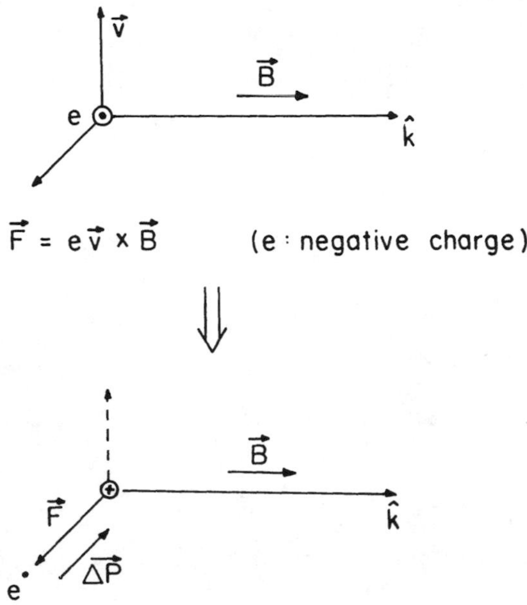

Fig. 10-2. Fundamental consideration of electron motion in a magnetic field leading to Faraday rotation.

which becomes, using eq. (10-7),

$$\Delta\vec{P} \propto -e(-i\vec{E}) \times \vec{B}$$

or

$$\Delta\vec{P} = i\,a\,e\vec{E} \times \vec{B} \qquad \text{a: proportionality constant}$$

$$= i\,a\,(-1)\,|e|\,(-\vec{B} \times \vec{E})$$

$$= i\,a\,|e|\,\vec{B} \times \vec{E} \tag{10-10}$$

Let

$$\epsilon_o\gamma\vec{G} \equiv a\,|e|\,\vec{B} \tag{10-11}$$

where $\gamma \equiv$ <u>magnetogyration coefficient</u>,

eq. (10-10) becomes

$$\Delta\vec{P} = i\gamma\epsilon_o G \times \vec{E} \tag{10-12}$$

The total field $\vec{D}$ induced in the medium is then

$$\vec{D} = \epsilon\vec{E} + \Delta\vec{P} = \epsilon\vec{E} + i\epsilon_o\gamma\vec{G} \times \vec{E} \tag{10-13}$$

§10.3 Discussion

Eq. (10-6) and eq. (10-13) both show that the field $\vec{D}$ is equal to the normal reaction $\epsilon\vec{E}$ plus a complex term. This complex term is due to the motion of the charges caused by magnetic fields. It is well known in electromagnetism that no work is done when charges are deviated by magnetic fields. This is also physically true in the above two cases and can be seen quantitatively as follows. The energy in the E - M field is proportional to $(\vec{D} \cdot \vec{E})$. From both eq. (10-6) and (10-13), the complex term's contribution is $(\vec{G} \times \vec{E}) \cdot \vec{E} = 0$ because $\vec{G} \times \vec{E} \perp \vec{E}$.

Thus, the magnetic term $(G \times \vec{E})$ doesn't contribute to any energy dissipation as in the case of the complex dielectric constant of metal optics. (G is called the Gyration Vector.)

One can now use the new expressions of $\vec{D}$ (eq. 10-6 and 13) and go through the same calculation as in §6.2, obtaining the Fresnel's equation of wave normals and the ray surfaces. The results show that the complex term $\pm\, i\vec{G} \times \vec{E}$ has the effect of transforming an incident linearly polarized E - M wave into two circularly polarized waves rotating in opposite directions. These two waves of counter rotating circular polarizations "see" two different indices of refraction in the medium. Let them be n_1 and n_2. Thus, when the waves pass through the medium, say, in the form of a plate of thickness ℓ, there is a phase difference between the two waves given by

$$\Delta\epsilon = (k_1 - k_2)\ell = \frac{\omega}{c}(n_1 - n_2)\ell \tag{10-14}$$

At the exit, the two circularly polarized waves recombine to form a linearly polarized wave again except that the latter has been rotated with respect to the incident linear polarization by an angle Θ where

$$\Theta = \frac{\ell}{2}(k_1 - k_2) = \frac{\ell\omega}{2c}(n_1 - n_2) \tag{10-15}$$

In the case of Faraday rotation, Θ is proportional to the external magnetic field B and the thickness ℓ and we have

$$\Theta \propto B\,\ell$$

or

$$\Theta = \frac{V}{\mu} B\ell \qquad (10\text{-}16)$$

where we have set the proportionality constant as V/μ , with μ the magnetic permeability of the medium. V is called the Verdet's constant. Combining eq. (10-15 and 16),

$$\frac{V}{\mu} B\ell = \frac{\ell\omega}{2c}(n_1 - n_2)$$

$$V = \frac{\mu\omega}{2cB}(n_1 - n_2) \qquad (10\text{-}17)$$

There are many examples of optically active materials, such as liquid crystals, quartz, cane sugar solution etc. In the case of quartz, which is also a uniaxial crystal, one might ask if the optical activity will influence the optical anisotropy of the medium. In principle, yes. But the effect is negligible if the wave is travelling in a direction significantly different from that of the optic axis. When the wave travels in or near the optic axis direction, the optical activity becomes more important. But then the anisotropy becomes isotropic or nearly so because the wave propagates in or near the direction of the optic axis.

Optical activity differs in practice from Faraday rotation. In the former case, the direction of rotation of linear polarization of an E - M wave propagating through the medium is the same when an observer looks into the wave, regardless of whether the wave propagates in one direction or in an opposite direction. Hence, if a linearly polarized wave passes through an optically active medium and is reflected back along the same path, the net rotation is null. In the case of Faraday rotation, a back reflection of the linearly polarized light wave (now in opposite direction to the externally applied magnetic field) will double the angle of rotation. Such Faraday rotation is used in laser amplifier chains to isolate reflected pulses and the device is called a Faraday rotator.

Detailed analyses of the magnetic rotation of polarization of an E - M wave can be found in Sommerfeld and Yariv and Yeh.

CHAPTER XI

IMPORTANCE OF ANISOTROPY IN SECOND HARMONIC GENERATION (SHG)

Since the invention of the laser, nonlinear optical phenomena were studied extensively resulting in the possibility of generating very intense coherent radiation from the infrared to the ultra-violet. The field is now so vast that there is no hope to cover everything even qualitatively in only one small chapter. Excellent texts exist in the market and the interested reader should consult any of these texts. The present chapter will only introduce the subject and show the importance of optical anisotropy in non-linear optics using second harmonic generation as an example.

§11.1 Introduction

In nonlinear optics as in all other electromagnetic phenomena, we normally ask what the material response is; i.e. what the induced polarization in the medium is. This is because the induced polarization is the source of the re-radiation by the medium described by the wave equation (assume a lossless medium)

$$\nabla^2\vec{E} - \frac{1}{v^2}\frac{\partial^2\vec{E}}{\partial t^2} = \mu_o \frac{\partial^2\vec{P}}{\partial t^2} \tag{11-1}$$

(cf. any text on electromagnetism) where $\vec{P}$ is the induced polarization in the medium and $\vec{E}$ the electric field vector of the E - M wave propagating in the medium.

The term on the right hand side of eq. (11-1) is the source of the re-radiation. Normally, using ordinary weak light, $\vec{P}$ is linear in $\vec{E}$ (or the medium is linear):

$$\vec{P} = \epsilon_o \chi \vec{E} \tag{11-2}$$

even if χ is a tensor; i.e. even if the medium is anisotropic. However, when an intense laser beam propagates in the medium, the medium becomes non-linear. That is, the laser induced polarization in the medium has the following general form (following Yariv and Yeh):

$$\begin{aligned} \vec{P} &\equiv P_i \qquad (i = 1,2,3) \\ &= \epsilon_o \chi_{ij} E_j + 2d_{ijk} E_j E_k + 4\chi_{ijk\ell} E_j E_k E_\ell + \cdots \end{aligned} \tag{11-3}$$

where repeated subscripts (j,j), (k,k) and (ℓ,ℓ) mean summation over (1,2,3), the three Cartesian coordinates. χ_{ij} , d_{ijk} , $\chi_{ijk\ell}$ etc are the linear, second, third etc. order susceptibilities, respectively. In principle, one substitutes the expression for $\vec{P}$ in eq. (11-3) into the wave equation (11-1), solves it and obtains the answer for the re-radiated field $\vec{E}$. The task is of course not trivial and we shall not do any such calculation here. We just remark that each of the non-linear terms in eq. (11-3) is proportional to the product of two or more fields. These electric fields could come from the same laser beam (thus same frequency) and we talk about harmonic generation. They could come from different laser beams of the same or different frequencies and we talk about sum and difference frequencies generation, parametric oscillation etc. Phase matching conditions relate the polarizations of different frequencies. For second harmonic generation, only the second term on the right hand side of eq. (11-3) is important and kept. (See below.)

§11.2 Second harmonic generation (SHG)

We now discuss the consequence of second harmonic generation in an anisotropic medium by skipping all the calculations under the following assumptions:

(1) The medium does not have an inverse symmetry (if so, the second order coefficients d_{ijk} will be identically zero).

(2) The medium is lossless

(3) Neglect all magnetic effect; i.e. $\mu \approx \mu_o$.

(4) Plane wave approximation i.e. the solutions of all the relevant fields are plane waves propagating in the z - direction; i.e. the incident laser beam's electric field is expressed as

$$E_i^{(\omega)}(z,t) = E_i^{(\omega)}(z)\cos(\omega t - kz)$$

$$= E_i^{(\omega)}(z)\mathrm{Re}\left\{ e^{\,i(\omega t - kz)} \right\}$$

$$= E_i^{(\omega)}(z)\left\{ e^{i(\omega t - kz)} + e^{-i(\omega t - kz)} \right\}$$

$$\equiv \frac{1}{2} E_i^{(\omega)}(z)\left\{ e^{i(\omega t - kz)} + \text{c.c.} \right\} \qquad (11\text{-}4)$$

where c.c. is the complex conjugate. The subscript i defines the polarization direction of $\vec{E}$ and the superscript denotes the frequency. Because the wave propagates in the z - direction, i = 1,2 or x,y only. Similarly, the field of the second harmonic and the polarization of the second harmonic are respectively

$$E_k^{(2\omega)}(z,t) = \frac{1}{2} E_k^{(2\omega)}(z)\left\{ e^{i(\omega t - kz)} + \text{c.c.} \right\} \qquad (11\text{-}5)$$

and $P_j^{(2\omega)}(z,t) = \frac{1}{2} P_j^{(2\omega)}(z)\left\{ \frac{1}{2} e^{i(\omega t - kz)} + \text{c.c.} \right\}$ (11-6)

Substituting eq. (11-4, 5, 6,) into the wave equation, with the dependence of $P^{(2\omega)}$ on the field (eq. 11-3) explicitly expressed keeping only the second term on the R.H.S., one obtains an expression for the intensity

$I_j^{(2\omega)}$ of the second harmonic radiation under an additional assumption of low conversion efficiency; i.e. $\frac{dE_i^{(\omega)}}{dz} \simeq 0$,

$$I_j^{(2\omega)} = f(\omega,n,d)\, I_i^{(\omega)} I_\ell^{(\omega)} L^2 \; \frac{\sin^2\left(\frac{1}{2}\Delta kL\right)}{\left(\frac{1}{2}\Delta kL\right)^2} \tag{11-7}$$

where $f(\omega,n,d)$ is a function of the frequency ω, index n of the second harmonic and second harmonic coefficient d. The subscripts j,i,ℓ of the I's are the polarization directions of the radiation at the appropriate frequencies (2ω) or (ω) denoted by the superscripts. L is the interaction length in the medium and Δk is given by

$$\Delta k \equiv k_j^{(2\omega)} - k_i^{(\omega)} - k_\ell^{(\omega)} \tag{11-8}$$

where $k_j^{(2\omega)}$ is the wave vector of the second harmonic wave polarized in the j - direction; $k_i^{(\omega)}$ and $k_\ell^{(\omega)}$ are the wave vectors of the fundamental frequencies polarized in the i - and ℓ - directions.

We analyse briefly eq. (11-7) keeping in mind that we wish to generate the highest possible $I_j^{(2\omega)}$. That is to say, we want to maximize the expression for $I_j^{(2\omega)}$. We note the following.

(1) The explicit expression for $f(\omega,n,d)$ is

$$f(\omega,n,d) = 8 \left(\frac{\mu_o}{\epsilon_o}\right)^{3/2} \frac{\omega^2}{n^3}\, d^2 \tag{11-9}$$

For a fixed frequency, we can choose materials of as large a d - value as possible while the index n of the second harmonic in the material cannot be changed

(2) One can increase either $I_i^{(\omega)}$ or $I_\ell^{(\omega)}$ or both to increase $I_j^{(2\omega)}$. Note that if $i = \ell$, i.e. if the second harmonic is produced by a single laser beam of frequency ω and linear polarization i ,

$$I_j^{(2\omega)} \propto \left[I_i^{(\omega)} \right]^2$$

which means that the second harmonic intensity is proportional to the square of the fundamental intensity.

(3) Apparently, from eq. (11-7), one would be able to increase the second harmonic intensity by increasing L because of the L^2 - dependence. But one should beware that the function

$$\frac{\sin^2 \left[\frac{1}{2} (\Delta k) L \right]}{\left[\frac{1}{2} (\Delta k) L \right]^2} \qquad (11\text{-}10)$$

overwhelms the L^2 - dependence as can be seen in Fig. 11-1 in which the function $\text{sinc}^2\, x$ is shown. Mathematically,

$$\text{sinc}^2\, x \equiv \frac{\sin^2 (\pi x/b)}{(\pi x/b)^2} \qquad (11\text{-}11)$$

so that expression (11-10) becomes eq. (11-11) if

$$\pi x/b = \frac{1}{2} (\Delta k) L$$

or

$$x = \frac{\Delta k}{2\pi} L\, b \qquad (11\text{-}12)$$

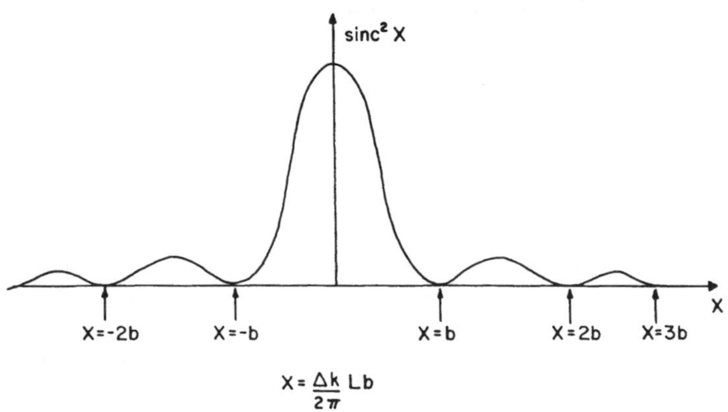

Fig. 11-1. Schematic representation of a $\text{sinc}^2 x$ function. The drawing is not to scale.

As shown in Fig. (11-1), $\text{sinc}^2\, x$ becomes zero periodically at

$$x = \pm\, mb \quad , \qquad m = 1,2,3,\ldots$$

Thus, the intensity of the second harmonic oscillates periodically. However, only the central peak (Fig. 11-1) is significant and we define the length L at which the first zero occurs as the coherence length ℓ_c

i.e. $\qquad L = \ell_c \qquad$ at $\qquad x = b$

i.e. $\qquad \frac{\Delta k}{2\pi}\, \ell_c\, b = b$

$$\ell_c = \frac{2\pi}{\Delta k} \tag{11-13}$$

i.e. the second harmonic intensity is significant only within the coherence length ℓ_c and is maximum

when $\qquad x = 0 \qquad$ (Fig. 11-1) ,

or when $\qquad \Delta k = 0 \qquad$ (11-14)

for a finite L (see eq. (11-12). Using eq. (11-8),

$$\Delta k \equiv k_j^{(2\omega)} - k_i^{(\omega)} - k_\ell^{(\omega)} = 0 \tag{11-15}$$

This is the very important phase matching condition.

§11.3 Phase matching

The phase matching condition (eq. 11-15) is usually the most critical one for ensuring an efficient second harmonic conversion. This is true not only in the low conversion efficiency approximation but also in the general case in which the fundamental laser beam is significantly depleted (turning into second harmonic). We examine this in more detail. Since in general

$$k = \frac{\omega}{c}\, n^{(\omega)}$$

eq. (11-15) becomes

$$\frac{2\omega}{c}\, n_j^{(2\omega)} - \frac{\omega}{c}\, n_i^{(\omega)} - \frac{\omega}{c}\, n_\ell^{(\omega)} = 0$$

where $n_k^{(\omega)}$ is the index of refraction of a plane wave of frequency ω and polarized in the k - direction, with k = i, j or ℓ .

i.e.
$$\left[n_j^{(2\omega)} - n_i^{(\omega)} \right] + \left[n_j^{(2\omega)} - n_\ell^{(\omega)} \right] = 0 \qquad (11\text{-}16)$$

This yields two solutions. The first is when $i = \ell$; i.e. the fundamental frequency comes from the same pump laser generating a second harmonic wave of polarization $j (j = 1,2)$. Thus, eq. (11-16) yields

$$n_j^{(2\omega)} - n_i^{(\omega)} = 0 \quad \text{(Type I phase matching)} \qquad (11\text{-}17)$$

The second is when $i \neq \ell$. i.e. one employs two laser beams at the same fundamental frequency but with orthogonal linear polarizations. Note: $i,\ell = 1,2$ so that when $i \neq \ell$, one of them must be equal to 1 polarized in the x - direction, and the other equal to 2 - polarized in the y - direction. Eq. (11-16) thus yields

$$n_j^{(2\omega)} = \frac{1}{2}\left[n_1^{(\omega)} + n_2^{(\omega)} \right] \quad \text{(Type II phase matching)} \qquad (11\text{-}18)$$

where we have explicitly expressed i and ℓ as 1 and 2.

Case 1: Type I phase matching

Using one linearly polarized laser beam to generate second harmonic radiation, eq. (11-17) is valid:

$$n_j^{(2\omega)} - n_i^{(\omega)} = 0 \qquad (11\text{-}19)$$

(a) If $j = i$, it means that the polarization of the second harmonic is parallel to that of the fundamental pump beam. If the medium is isotropic, because of the normal dispersion of the medium,

$$n_i^{(2\omega)} \neq n_i^{(\omega)}$$

For example, most types of glass exhibit

$$n_i^{(2\omega)} > n_i^{(\omega)} \qquad (11\text{-}20)$$

in the near i.r. to the near u.v. region. Thus, it is almost impossible to satisfy eq. (11-19) in an isotropic medium.

If the medium is anisotropic, j = i means that the second harmonic and the fundamental are either both ordinary waves or both extraordinary waves in the medium (using uniaxial crystal as an example). If they are both ordinary waves, their index ellipsoids are concentric spheres of radii $n^{(\omega)}$ and $n^{(2\omega)}$ and they will never be equal in general because of the dispersion (eq. 11-20). Thus, eq. (11-19) cannot be satisfied. If the two waves are both extraordinary waves, their index ellipsoids are two concentric ellipsoids without any intersection between them. Again, $n^{(2\omega)}$ cannot be made equal to $n^{(\omega)}$ due to dispersion (eq. 11-20).

In conclusion, if j = i, the index matching condition (11-19) cannot be satisfied in both isotropic and anisotropic media.

(b) If j ≠ i, eq. (11-19) still cannot be satisfied in an isotropic medium because

$$n_j^{(\omega)} \equiv n_i^{(\omega)} \neq n_j^{(2\omega)} \tag{11-21}$$

We are left with the possibility of using an anisotropic medium. Using a uniaxial medium as an example, one can choose a direction in the medium such that the second harmonic propagates as an ordinary wave and the fundamental as an extraordinary wave (or vice versa) with equal velocity (hence equal indices, satisfying eq. (11-19)). Fig. (11-2 and 3) illustrate the above statement.

Fig. (11-2) shows the surfaces of wave normals of a negative uniaxial medium. (a) corresponds to the surfaces of the wave normals at frequency 2ω and (b) at ω. In (c), the two are superimposed and we see that in the

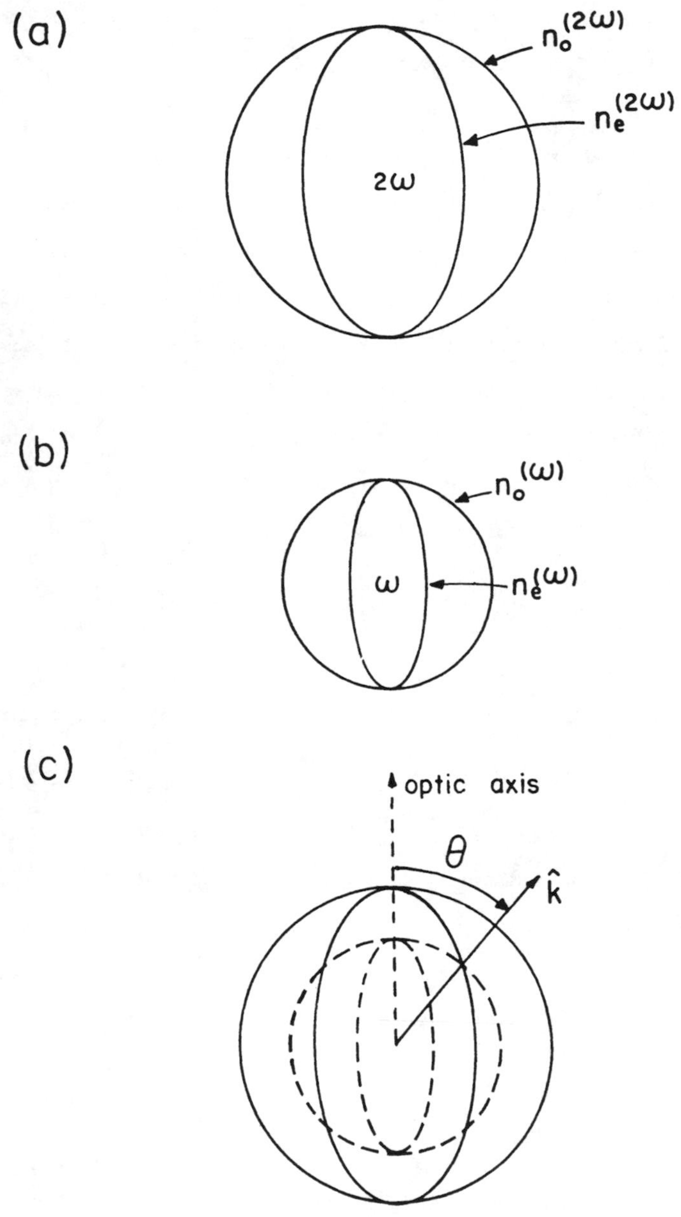

Fig. 11-2. Phase matching in a negative uniaxial crystal.

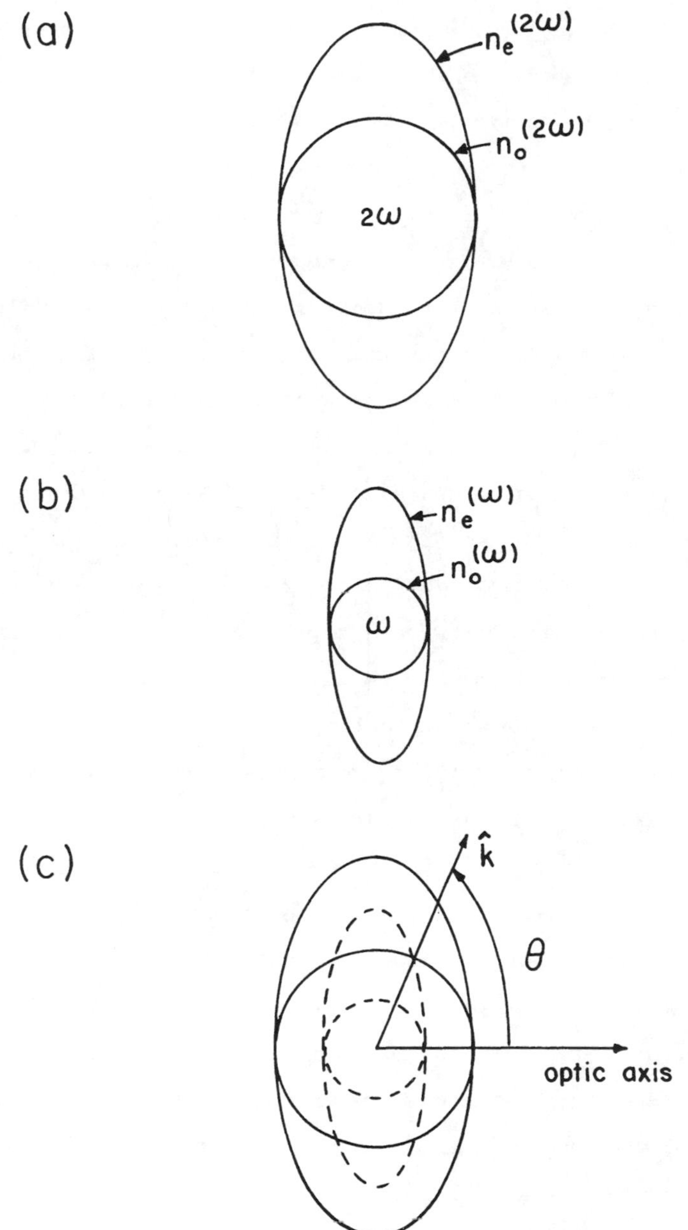

Fig. 11-3. Phase matching in a positive uniaxial crystal.

direction $\hat{k}$, the spherical surface of radius $n_o^{(\omega)}$ of the ordinary wave of frequency ω intersects the ellipsoid of index $n_e^{(2\omega)}(\Theta)$ (Θ is the angle between the optic axis and $\hat{k}$) of the extraordinary wave of frequency 2ω. At the intersection point, $n_o^{(\omega)} = n_e^{(2\omega)}(\Theta)$, thus satisfying eq. (11-19). Physically speaking, it means that the second harmonic wave generated inside the anisotropic medium will have an ordinary and an extraordinary component both propagating in the $\hat{k}$ - direction. The ordinary component will "see" an index $n_o^{(2\omega)}$ which will never be equal to $n_o^{(\omega)}$ or $n_e^{(\omega)}$ of the fundamental wave of frequency ω. Hence, the ordinary component of the second harmonic wave will quickly die out (or not be amplified) because of the phase mismatch (i.e. not satisfying eq. (11-19)). The extraordinary component "sees" an index $n_e^{(2\omega)}(\Theta)$ equal to $n_o^{(\omega)}$ of the fundamental. Note that their polarizations are orthogonal. The phase matching condition is thus satisfied and the second harmonic is amplified as it propagates at the same velocity as the pump beam (fundamental). Experimentally, one should thus align the fundamental laser beam in the direction of $\hat{k}$, i.e. the anisotropic crystal should be properly cut to permit an easy alignment. The angle Θ can be calculated. We shall come back to this.

Similarly, Fig. (11-3) shows the surfaces of wave normals of a positive uniaxial medium. Again, (a) represents the surfaces of wave normals at 2ω and (b) at ω. The superposition of the two shows that in the direction $\hat{k}$, the propagation velocity (and hence indices) of the ordinary wave at 2ω (index $n_o^{(2\omega)}$) and the extraordinary wave at ω (index $n_e^{(\omega)}(\Theta)$) are equal, satisfying the phase matching condition (eq. 11-19).

$n_e^{(\omega)}(\Theta)$ (Fig. 11-3) or $n_e^{(2\omega)}(\Theta)$ (Fig. 11-2) can be calculated easily in the case of a uniaxial crystal. Referring back to §6.3 and Fig. 6-12, a plane

E - M wave of frequency ω propagating in the direction $\hat{k}$ inside a uniaxial crystal generates two waves, ordinary and extraordinary, travelling in the $\hat{k}$ - direction, with electric vectors $\vec{E}_o$ and $\vec{E}_e$ ($\vec{E}_o \perp \vec{E}_e$) and indices n_o and $n_e(\Theta)$ respectively. Note that $\hat{k}$ is constrained in the y - z plane without loss of generality and $n_e(\Theta)$ changes from n_o at $\Theta = 0$ to n_e at $\Theta = \frac{\pi}{2}$. From Fig. 6-12, one sees that

$$\frac{1}{n_e^2(\Theta)} = \frac{\cos^2\Theta}{n_o^2} + \frac{\sin^2\Theta}{n_e^2} \tag{11-22}$$

the derivation is left as an <u>exercise</u> to the reader.

In the case of a negative uniaxial crystal, one needs to calculate $n_e^{(2\omega)}(\Theta)$. Eq. (11-22) becomes

$$\frac{1}{\left[n_e^{(2\omega)}(\Theta)\right]^2} = \frac{\cos^2\Theta}{\left[n_o^{(2\omega)}\right]^2} + \frac{\sin^2\Theta}{\left[n_e^{(2\omega)}\right]^2} \tag{11-23}$$

At $\Theta = \Theta_{pm}$, the phase matching condition is assumed to be valid, and eq. (11-19) becomes

$$n_o^{(\omega)} = n_e^{(2\omega)}(\Theta_{pm}) \ . \tag{11-24}$$

Θ_{pm} is the phase matching angle. Substituting (11-24) into (11-23), we obtain

$$\frac{1}{\left[n_o^{(\omega)}\right]^2} = \frac{\cos^2\Theta_{pm}}{\left[n_o^{(2\omega)}\right]^2} + \frac{\sin^2\Theta_{pm}}{\left[n_e^{(2\omega)}\right]^2} \tag{11-25}$$

With $n_o^{(\omega)}$, $n_o^{(2\omega)}$ and $n_e^{(2\omega)}$ known, eq. (11-25) can be solved for the angle Θ_{pm} (<u>exercise</u>):

$$\sin^2\Theta_{pm} = \frac{\left[n_o^{(\omega)}\right]^{-2} - \left[n_o^{(2\omega)}\right]^{-2}}{\left[n_e^{(2\omega)}\right]^{-2} - \left[n_o^{(2\omega)}\right]^{-2}} \quad \text{(negative uniaxial crystal)} \tag{11-26}$$

Similarly, in the case of a positive uniaxial crystal, we need to calculate $n_e^{(\omega)}(\Theta)$. Eq. (11-22) becomes

$$\frac{1}{\left[n_e^{(\omega)}(\Theta)\right]^2} = \frac{\cos^2\Theta}{\left[n_o^{(\omega)}\right]^2} + \frac{\sin^2\Theta}{\left[n_e^{(\omega)}\right]^2} \tag{11-27}$$

Using the phase matching condition at $\Theta = \Theta_{pm}$,

$$n_o^{(2\omega)} = n_e^{(\omega)}(\Theta_{pm}) \tag{11-28}$$

one obtains (exercise)

$$\sin^2\Theta_{pm} = \frac{\left[n_o^{(2\omega)}\right]^{-2} - \left[n_o^{(\omega)}\right]^{-2}}{\left[n_e^{(\omega)}\right]^{-2} - \left[n_o^{(\omega)}\right]^{-2}} \quad \text{(positive uniaxial crystal)} \tag{11-29}$$

Case 2 Type II phase matching

Using two parallel pump beams with orthogonal linear polarizations to generate second harmonic radiation, eq. (11-18) is valid:

$$n_j^{(2\omega)} = \frac{1}{2}\left[n_1^{(\omega)} + n_2^{(\omega)}\right] \tag{11-30}$$

The situation becomes much more complicated. We shall consider simple situations only. Again, using a uniaxial crystal as an example, the polarization directions 1 and 2 in eq. (11-30) are assumed to be those of the ordinary and extraordinary waves for the sake of simplicity. Thus, eq. (11-30) becomes

$$n_j^{(2\omega)} = \frac{1}{2}\left[n_o^{(\omega)} + n_e^{(\omega)}\right] \tag{11-31}$$

The propagation direction $\hat{k}$ of the three waves (two pump beams and the second harmonic) are assumed collinear. $\hat{k}$ makes angle Θ, in general, with the optic axis, (see Fig. 11-2, 11-3 and 6-12). There could now be several ways of fulfilling eq. (11-31) depending on the indices of the anisotropic material. First of all, one have, in general,

$$n_j^{(2\omega)}(\Theta_{pm}) = \frac{1}{2}\left[n_o^{(\omega)} + n_e^{(\omega)}(\Theta_{pm}) \right] \tag{11-32}$$

where Θ_{pm} is the phase matching angle. It is easy to imagine that some material could satisfy eq. (11-32) at $\Theta_{pm} = 0$. Under such a condition, all the waves become ordinary and

$$n_j^{(2\omega)}(\Theta_{pm} = 0) = n_o^{(2\omega)}$$

$$n_e^{(\omega)}(\Theta_{pm} = 0) = n_o^{(\omega)}$$

and eq. (11-32) becomes

$$n_o^{(2\omega)} = \frac{1}{2}\left[n_o^{(\omega)} + n_e^{(\omega)} \right] = n_o^{(\omega)} \tag{11-33}$$

regardless of polarization directions. But this condition is impossible due to dispersion (c.f. eq. 11-20). Thus, we are left with two more possibilities for $\Theta_{pm} \neq 0$, namely

$$n_j^{(2\omega)}(\Theta_{pm}) = \begin{cases} n_o^{(2\omega)}(\Theta_{pm}) \\ n_e^{(2\omega)}(\Theta_{pm}) \end{cases} \tag{11-34}$$

i.e. the second harmonic could be an ordinary ($n_o^{(2\omega)}(\Theta_{pm})$ or an extraordinary ($n_e^{(2\omega)}(\Theta_{pm})$ wave.

If the second harmonic is an ordinary wave, we have, from eq. (11-32),

$$n_o^{(2\omega)}(\Theta_{pm}) = \frac{1}{2}\left[n_o^{(\omega)} + n_e^{(\omega)}(\Theta_{pm}) \right] \tag{11-35}$$

But $n_o^{(2\omega)}(\Theta_{pm})$ is constant, independent of Θ_{pm} because, by definition, the surface of wave normal of the ordinary wave is a sphere and thus the wave propagates at the same velocity in all directions, i.e. constant index in all directions. Eq. (11-35) is thus not satisfied because $n_e^{(\omega)}(\Theta_{pm})$ varies with Θ_{pm} . We are left with the last possibility, namely, the second harmonic must be an extraordinary wave in order to satisfy the phase matching condition. Eq. (11-32) becomes

$$n_e^{(2\omega)}(\Theta_{pm}) = \frac{1}{2}\left[n_o^{(\omega)} + n_e^{(\omega)}(\Theta_{pm}) \right] \tag{11-36}$$

To solve for Θ_{pm}, we use eq. (11-22) for both ω and 2ω:

$$\frac{1}{\left[n_e^{(2\omega)}(\Theta_{pm}) \right]^2} = \frac{\cos^2\Theta_{pm}}{\left[n_o^{(2\omega)} \right]^2} + \frac{\sin^2\Theta_{pm}}{\left[n_e^{(2\omega)} \right]^2} \tag{11-37}$$

$$\frac{1}{\left[n_e^{(\omega)}(\Theta_{pm}) \right]^2} = \frac{\cos^2\Theta_{pm}}{\left[n_o^{(\omega)} \right]^2} + \frac{\sin^2\Theta_{pm}}{\left[n_e^{(\omega)} \right]^2} \tag{11-38}$$

By substituting eq. (11-37) and (11-38) into (11-36), one obtains an equation with Θ_{pm} as the only unknown variable. We shall not go into the details here.

<u>Summary (1)</u>

<u>Type I phase matching condition</u>

Both the linearly polarized pump and second harmonic waves propagate in the same direction $\hat{k}$ that makes an angle Θ_{pm} (the phase matching angle) with the optic axis. For a negative uniaxial medium,

$$n_o^{(\omega)} = n_e^{(2\omega)}(\Theta_{pm}) \qquad \text{(eq. 11-24)}$$

i.e. the pump beam is an ordinary wave and the second harmonic is an extraordinary wave both propagating in the direction $\hat{k}$ at the same velocity (in phase). The experimental schematic is given in Fig. (11-4)(a). The medium is cut such that Θ_{pm} can be easily aligned. For a positive uniaxial medium,

$$n_o^{(2\omega)} = n_e^{(\omega)}(\Theta_{pm}) \qquad \text{(eq. 11-28)}$$

i.e. the pump beam is an extraordinary wave and the second harmonic, an ordinary wave both propagating in the direction $\hat{k}$ at the same velocity (in phase). The angle Θ_{pm} is given by eq. (11-29). The experimental schematic is shown in Fig. (11-4)(b).

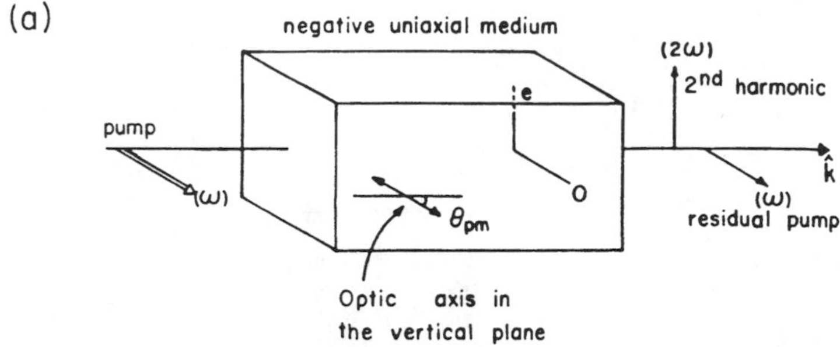

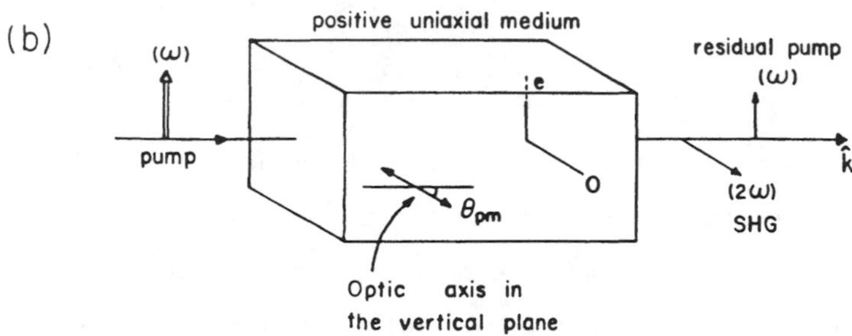

Fig. 11-4. Type I phase matching in the SHG in (a) negative (b) positive uniaxial media.

Summary (2)

Type II phase matching condition

Two pump beams with orthogonal linear polarizations are used; one is an ordinary wave, the other an extraordinary wave. The generated second harmonic is an extraordinary wave. All the waves propagate in the same direction $\hat{k}$ making an angle Θ_{pm} with respect to the optic axis. Eq. (11-36) has to be satisfied. This means that the "mean" velocity of the combined pump waves is equal to the velocity of the second harmonic wave. Fig. (11-5) shows one practical way of such phase matching starting from one pump beam. The pump beam first passes through a half-wave plate such that the incident linearly (vertically) polarized pump beam becomes polarized at

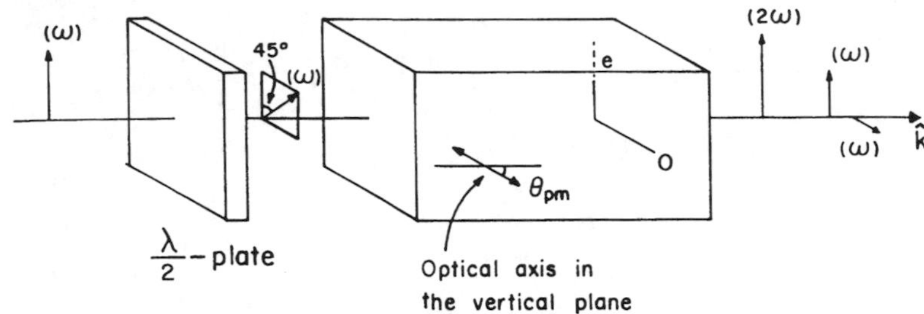

Fig. 11-5. One practical way for type II phase matching in the SHG in a uniaxial crystal.

45° with respect to the vertical axis (e - axis in the crystal) (see Chapter VII). In entering the second harmonic medium the incident wave is decomposed into the ordinary and extraordinary waves. These two waves thus fulfills the two pump beams condition. At the output, there are three waves, the two pump waves (now phase shifted with each other) and the second harmonic wave polarized vertically. The two pump waves combine forming a resultant wave of elliptic polarization in general.

The output waves of frequencies 2ω and ω in Fig. (11-4) and (11-5) can be separated by various means. For instance, one can use a prism to separate them or a thin film dichroic beam splitter that transmits preferentially one frequency and reflects the other.

CHAPTER XII

SHORT LASER PULSE GENERATION – A REVIEW

We arbitrarily define a short laser pulse as one whose duration (full width at half maximum or FWHM) is of the order of a few tens of nanoseconds and shorter. The lower limit depends on the wavelength of the laser and the shortest laser pulse to date is a few femtoseconds in duration in the visible region. In terms of optical cycle, it lasts only a few optical cycles. We have already qualitatively described Q-switching and mode-locking in chapters VIII and IX. The present chapter presents reprints of the author's previous publications reviewing different methods of producing short laser pulses. Although femtosecond pulse generation is not included, the general principles are similar. After understanding the general principles presented in this chapter, the reader should have little difficulty in understanding the most recent research papers on femtosecond laser pulse generation.

Short laser pulse generation: part one

D. FAUBERT, S.L. CHIN

The status of short laser pulse generation is reviewed. Achievements, limitations and potential of techniques producing pulses shorter than 50 ns, from the ultra-violet to the infra-red, are examined.

In this half of the paper the physical principles underlying the generation of ultra-short pulses are introduced and the methods of Q-switching, gain switching, cavity dumping and mode-locking are considered.

KEYWORDS: lasers, short pulse generation, Q-switching, gain switching, cavity dumping, mode-locking

List of symbols

c	speed of light in a vacuum
f	frequency of the acoustic wave travelling in an acousto-optic crystal
$I(t)$	intensity of a laser pulse
k	constant relating T and $\Delta\omega$
l	length between the mirror and the Pockel cell in the electro-optical intra-cavity gating scheme ('Gating', 4.f.)
L	length of a laser resonator
l_a	length of the active medium of a laser
l_1	length of the optical path in the two electro-optical crystals gating scheme ('Gating', 4.e.)
L_T	length of a charged transmission line in electro-optical gating
$l_{1,2}$	length of transmission lines 1 and 2 in electro-optical gating
$n(n')$	index of refraction of the medium outside (inside) a laser
Q	quality factor of a resonator
$R_{1,2}$	reflectivities of the two mirrors of a laser
R_E	$= (R_1R_2)^{1/2}$
t	time
T	full width at half maximum of a laser pulse
T_c	cutting time of a laser pulse
T_F	total duration of a laser pulse
T_2	transverse relaxation time
Δt	time interval
t_v	photon lifetime within a laser resonator
t_{rec}	recovery time of a saturable absorber
$V(t)$	a high voltage pulse
$V_{\lambda/2}(V_{\lambda/4})$	half-wave (quarter-wave) voltage of an electro-optical device
V_0	voltage to which a transmission line is charged
v	speed at which electrical current flows in a transmission line
ω_0	carrier frequency of an electromagnetic wave
$\Delta\omega$	spectrum of an electromagnetic wave

Introduction

Laser technology opened the way to many new experiments and very often gave new impulses to older ones. In particular, short laser pulses have found a wide range of applications, such as the study of ultra-fast mechanisms within molecular systems or others, controlled thermonuclear fusion, laser induced chemistry (laser isotope separation for instance), high data rate communication systems, ultra-high resolution radars, short wavelength laser pumping etc.

DF is at the Defence Research Establishment Valcartier, PO Box 8800, Courcelette, Québec, Canada. SLC is in the Laboratoire de Recherches en Optique et Laser, Université Laval, Québec, Canada.
Received 9 December 1981.

As a result, a myriad of techniques have been developed to yield pulses almost anywhere between 200 fs (2×10^{-13} s)[1] to continuous wave operation, covering parts of the electromagnetic spectrum from the ultra-violet to the infra-red.

The aim of this paper is to examine the status of short laser pulse generation today. We will limit our attention to the techniques producing pulses 50 ns and shorter since this is the range in which most of the work has been done. It is not our purpose to look at each individual contribution to a given method. Rather, we will outline the fundamental principles of each technique together with its performances, potential and the actual technical set-up involved. Therefore, the discussion within each section is, whenever possible, arranged according to the following pattern:

0030-3992/82/040197-10/$03.00

1. Description of the physical principles underlying the technique.

2. Recorded performances, potential and limitations regarding the pulse width, frequency and power.

3. Standard experimental arrangements, to substantiate the physical principles.

4. Significant improvements or variations to the classical set-up.

The discussion has been divided into eight main sections: Q-switching, gain switching, cavity dumping, mode-locking, gating, interferometric techniques, compression, pulse shaping. The first four sections will be considered here, the latter four in part two.

For each item discussed, an attempt has been made to give reference to the most recent publication and/or important contributors on the subject. Unfortunately, references to earlier pioneering or fundamental contributions are sometimes omitted, but we felt it was the best approach to minimize the number of references contained in this paper and to give access to the latest information on the subject. Very often reference to other work is made in the following format: (wavelength; pulse width) reference number. For example: HF and DF lasers (3-5 μm; 3-25 ns)[31].

Before describing the peculiarities of the different short laser pulse generation techniques, let us discuss very briefly a few relationships common to all of them.

The spectrum of a laser pulse and its temporal shape are related by a Fourier transform. If $f(t)$ is the envelope of the pulse and $F(\omega)$ its spectrum, then

$$F(\omega) = \int_{-\infty}^{\infty} f(t)\, e^{-i\omega t} dt \qquad (1)$$

$$f(t) = \frac{1}{2\pi} \int_{-\infty}^{\infty} F(\omega)\, e^{i\omega t} d\omega \qquad (2)$$

When the full-width at half-maximum of the spectrum, $\Delta\omega$, and the envelope T are related by

$$T = \frac{2\pi k}{\Delta\omega} \qquad (3)$$

where k is a constant depending on the shape of the pulse (k = [2 ln 2] / π = 0.441 for a Gaussian), the pulse is said to be bandwidth limited.

The single most fundamental consideration behind short pulse generation is given by (3), that is, the wider the spectrum radiated by an optical source, the shorter the pulse it can sustain. To generate the shortest possible laser pulses, we must therefore use laser systems with the widest possible gain width $\delta\omega$. But this ultimate duration is not always achieved (in fact it seldom is) so that in practice

$$T \gtrsim \frac{2\pi k}{\Delta\omega} \qquad (4)$$

The actual pulse width a given technique will achieve results from the dynamics of the laser action. It depends on many parameters, like the active medium (its gain width, upper and lower level lifetimes, quenching mechanisms etc), the pump source (its rise-time, duration, intensity, spectrum etc), the resonator (the coupling, losses etc), as well as modulators or some other accessories which control the laser action in a predetermined manner. Non-linear interactions, because of the high optical fields involved, often play determinant roles also.

An evaluation of the laser pulse parameters (shape, width and intensity) for a given laser system is often an involved task, often requiring the help of numerical analysis and a computer. For this reason the influence of the key system elements on the pulse parameters has been discussed qualitatively, avoiding tedious theoretical considerations whenever possible. This enabled us to explain the mechanisms and performances of each method in a concise manner, making it easier to survey the subject comprehensively.

Short pulse generation methods

Q-switching

1. Q-switching prevents laser action (low resonator quality factor Q) while the active medium is being pumped. The population inversion can then build up to a high level since it is not continuously being depleted by stimulated emission. Since the gain is proportional to the population inversion, it reaches a high value (the longer the lifetime of the upper laser level, the higher it is). When the gain peaks, a high Q condition is suddenly switched on. The laser pulse then builds up rapidly because the gain is high, quickly depleting the population inversion. Laser action terminates rapidly. The more intense the output burst of light, the shorter it is. This mode of operation has also been referred to as Q-spoiling, pulse reflection mode (PRM) or giant pulse laser. The first publication on the subject was by McClung and Hellwarth[2] in 1962. Refs 3 and 4 are review articles.

2. The pulse width is influenced by many factors among which the two most important are the speed with which the high Q condition is switched on and the photon lifetime t_v within the resonator, given by

$$t_v = [n' l_a + n(L - l_a)] / c\,(1 - R_E) \qquad (5)$$

where $n(n')$ is the refractive index outside (within) the active medium, $l_a(L)$ the length of the active medium (cavity), c the speed of light in the vacuum, and $R_E \equiv (R_1 R_2)^{1/2}$ where R_1 and R_2 are the reflectivities of the two mirrors.

The exact pulse width generally has to be evaluated with rate equations, but for a typical 30 cm resonator pulses shorter than 1 ns cannot be achieved.

The pulse tends to be slightly asymmetric as shown in Fig. 1. The leading edge is sharper and closely linked to the speed of Q-switching while the trailing edge is mainly the decay of the energy within the cavity, when laser action has terminated after exhausting the population inversion.

Q-switching is widely used with the neodymium (1.06 μm) and ruby (0.6943 μm) lasers for which the upper laser level lifetimes are long. Typical pulse widths achieved are around 10 ns, but with short cavities and hard pumping, pulses as short as 2 ns have been produced[5]. In fact, damage to the windows or to the active medium itself is a serious limiting factor on the shortest pulse achievable[6]. We can exercise a certain control over the pulse width (from 2 ns to 17 ns in Ref. 5) by varying the parameters of the resonator (length, losses etc). Output power around 10 MW (100 mJ per pulse) is easily obtainable. Q-switching has also been applied to CO_2 (10.6 μm; T ⩾ 50 ns)[7,8] and iodine lasers (1.35 μm;

5 ns)[9]. A 300 μm long rhodamine 6 G dye laser was also Q-switched (0.6 μm; 350 ps)[10], but since the fluorescence lifetime of these materials is so short (a few nanoseconds) it is not a widespread practice.

3. Under certain conditions, some lasers have been observed to Q-switch on their own[11], mostly when they are misaligned, but not necessarily. Since the reasons for this happening are still rather vague, other means of achieving Q-switching are more reliable.

The addition of a variable loss element within the cavity (as shown in Fig. 1) is the standard way to Q-switch a laser. At the proper time, the transmission of the switch is changed from a low to a high value, triggering the laser action. The most common switches are:

a. Electro-optic (active): A Kerr or Pockel cell is placed between two crossed polarizers (Fig. 1). As long as the electro-optical element (EO) is not activated, light cannot go through this system and losses are very high. But as soon as a voltage $V_{\lambda/2}$ is applied to EO, light going through it has its polarization vector turned by 90° and can cross both polarizers, initiating laser action. This method has been applied to ruby and neodymium[2,3,4], CO_2[8] and I_2[9] lasers.

b. Electro-optic (passive): A saturable absorber having a small low-level signal transmission (typically around 5%) is placed within the resonator. The initially high absorption impedes the growth of fluorescence, but as it gets stronger it partially bleaches the absorber, that is, the transmission of the saturable absorber increases. Soon after the onset of this non-linear behaviour, the laser becomes strong enough to bleach fully or 'open' this passive switch. This approach tends to be less reproducible, but being very simple, inexpensive and easy to implement, it is widely utilized, mainly with ruby and neodymium[3,4,5,6,12]. With these systems, a dye solution is usually used as the saturable absorber. Changing its concentration is the way in which the 2 to 17 ns tuning range has been achieved[5]. Iodine vapour was also used for ruby[13]. A CO_2 laser was also Q-switched by this method[7].

4. A few other Q-switching methods exist, but being less frequently utilized no detailed description will be given. They use:

(a) A rotating mirror[3,4], grating[14], chopper[15], or total internal reflection prism[9].

(b) Frustrated total internal reflection[3,4].

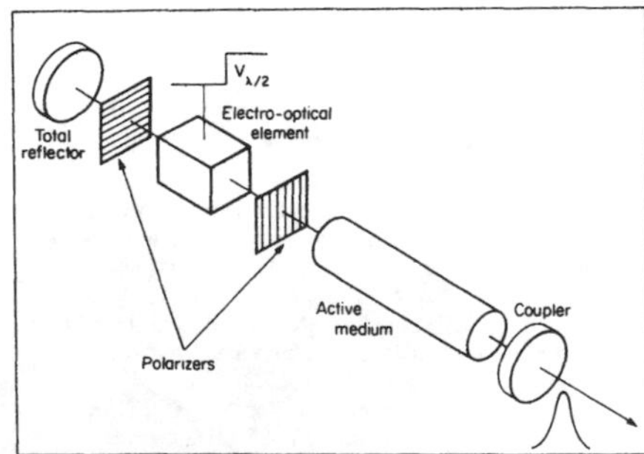

Fig. 1 Q-switching using an electro-optical device. Note the slightly asymmetric output pulse shape

(c) An acousto-optic modulator[3,4].

(d) An exploding film. A PbS film inside a Nd laser resonator (1.06 μm; 40 ns)[16].

(e) A moving mirror (reactive Q-switching) (10.6 μm; $T \geqslant 300$ ns)[17].

(f) The Stark effect (CO_2 laser)[18].

(g) An open-air spark (I_2 laser) (1.3 μm; > 200 ns)[19].

(h) Theoretical investigations of a Q-switched semiconductor laser were also reported[20] but it was not realized in practice. Due to their very small dimension (L = 10 μm) they could produce very short pulses (~50 ps).

(i) Several ways to increase the pulse width of Q-switched devices exist, but since they produce pulses whose duration are outside our range of interest (> 50 ns) they will not be described. Ref. 3 covers these schemes in fairly great detail.

Recent work in this area is directed towards improving the stability of the output. For instance, fluorescence from the active medium is monitored using a photodiode.[21] Being a direct indication of the gain in the system, the photodiode signal, when it reaches a predetermined level, is used to trigger the opening of the switch. The laser action is thus initiated under almost identical conditions from one shot to the next, yielding more reproducible results.

We should also mention that Q-switching has been applied to unstable resonators, mainly to increase the output power[22].

Gain switching

1. The object of this technique is to pump the active medium so fast that the population inversion builds up more rapidly than the photon field inside the resonator, which grows due to fluorescence. Under this condition, the final growth of the laser pulse takes place when the gain is high, resulting in a short and powerful burst of light. This is much like Q-switching except that there is no switch.

The dynamics of the laser itself are responsible for the giant pulse. If the pumping mechanism is still powerful enough after the first spike, the process will repeat. The output then usually goes through a series of spikes, unless the pump is adjusted to go above threshold only once.

To favour spiking, one tries to restrain the photon field from building up too fast due to noise. A low Q cavity is used, that is, the shortest resonator possible with low reflectivity mirrors. The photon lifetime within the resonator then being very small (1), the field within the cavity takes longer to grow. This technique has also been called 'controlled resonator transients' and was first proposed by Roess in 1966[23].

2. This method can yield ultra-short pulses with high gain lasers that can be made very small, like dye and semiconductor lasers. For instance, a 50 μm dye laser produced 5 ps pulses[24]. Under this condition a complete round trip within the resonator takes only 0.33 ps. Similarly, a 5–7 μm GaAs laser was optically pumped by a single 1 ps dye laser pulse to yield a 1 ps output at 0.84 μm wavelength[25]. We can vary the pulse width to a considerable extent by changing the length of the resonator[26]. The power generated is usually small because the active medium is thin and it is pumped just slightly above threshold. For example, 50–100 W peak power for ~600 ps has been reported[26] and ~1 kW for

~200 ps[27]. In these examples both active media were dyes operating in the 0.4 to 0.6 μm wavelength range. This technique has mostly been applied to dye lasers[24,26–28] for which pulses can be produced almost anywhere between a few ps and 1 ns. This scheme was also applied to the CO_2 laser (10–20 ns[29] and ~200 ns[30]), the HF and DF lasers (3–5 μm; 3–25 ns)[3] and the I_2 laser (1.31 μm; 1 ns)[30,32].

3. As already discussed in section 2 above, no special element is required within the cavity. We have only to pump the active medium as fast as possible in a cavity as short as possible.

4. A few other interesting versions of this technique exist:

a. The use of a grating[33], or injection techniques[34], to reduce the lasing bandwidth with dye lasers.

b. Mixing two dyes together. The first one (rhodamine 6 G[35]) absorbs the pumping energy (a N_2 laser) and undergoes spiking. Part of the pulse emitted is absorbed by the second dye (DODCI) which also emits a burst of light about the same width (600 ps) but 1.5 ns later[35]. We thus have two short pulses of different wavelength.

c. The use of a distributed feedback dye laser together with a conventional resonator (0.6 μm; 36 ps)[36].

d. The use of a high pressure CO_2 waveguide laser to increase the pulse energy (10.6 μm; 20 ns)[37].

e. The use of fast electrical pumping as well as optical pumping, mentioned above for semiconductor lasers (0.82 μm; 40 ps)[38] and (0.82 μm; ⩽ 100 ps)[39].

Cavity dumping

1. Cavity dumping takes place in two stages. First we let the electromagnetic field build up inside the cavity. This is usually done with very little coupling of the resonator (high reflectivity mirrors). Then, when the internal optical field reaches its maximum value, *all* the energy contained in the resonator is extracted. The cavity is said to have 'dumped' the energy it contained. The first work on the subject was done by Vuylsteke in 1963[40]. This technique is also called pulse transmission mode.

2. Since all the energy is taken out of the cavity, the pulse length (full width T_f) is given by

$$T_f = \frac{2n'l_a}{c} + \frac{2n(L-l_a)}{c} \simeq \frac{2L}{c} \qquad (6)$$

Typical pulse width is around 3 ns, but shorter durations are possible if more compact optical components are used (for a 1 ns pulse the cavity should measure ~15 cm). We see from (6) that the pulse width is adjustable simply by varying the length L of the resonator. But we should mention that when this is done, changes in the pulse shape, peak power and intensity distribution are often observed. The power obtained is comparable to that of Q-switched lasers and even tends to be slightly higher because approximately the same energy is given out but in a shorter time. This technique has been applied to Nd (1.06 μm; 4 ns)[41], ruby (0.6943 μm; 3 ns)[42] and dye lasers (0.49 μm; 20 ns)[43].

3. The only way this has been accomplished up to now is by again inserting a Kerr[41,42] or Pockel[43] cell inside the resonator together with a polarizer/extractor element such as a Glan-Thompson (GT) prism (Fig. 2). These prisms are made of birefringent material and are arranged so that the two polarization components are separated. Under this condition, when the laser is turned on, a polarized optical field builds up in the cavity. At the proper time, a voltage $V_{\lambda/4}$ is applied to the electro-optic element (EO) (45° rotation of the polarization in one pass through EO). After two passes inside EO, the light-wave has its polarization flipped by 90° and is deflected out of the resonator by GT. For this to work, the high voltage pulse applied to EO must be longer than $2L/c$ (a cavity round-trip time) and its rise-time should be as short as possible.

4. This technique enjoyed more popularity in the early days of quantum electronics, but now is seldom used. Instead, with the same kind of set-up, a voltage pulse shorter than $2L/c$ is applied to the EO. A shorter laser pulse is produced but not all the energy is extracted from the resonator. Such a method is not the cavity dumping defined above and will be discussed under 'gating'.

Mode-locking

1. This technique, because of its ability to produce the shortest light-pulse to date, is widely used. It can be understood as follows. The electromagnetic wave radiated by a laser is made up of a set of longitudinal modes, spaced by a value of $c/2L$ in the frequency spectrum (we assume that a single transverse mode is lasing), which, if undisturbed, oscillate with random phase relative to each other. The result is a field whose amplitude also varies randomly in time. The idea of mode-locking is to force the modes to keep a fixed phase relationship relative to each other. This modal configuration corresponds in the time domain to a short packet of light going back and forth within the resonator. The output of the laser through the partially transparent mirror is then a train of pulses (Fig. 3) spaced in time by $2L/c$.

This technique is also referred to as phase-locking or mode-coupling. The first paper clearly dealing with this subject was published in 1964 by Didomencio[44]. A whole book has been devoted to the subject[45] as well as numerous review articles[45].

2. The pulse width of perfect mode-locking is approximately equal to the inverse of the bandwidth $\Delta\omega$ of the lasing transition (bandwidth limited). This is why dye lasers, with their large bandwidth (a few hundred angstroms), have yielded the shortest pulses to date (200 fs[1]). The extinction

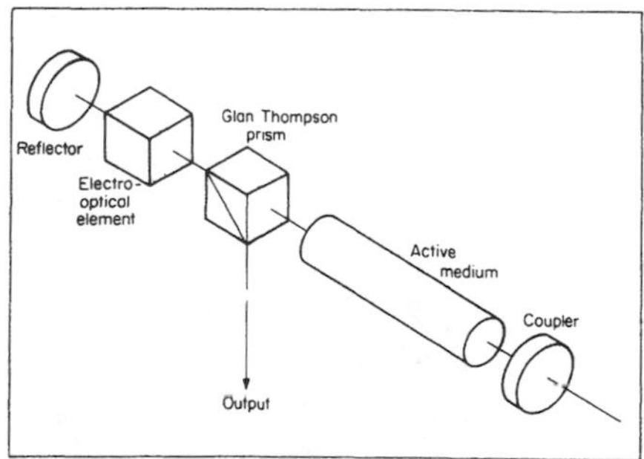

Fig. 2 Cavity dumping with an electro-optical element and a Glan-Thompson prism

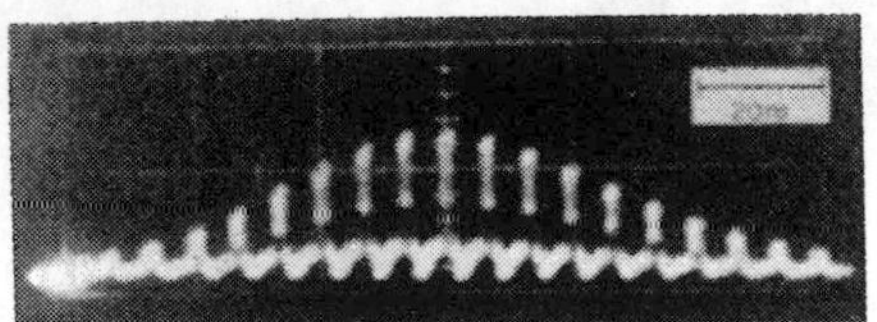

Fig. 3 A typical mode-locked laser output. The pulse separation is 6 ns and the pulse width is around 800 ps[92]

coefficient (the ratio of the intensity of the pulses to that of the background) for perfect mode-locking is proportional to the number of longitudinal modes locked together. Peak power around 1 MW is easily achieved. This technique has been applied to a multitude of laser systems, both cw and pulsed, extending from the ultra-violet to the far infra-red. A list of these is given in Table 1.

The biggest drawback of mode-locking is that it produces a train of pulses rather than a single one. Most of the time, a switch-out apparatus has to be used to extract a pulse from the train to perform a given experiment.

3. To lock together the modes of a laser it suffices to modulate either the losses, the gain or the phase (they are all equivalent) of the intra-cavity field at a frequency of $c/2L$, that is, once every round trip of the cavity. It is not evident *a priori* how this action will mode-lock the laser. The following is an explanation. Let us consider a modulator opening once every round trip. The modal configuration which is going to be least affected by the modulator is the one corresponding to a short pulse which will go through the modulator when it is open. Any other arrangement of the modes will cross the modulator in a closed state at one time or another. Then only the short wave-packet or the mode-locked condition will lase because its losses are lowest.

Many laser systems have been observed to self-lock, that is, to mode-lock on their own[51,66]. However, this effect is not well understood and is rather unreliable. Many artificial ways have been devised to achieve mode-locking. They can be divided into two broad categories; active (see Ref. 67 for a review) and passive modulation schemes.

a. Active approaches.

(i) Acousto-optic: This is one of the most popular active approaches. An acoustic wave of frequency f is launched in a crystal. A standing wave is created which acts as a diffraction grating producing diffraction orders which fluctuate in time at a frequency $2f$. This is because a medium in which a standing wave has been set up becomes homogeneous twice per period. We thus arrange so that $2f$ equals $c/2L$. This method has been applied to XeF[46], HeCd[48], Ar and Kr[50], HeNe[51], Nd[58] and CO_2[62,68,69] lasers. The acousto-optic modulator (like all other active schemes) will not generate pulses shorter than approximately 50 ps because of its finite modulation depth[67].

(ii) Electro-optic: As in the case of Q-switching (Fig. 1), an EO is placed between two crossed polarizers, but for mode-locking a voltage pulse $V_{\lambda/2}$ is applied to EO once every $2L/c$ seconds (visibles[51,70] and I_2[58]).

The greatest advantages of the two active methods discussed so far are reproducibility and a short pulse well synchronized to the modulating voltage. In turn it can then be easily synchronized to other external experimental events.

(iii) Quartz modulator: Applied once to a dye laser ($\sim 0.6\ \mu$m; 50 ps)[71], and will not be discussed in detail here.

(iv) Phase (or frequency) modulation: An electro-optical phase modulator (EPM) is placed within the resonator and will shift the frequency of the field up or down each time it passes through (once every $2L/c$ seconds) except if the

Table 1. Laser systems that have been mode-locked

Laser	Wavelength [μm]	Pulse width	Reference
KrF	0.2485	$\lesssim 2$ ns	46
XeF	0.351	$\lesssim 2$ ns	47
HeCd	0.3250 and 0.4416	~ 400 ps	48
Dye	Visible to near infra-red	0.2 ps to 10 ps	1, 45, 49
Ar	0.5140	130 ps	50
Kr	Six lines from 0.5208 to 0.7525	50–75 ps	50
HeNe	0.6328 and 3.39	~300 ps	51, 52
Ruby	0.6943	25 ps	53
GaAlAs	0.86	16 ps	54
GaInAsP	1.21	18 ps	55
Nd	1.06 and 1.31	5.25 ps	53, 56, 57
I_2	1.31	500 ps	58
Colour centre	0.8 to 2.2	4 ps	59
HF	~ 2.7	5–20 ns	60
CO_2	10.6	$\lesssim 1$ ns	7, 61–64
D_2O	385	$\lesssim 1$ ns	65
$C^{12}H_3$	496	$\lesssim 1$ ns	65

light happens to cross it when the voltage applied to the EPM is zero. A wave travelling through the EPM at this voltage is frequency shifted away from the gain curve of the active medium and is attenuated. Only small wave-packets crossing the EPM at the right time are amplified[51,67].

b. Passive mode-locking. Much as in Q-switching, a thin cell containing a saturable absorber is placed within the cavity, in contact with the totally reflecting mirror. The way it achieves mode-locking is of a statistical nature and because of that is somewhat unreliable. However, being extremely simple, inexpensive and efficient (it has produced the shortest pulses to date[1]) it is widely used.

Without going into detail, here is how it works. When the pumping mechanism is activated, fluorescence sets in. Since the unbleached transmittivity of the non-linear absorber is quite high (~ 70%), the fluorescence noise travels back and forth in the resonator and grows in intensity. The intensity distribution of the noise is random and contains many sharp peaks. When their intensity becomes strong enough, they will start to bleach, totally or partially, the saturable absorber. The lower intensity portions of the noise are absorbed and eventually disappear. Slowly, one peak emerges (not necessarily the one that was the most powerful to start with because of the relative timing with respect to bleaching) and eventually is the only one to lase.

Besides discriminating in favour of one peak, the dye also has another very important function; it compresses the sharp peak. Part of its leading edge is used up or absorbed by the molecule to bleach it, but its tail goes through almost intact. The result is a shorter pulse whose length is ultimately approximately equal to t_{rec}, the recovery time of the absorber. This is so because after being bleached the absorption recovers after a time t_{rec} and cuts the remainder of the tail.

Up to now we have assumed that the gain was linear, which is almost the case with laser systems whose upper-level lifetimes are long (ruby, Nd). But for dye lasers for instance, the active medium itself behaves non-linearly and contributes as much as the saturable absorber to the ultra-short pulse generation process. It does so by sharpening the small pulse to a value much shorter than t_{rec}. The leading edge of the pulse saturates the gain in the active medium so that the tail is amplified less than the front, resulting in a shorter pulse. To summarize, we can distinguish two classes of passively mode-locked lasers: class I (ruby, neodymium) where the pulse width equals t_{rec}, and class II (dyes) for which it is not limited by t_{rec}.

The biggest drawback of passive mode-locking is its lack of reliability. For instance, two short pulses can survive the critical action of the saturable absorber; the intensity fluctuates from shot to shot, as does as the extinction ratio; the short pulse is not synchronized, making it extremely difficult to synchronize with other events. Typically, proper mode-locking is obtained only 80% of the time. However, because of the advantages mentioned above, it is still very popular.

The theory just summarized is the result of many investigations, but it is generally recognized that the most fundamental work has been done by Letokhov[72] and Fleck[73] for class I lasers. New[74] also contributed much to the understanding of the class II type. See the book by Shapiro[45] for the complete story.

The technique is widely applied in the visible spectrum using dyes as the saturable absorber. See, for instance, dye[1,45,49], ruby[53], Nd lasers[53,56,57]. It is also applied to the KrF laser using a dye[46], to the HF laser using a gas[60] and to the CO_2 laser, using a gas[7] or a semiconductor[61]. An optical Kerr modulator was also used with the Nd laser (1.06 μm; 20 ps)[75] and a simple glass filter with a ruby laser[76]. A plasma acting as a passive switch has also been investigated for the CO_2 laser (10.6 μm; 1 ns)[77].

c. Synchronous mode-locking. In this scheme, the active medium of the laser is optically pumped by another mode-locked laser (Fig. 4). The gain is modulated every $2L/c$ seconds. It is critical[78] that the length of the resonators of the two lasers be matched, otherwise the gain of the laser being pumped will not be modulated precisely every $2L/c$ seconds. This scheme offers certain unique advantages, like simplicity and two well synchronized trains of short pulses at different wavelengths (the pump and the laser itself). It is worth noting that the optical pumping does not have to be coaxial with the optical axis of the laser being pumped[45,78]. This scheme has been applied to dye lasers[45,78-80], colour centre[60] and far infra-red lasers; D_2O and $C^{12}H_3$[65]. It is reliable and yields pulses almost as short as passive mode-locking provided that the pumping pulses are spaced precisely by $c/2L$ (or a multiple). This is the most critical adjustment to perform.

d. Injection mode-locking. This scheme requires a master oscillator and a slave amplifier (Fig. 5). The master is mode-locked using one of the techniques described above. A single pulse is switched out of the train and injected into the slave where it undergoes regenerative amplification. That is, the injected pulse, bouncing back and forth within the slave, is amplified each time it goes through the active medium and gives rise to an output pulse each time it strikes the coupler of the slave. In fact, the master takes care of locking the modes together and the slave amplifies them. The master pulse has to be injected at the right time to dominate and grow more rapidly than the noise in the slave.

This technique is extremely simple, reliable, and being scalable, can produce extremely high peak powers. Originally named RAAT, standing for regenerative amplification above threshold[63], it has been developed for CO_2 lasers[63,64,81,82] and is also applied to dye and Nd lasers[81].

4. By modifying and/or combining the basic schemes

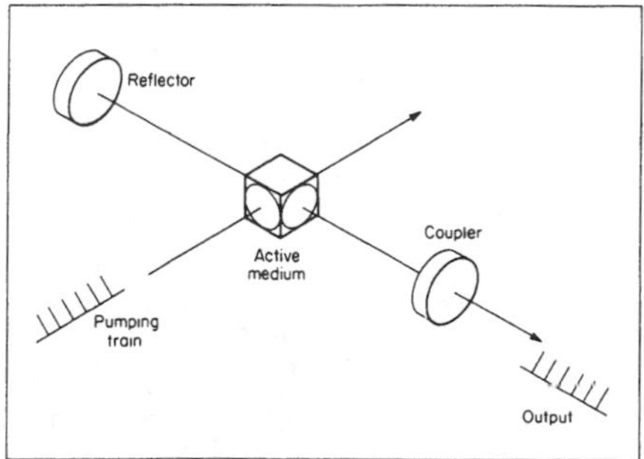

Fig. 4 Synchronous mode-locking. The optical pumping beam can also be coaxial with the optical axis of the laser being pumped

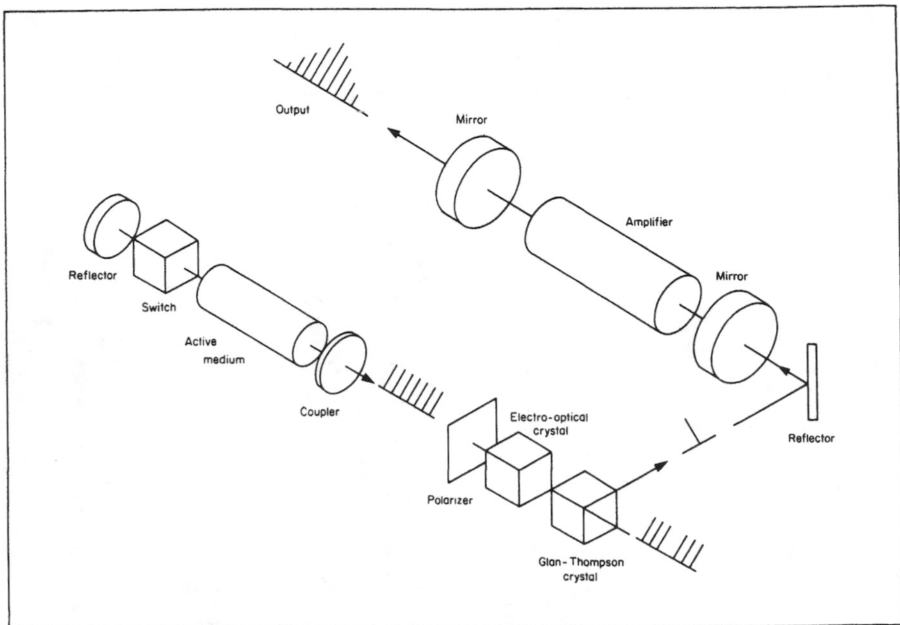

Fig. 5 Injection mode-locking. One short pulse is selected from the train of the master oscillator and is injected into the slave amplifier

described above, many different modifications have been tried successfully. They are briefly summarized as follows:

a. Combination of the active and passive schemes. This combines the ultra-short pulse capability of the passive method with the reliability and synchronizability of the active method. Pulses obtained are not as short as with passive mode-locking alone, but reliability is greatly enhanced[57,83,84]. In Ref. 84, the introduction of the saturable absorber decreased the pulse width of a ruby laser from 100 ps to 5 ps.

b. Combination of the synchronous and passive schemes. This, again, is to take advantage of both schemes, or in other words to further shorten the pulses from the synchronously mode-locked laser. Pulses as short as 350 fs have been obtained[95].

c. Combination of the injection and passive schemes. Incorporating a saturable absorber in the slave of an injection mode-locked laser increases the signal-to-noise ratio and compresses the short pulse (from ~ 100 ps to ~ 15 ps[81]).

d. Combination of active Q-switching and active mode-locking[56,57]. This technique will produce reliable high peak power pulses if one small difficulty can be overcome. This is that active Q-switched lasers are above threshold for only a very short time (< 100 ns), which is not long enough for the active modulator to lock the modes together initially. One way to get around the problem is to open the Q-switching switch just slightly so that fluorescence can build up and the modulator can work. Then, around at least 1 μs later, the full Q of the cavity is switched on and a high power train of pulses is produced. Their widths, as in the pure active scheme, are not very small ($\gtrsim$ 50 ps). This method has been, up to now, applied to Nd (1.06 μm; 37 ps)[57] lasers.

This active Q-control of the resonator has the added advantage that it facilitates the synchronization of two lasers together. As an example, an ultra-short Nd pulse was synchronized to a CO_2 laser pulse[57]. Both their active modulators were driven from the same reference signal, and active Q-control of the Nd laser allowed time for the CO_2 pulse to build up.

e. Synchronous mode-locking (pump laser not mode-locked). It is possible to use a cw laser, whose intensity is modulated by ~ 30% at a rate of $2L/c$ seconds, to pump another laser and still achieve mode-locking in the latter[85]. This technique is simpler than a completely mode-locked pump laser.

f. Double mode-locking. Two dyes, A and B, are placed within the same resonator (Fig. 6). A is pumped by a cw[86] or mode-locked[87] laser. According to the normal passive mode-locking process described above, B is the saturable absorber and it locks the modes of A. A short pulse at the frequency of A is thus produced which synchronously pumps the dye B. Under proper conditions it will also produce a second short pulse at frequency B. We thus have two short pulses of different wavelengths which are perfectly synchronized (three pulses if the pumping laser is mode-locked).

g. External modulation of the saturable absorber in a passive mode-locking scheme. A mode-locked laser shines on the intra-cavity saturable absorber (Fig. 7), opening it at a frequency of $c/2L$. Again, we have two short pulses at

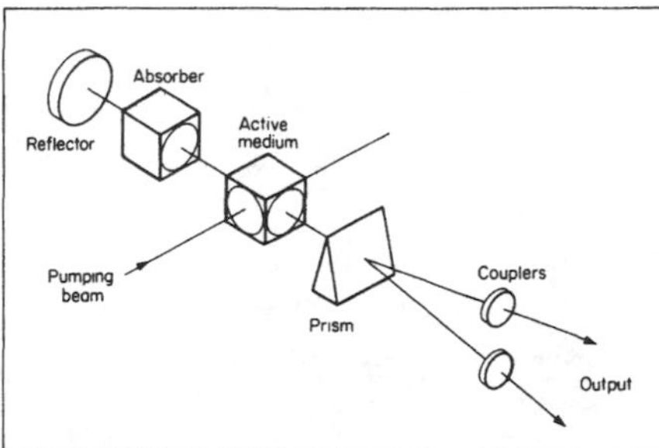

Fig. 6 Double mode-locking. The output beams are produced by the active medium and the saturable absorber

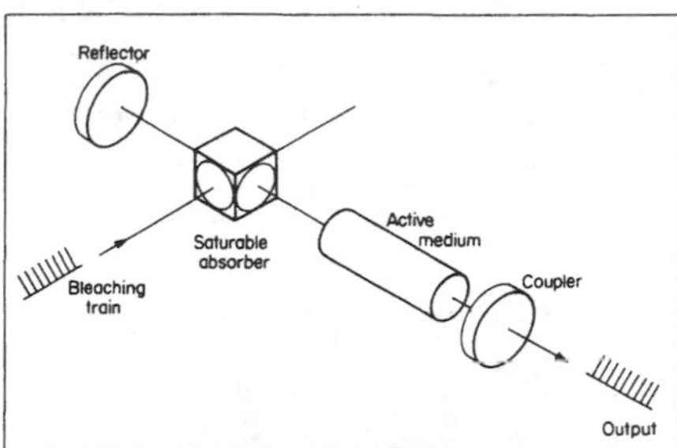

Fig. 7 Mode-locking by an external modulation of the transmittivity of the saturable absorber

different wavelengths, perfectly synchronized[88].

h. Use of a second cavity to enable the saturable absorber of a passive mode-locking scheme to lase on its own (Fig. 8). This produces an artificial decrease in the recovery time t_{rec} of the passive switch, thus producing shorter pulses for class I lasers[89,90].

i. The control of t_{rec}. For class I lasers, if we can control the value of t_{rec} for a given saturable absorber, we will be able to generate pulses whose widths will be tunable (because $T = t_{rec}$). This can be done using different solvents[91].

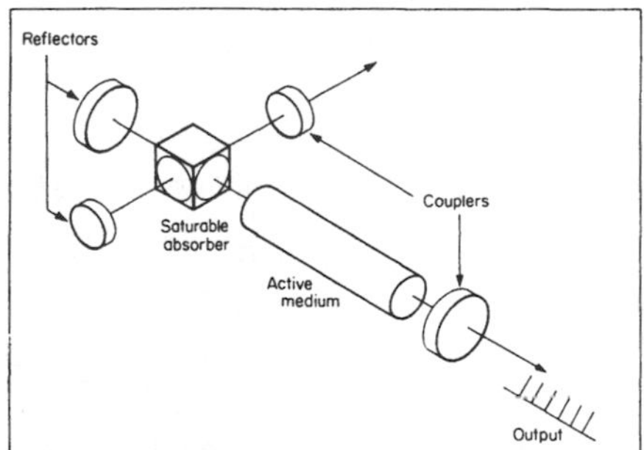

Fig. 8 Passive mode-locking enhanced through stimulated emission of the saturable absorber

The use of viscous solvents usually dramatically increases t_{rec}. Pulses up to 800 ps have been produced[92] out of a passively mode-locked ruby laser which gives ~ 20 ps pulses in low viscosity solvents.

j. Insertion of a Fabry-Perot interferometer within the resonator. This reduces the lasing bandwidth and thus increases the pulse width. The pulse width can be varied (25 - 550 ps[83] and 80-600 ps[93]) by changing the parameters of the interferometer[83,93].

k. Insertion of a second harmonic crystal within the resonator[94]. By varying the conversion efficiency to the second harmonic, the pulse width of a Nd laser has been tuned from 50 to 900 ps by varying the temperature of the crystal.

l. Use of an internal Fabry-Perot etalon to eliminate a certain number of longitudinal modes. This simulates a shorter resonator (since the spacing of the modes increases) and thus decreases the time interval between each small pulse of the train[96].

m. Use of an unstable resonator to increase the pulse energy[97].

n. Use of a hybrid laser. A hybrid laser (a cw gain section within the cavity in addition to the normal active medium) has been mode-locked using an acousto-optic device. It considerably increased the stability of the pulse train as well as extended the range over which the pulse width could be varied[98].

o. Optoelectronic regenerative pulser[99]. The pumping mechanism of a semiconductor laser is part of a feedback loop including a photodetector, an amplifier and a delay line. Initially, a dc bias slightly above threshold is applied to the laser. Part of the output is detected by the photodetector. After amplification this electrical signal is fed back to the pumping circuit. The laser is observed to pulse itself ($\lesssim$ 100 ps) because this laser tends to emit a light pulse whose duration is smaller than the pumping pulse. The system thus evolves rapidly towards the shortest possible pulse that the electronics can process.

p. Excitation of longitudinal modes which fall outside the gain curve of the laser. Energy is transferred from the lasing modes to some adjacent passive modes of the cavity lying outside the gain curve. This constitutes a loss for the lasing modes that can be overcome with a high gain system under proper conditions. It has the advantage of generating a pulse shorter than one which the bandwidth of the lasing transition alone could sustain. It has been applied to a HeNe laser on the 3.39 μm transition, yielding pulses 170 ps long[100].

q. Use of many lasers whose frequency separations and phases are precisely adjusted. Combining the output of all these is equivalent to mode-locking. Pulses 2 ns wide at the CO_2 frequency have been obtained in this way[101].

r. Feedback controlled self-excitation. A fraction of the wave propagating in one direction, V_+, of a ring laser is fed back into the resonator with the aid of a partially reflecting mirror located outside the cavity (Fig. 9). Thus part of the V_+ wave is transformed into a V_- wave. This power transposition from V_+ to V_- is equivalent to an amplitude modulation[102], and the output of the laser is observed to be strongly modulated (Fig. 9). This approach is not, strictly speaking, mode-locking, but it has produced a train of pulses with a rhodamine dye laser (0.6 μm; 5 ns)[102].

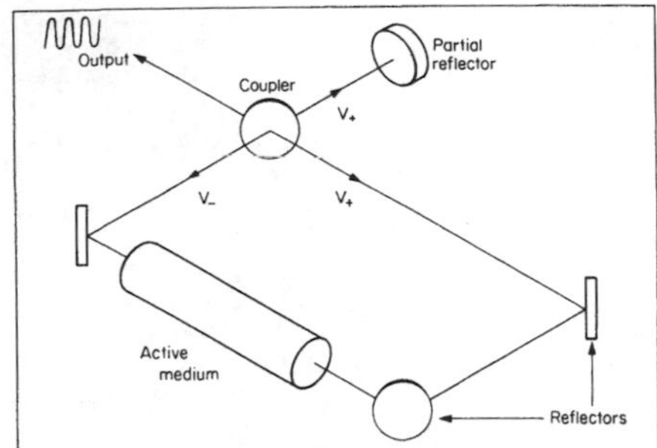

Fig. 9 Feedback controlled self-excitation[102]

References

1 Diels, J.C., Van Stryland, E., Benedict, G. *Opt Comm* 25 (1978) 93
2 McClung, F.J., Hellwarth, R.W. *J Appl Phys* 33 (1962) 828
3 Arsenev, V.A., Matveev, I.N., Ustinov, N.D. *Sov J Quant Electr* 7 (1977) 1321
4 Hudspeth, W.S. EOSD (Aug. 1977) 40
5 Jackel, S., Loebenstein, H.M., Zigler, A., Zmora, H., Zweignebaum, S. *J Phys E : Appl Phys* 13 (1980) 995
6 Vroba, M. *IEEE J Quant Electr* QE-14 (1978) 596
7 Izatt, J.R., Caudle, G.F., Bean, B.L. *Appl Phys Lett* 25 (1974) 446
8 Day, G.W., Gaddy, O.L., Jungling, K.C. *IEEE J Quant Electr* QE-6 (1970) 553
9 Katulin, V.A., Yu Nosach, V., Petrov, A.L. *Sov J Quant Electr* 6 (1976) 205
10 Braverman, L.W. *Appl Phys Lett* 27 (1975) 602
11 Szabo, A. *J Appl Phys* 49 (1978) 533
12 Sinha, B.B., Gopi, N. *IEEE J Quant Electr* QE-17 (1981) 92
13 Agee, B.G., Yu Nechaev, S., Yu Ponomarev, N. *Sov J Quant Electr* 7 (1977) 368
14 Lee, L.C., Gundersen, M., Faust, W.L. *Opt Comm* 1 (1970) 291
15 Meyerhofer, D. *IEEE J Quant Electr* QE-4 (1968) 762
16 Landry, M.J. *Appl Opt* 17 (1978) 635
17 Pardue Jr. A.L., McDuff, O.D. *IEEE J Quant Electr* QE-6 (1970) 753
18 Hall, D.R., Yoh-Han Pao. *IEEE J Quant Electr* QE-7 (1971) 427
19 Ishii, S., Ahlborn, B. *Rev Sci Instr* 46 (1975) 1287
20 Ito, H., Gen-Ei, K., Inaba, H. *Electr Lett* 16 (1980) 846
21 Downs, D.C., Murray, J.E., Lowdermilk, W.H. *IEEE J Quant Electr* QE-14 (1978) 571
22 Brown, A.D.E. *Opt Comm* 27 (1978) 253
23 Roess, D. *J Appl Phys* 37 (1966) 2004
24 Fan, B., Gustafson, T.K. *Appl Phys Lett* 28 (1976) 202
25 Duguay, M.A., Damen, T.C., Stone, J., Wiesenfeld, J.M., Burrus, C.A. *Appl Phys Lett* 37 (1980) 369
26 Lin, C., Shank, C.V., *Appl Phys Lett* 26 (1975) 389
27 Walden, G.L., Bradshaw, J.D., Winefordner, J.D. *Appl Spectr* 34 (1980) 18
28 Salzman, H., Strohwald, H. *Phys Lett* 57A (1976) 41
29 Bychkov, Yu I., Kudryashov, V.P., Kurbatov, Yu A., Orlovskii, V.M., Osipov, V.V., Savin, V.V. *Sov Phys Tech Phys* 24 (1979) 875
30 Gilbert, J., Lachambre, J.L., Rheault, F., Fortin, R. *Can J Phys* 50 (1972) 2523
31 Schilling, P., Decker, G. *Inf Phys* 16 (1976) 14
32 Brederlow, G., Fill, E., Fuss, W., Hohla, K., Volk, R., Witte, K.J. *Sov J Quant Electr* 6 (1976) 491
33 Wyatt, R. *Opt Comm* 26 (1978) 429
34 Ganiel, V., Hardy, A. *Opt Comm* 19 (1976) 14
35 Buck, J.A., Yasa, Z.A. *IEEE J Quant Electr* QE-13 (1977) 935
36 Bor, Z. *IEEE J Quant Electr* QE-16 (1980) 517
37 Brink, D.J., Hasson, V. *J Appl Phys* 49 (1978) 2250
38 Lin, C., Liu, P.L., Damen, T.C., Eilenburger, D.J., Hartman, R.L. *Electr Lett* 16 (1980) 600
39 Torphammar, P., Eng, S.T. *Electr Lett* 16 (1980) 587
40 Vuglsteke, A.A. *J Appl Phys* 34 (1963) 1615
41 Hook, W.R., Dishington, R.H., Hilberg, R.P. *Appl Phys Lett* 9 (1966) 125
42 Ernest, J., Michon, M., Debric, J. *Phys Lett* 22 (1966) 147
43 Morton, R.G., Mack, M.E., Itzkan, I. *Appl Opt* 17 (1978) 3268
44 Didomencio, M. *J Appl Phys* 35 (1964) 2870
45 Ultrashort light pulses, Ed. S.L. Shapiro, Springer-Verlag (1977) 389
46 Efthimiopoulos, T., Banic, J., Stoicheff, B.P. *Can J Phys* 57 (1979) 1437
47 Christensen, C.P., Braverman, L.W., Steier, W.H., Wittig, C. *Appl Phys Lett* 29 (1976) 424
48 Selfvast, W.T., Smith, P.W. *Appl Phys Lett* 17 (1970) 70
49 Azuma, K., Nakagawa, O., Segawa, Y., Aoyagi, Y., Namba, S. *Jap J Appl Phys* 18 (1979) 209
50 Steinmetz, L.L., Bookless, W.A.S., Richardson, J.H. *Appl Opt* 19 (1980) 2663
51 Smith, P.W. *Proc IEEE* 58 (1970) 1342
52 Kobayashi, T., Konishi, S., Sueta, T. *Opt and Quant Electr* 7 (1975) 229
53 DeMaria, A.J., Glenn, W.H. Jr, Brienza, M.J., Mark, M.E. *Proc IEEE* 57 (1969) 2
54 Holbrook, M.B., Sleat, W.E., Bradley, D.J. *Appl Phys Lett* 37 (1980) 59
55 Glasser, L.A. *Electr Lett* 14 (1978) 725
56 Jones, E.D., Palmer, M.A. *Opt Quant Electr* 7 (1975) 520
57 Tomov, I.V., Fedosejevs, R., Richardson, M.C. *Sov J Quant Electr* 10 (1980) 797
58 Baker, H.J., King, T.A. *J Phys E: Sci Instr* 9 (1976) 287
59 *Laser Focus* (Aug. 1978) 23
60 Simonis, G.J. *Appl Phys Lett* 29 (1976) 42
61 Taylor, R.S., Garside, B.K., Ballik, E.A. *IEEE J Quant Electr* QE-14 (1978) 532
62 Davis, D.T., Smith, D.L., Koval, J.S. *IEEE J Quant Electr* QE-8, (1972) 846
63 Bélanger, P.A., Boivin, J. *Can J Phys* 54 (1976) 720
64 Bernard, P., Mathieu, P., Bélanger, P.A. *Opt Comm* 34 (1980) 101
65 Lee, S.H., Petuchowski, S.J., Rosenberger, A.T., Detemple, T.A. *Opt Lett* 4 (1979) 6
66 Haken, H., Ohno, H. *Opt Comm* 16 (1976) 205
67 Siegman, A.E., Kuizenga, D.J. *Opto Electr* 6 (1974) 43
68 Karapuzikov, A.I., Troshin, B.I. *Sov J Quant Electr* 10 (1980) 816
69 Smith, P.W., Bridges, T.J., Burkhardt, E.G., Wood, O.R. *Appl Phys Lett* 21 (1972) 471
70 Krivoshchekov, G.V., Kulevskii, L.A., Nikulin, N.G., Semibalamut, V.M., Smirnov, V.A., Smirnov, V.V. *Sov Phys JETP* 37 (1973) 1007
71 Marinero, E.E., Jasny, J. *Opt Comm* 36 (1981) 69
72 Kryukov, P.G., Letokhov, V.S. *IEEE J Quant Electr* QE-8 (1972) 766
73 Fleck, J.A. *Phys Rev B* 1 (1970) 84
74 New, G.H.C. *IEEE J Quant Electr* QE-10 (1974) 115
75 Sala, K., Richardson, M.C., Isenor, N.R. *IEEE J Quant Electr* QE-13 (1977) 915
76 Winer, I.M. *Appl Opt* 12 (1973) 2809
77 Pichi, M., Bélanger, P.A. *Opt Comm* 24 (1978) 158
78 Ferguson, A.I., Eckstein, J.N., Hansh, T.W. *J Appl Phys* 49 (1978) 5389
79 Azuma, K., Nakagawa, O., Segawa, Y., Aoyagi, Y., Namba, S. *Jap J Appl Phys* 18 (1975) 209
80 Frigo, N.J., Daly, T., Mahr, H. *IEEE J Quant Electr* QE-13 (1977) 101
81 Corkum, P.B. *Laser Focus* (June 1979) 80
82 Dyer, P.E., Perera, I.K. *Appl Phys* 23 (1980) 245
83 Seka, W., Bunkenburg, J. *J Appl Phys* 49 (1978) 2277
84 Krivoshchekov, G.V., Nikulin, N.G., Smirnov, V.A. *Sov J Quant Electr* 5 (1975) 1096
85 Kobayashi, T., Hosokawa, T., Sueta, T. *Opt Comm* 27 (1978) 431
86 Bouzkoff, E., Winnely, J.R., Dienes, A. *Opt Lett* 4 (1979) 179
87 Mahr, H. *IEEE J Quant Electr* QE-12 (1976) 554
88 Lill, E., Schneider, S., Dörr, F. *Opt Comm* 22 (1977) 107
89 Abakumov, G.A., Antipov, A.I., Simonov, A.P., Sinitsyn, A.B., Fadew, V.J. *Sov J Quant Electr* 7 (1978) 1314

90 **Stappaerts, E.A., Long Jr, W.H.** *Opt Lett* **3** (1978) 226
91 **AL-Obaidi, H., Dewhurst, R.J., Jacoby, D., Oldershaw, G.A., Ramsden, S.A.** *Opt Comm* **14** (1975) 219
92 **Faubert, D., Chin, S.L.** *Appl Opt* **18** (1979) 2347
93 **Soures, J., Kumpa, S., Hoose, J.** *Appl Opt* **13** (1974) 2081
94 **Falk, J., Hitz, C.B.** *IEEE J Quant Electr* **QE-11** (1977) 365
95 **Kuhl, J., Klingenberg, H., Von der Linde, D.** *Appl Phys* **18** (1979) 279
96 **Berry, J.A., Salomo, G.J.** Opt Soc Am Ann Meet (1978) Communication W15
97 **Reali, G.C.** *Opt Comm* **35** (1980) 264
98 **Bernard, P., Bélanger, P.A.** *Opt Lett* **4** (1979) 196
99 **Damen, T.C., Duguay, M.A.** *Electr Lett* **16** (1980) 166
100 **Kobayashi, T., Konishi, S., Sueta, T.** *Opt and Quant Electr* **7** (1975) 229
101 **Hayes, C.L., Laughman, L.M.** *Appl Opt* **16** (1977) 263
102 **Kae, D.M., Marosky, G., Tittel, F.K.** *Appl Phys* **15** (1978) 59

Short laser pulse generation: part two

D. FAUBERT, S.L. CHIN

The status of short laser pulse generation is reviewed. Achievements, limitations and potential of techniques producing pulses shorter than 50 ns, from the ultra-violet to the infra-red, are examined.

In this second half of the paper the physical principles underlying the generation of ultra-short pulses by gating, interferometric techniques, pulse compression and pulse shaping are considered.

KEYWORDS: lasers, short pulse generation, gating, interferometric techniques, pulse compression, pulse shaping

List of symbols

c	speed of light in a vacuum
f	frequency of the acoustic wave travelling in an acousto-optic crystal
$I(t)$	intensity of a laser pulse
k	constant relating T and $\Delta\omega$
l	length between the mirror and the Pockel cell in the electro-optical intra-cavity gating scheme ('Gating', **4**.f.)
L	length of a laser resonator
l_a	length of the active medium of a laser
l_l	length of the optical path in the two electro-optical crystals gating scheme ('Gating', **4**.e.)
L_T	length of a charged transmission line in electro-optical gating
$l_{1,2}$	length of transmission lines 1 and 2 in electro-optical gating
$n(n')$	index of refraction of the medium outside (inside) a laser
Q	quality factor of a resonator
$R_{1,2}$	reflectivities of the two mirrors of a laser
R_E	$= (R_1R_2)^{1/2}$
t	time
T	full width at half maximum of a laser pulse
T_c	cutting time of a laser pulse
T_F	total duration of a laser pulse
T_2	transverse relaxation time
Δt	time interval
t_v	photon lifetime within a laser resonator
t_{rec}	recovery time of a saturable absorber
$V(t)$	a high voltage pulse
$V_{\lambda/2}(V_{\lambda/4})$	half-wave (quarter-wave) voltage of an electro-optical device
V_0	voltage to which a transmission line is charged
v	speed at which electrical current flows in a transmission line
ω_0	carrier frequency of an electromagnetic wave
$\Delta\omega$	spectrum of an electromagnetic wave

Gating

1. Under the single heading of gating we have gathered all the techniques which in some way or another gate or extract a short pulse of light from a longer one. We do not pay too much attention to the laser itself, but arrange, most often outside the cavity, a scheme to isolate a small fraction of the laser output. Fig. 1 shows a Q-switched ruby laser pulse that was gated for 1.0 ns.

2. The width of the small pulse produced is obviously equal to the opening time of the gating mechanism, and differs from one scheme to the other. Section **3** summarizes performances as it describes each system. Very often the gating time can be relatively easily controlled to produce pulses of

DF is at the Defence Research Establishment Valcartier, PO Box 8800, Courcelette, Québec, Canada. SLC is in the Laboratoire de Recherches en Optique et Laser, Université Laval, Québec, Canada.
Received 9 December 1981.

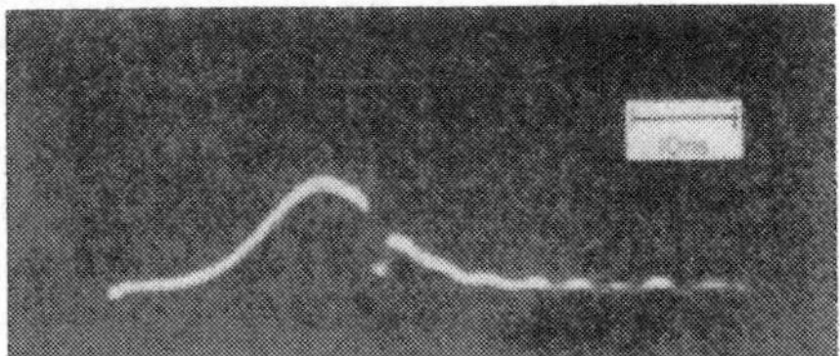

Fig. 1 A 1 ns ruby laser pulse gated out of a 10 ns Q-switched pulse

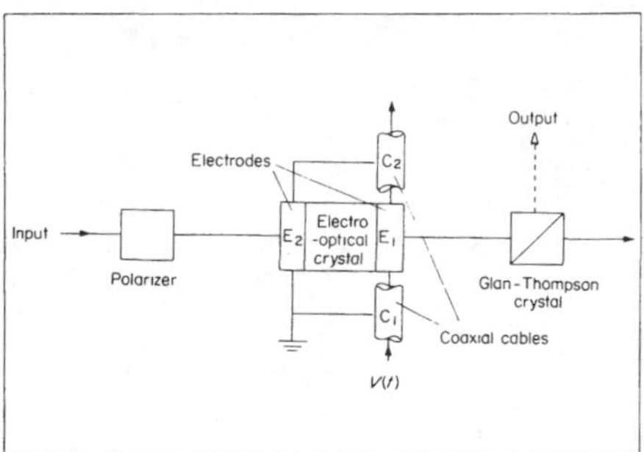

Fig. 2 Conventional electro-optical gating. The dotted line is the useful short laser pulse output

adjustable duration. Furthermore, in most cases the output power is limited only by the power density the optical elements can sustain.

3. We will discuss six different extraction schemes.

a. Electro-optic. Certainly the most popular, it can be understood as follows. As depicted in Fig. 2, a long and polarized (say horizontally) light pulse is directed into an electro-optical element followed by a birefringent analyser (a device such as a Glan-Thompson prism which can split the two orthogonal polarizations). As long as the electro-optical element is not activated, the incident light is not deflected by the prism, but if a voltage $V_{\lambda/2}$ is applied to the element for a short period of time, the portion of the incident laser pulse that happens to cross it during that time has its polarization shifted by 90°. This slice of the incident pulse will then be reflected by the prism. In Fig. 2 an extra polarizer was placed before the electro-optical element to make sure the incident light was well polarized. This method has been applied to XeCl lasers (3080 Å; 1 ns)[1], ruby or Nd lasers (6943 Å or 10600 Å; 70 ps-5 ns)[2-4], HF lasers (2.7 μm; 1-5 ns)[5] and CO_2 lasers (10.6 μm; 600 ps)[6,7].

The generation of the shortest possible high voltage pulse (usually around 5 kV or more) is the main feature of this technique. The widespread laser-triggered spark gaps[5,6] or the newer semiconductor switches[8-10] coupled to low impedance transmission lines, carefully matched to a wide bandwidth Pockel[1-10] or Kerr cell[11,12], are the building blocks of the method. The spark gap (Fig. 3a) utilizes a laser-produced plasma to create a short circuit between two electrodes connected to a transmission line of length L_T charged to a voltage V_0. This produces a voltage pulse of amplitude $V_0/2$ and width $2\,L_T/v$ where v is the speed of propagation of an electrical signal in the transmission line[13]. The width (FWHM) of the light pulse produced is equal to that of $V_{\lambda/2}$. We thus have to arrange that $V_0/2$ equals $V_{\lambda/2}$ and that the transmission line is the right length for the laser pulse width we want to produce. We see that the pulse width can be varied by changing L_T, but much more flexible ways to do that will be discussed in section **4**. The fastest high voltage pulses produced this way had a rise-time of around 300 ps.

Faster voltage pulses (rise-time < 50 ps) can be obtained with semiconductor switches (Si, GaAs or CdS). In this case, the charged transmission line (Fig. 3b) is joined to the electro-optical circuit by a semiconductor. A powerful laser light pulse creates free carriers within the semiconductor, thus increasing its conductivity and closing the circuit. The disadvantage of these switches is that they require two independent, well synchronized laser pulses, one that will be gated and the other to open the gate. Besides spark gaps and semiconductor switches, krytrons[14], thyratrons[1] or avalanche transistors[3] can be used, but they produce longer voltage pulses ($\geqslant$ 1 ns). Typical pulse widths achieved are a few hundred picoseconds to a few nanoseconds, but pulses as short as 70 ps[4] have been produced. The peak power of the gated pulse at most equals that of the input, and megawatts are common.

This versatile and straightforward scheme can be applied at any wavelength, provided optical components for these frequencies can be found. The most common electro-optical materials used are KDP and $LiNbO_3$ (visible), and CdTe and GaAs (infra-red). Extinction ratios (signal-to-noise

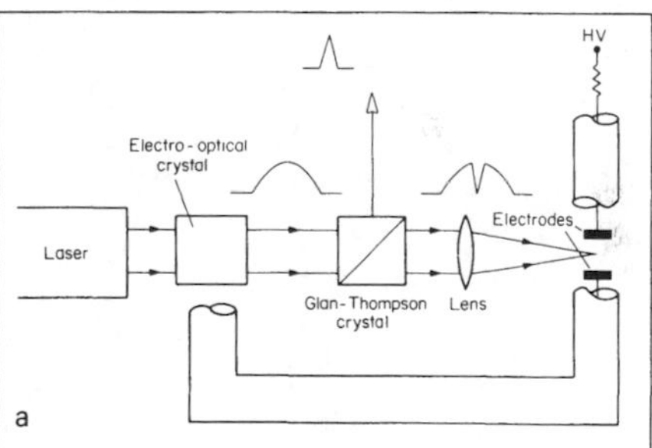

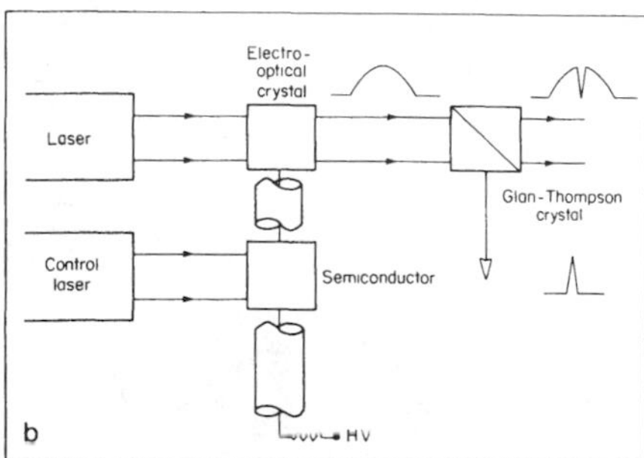

Fig. 3 An ultra-fast electrical switch launching the high voltage pulse in the circuit of an electro-optical element: a — a spark gap; b — a semiconductor switch

ratios) between 10^3 and 10^4 are possible[15]. Kerr cells, even if they are cheap, have very good optical quality, a high damage threshold and a greater extinction ratio (10^5, because there is no residual birefringence due to stresses in the material), are less popular. This is because their high dielectric constant increases the capacitance of the cell, thus slowing its response unless very low impedance transmission lines (1.5 or 2.5 Ω) are used[11].

b. Acousto-optic deflector[16]. An acousto-optic modulator similar to the one used for mode-locking (see part one, mode locking, 3a(i)) is used to produce bursts of power from the laser cavity. To our knowledge, it has only been used with a Nd laser and yielded pulses a few ns wide. The problem it suffers compared to the electro-optic modulator is that the acoustic grating cannot be set up very quickly in the crystal because of the relatively slow speed of sound. It must, therefore, be used in a standing-wave mode, thus giving a high pulse-repetition frequency output rather than a single pulse.

c. Semiconductor switching (CO_2 laser). The idea here is to modulate the reflectivity (or transmittivity) of a semiconductor material by illuminating it with a light source (a short pulse visible laser for instance) and thereby creating free carriers at its surface (see Fig. 4a). The free carriers are generated when electrons are excited to the conduction band by absorbing photons whose energy is greater than the band-gap energy. At first the semiconductor is normally transparent to the incoming CO_2 long laser pulse, but when the shorter control pulse produces surface charges, the CO_2 laser light will be reflected (or the transmission will terminate due to free carrier absorption) for a time equal to the duration of the control pulse plus the diffusion time of the carriers.

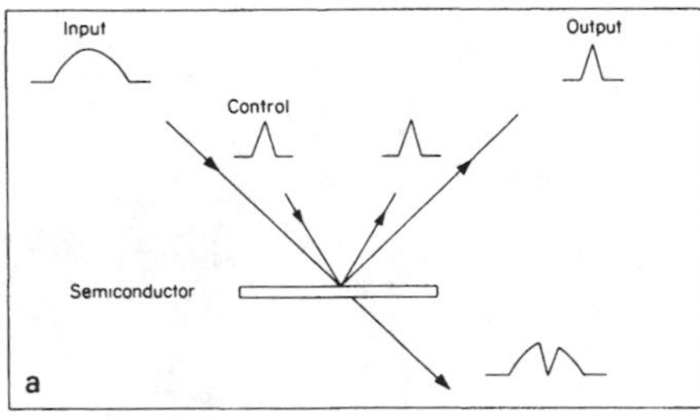

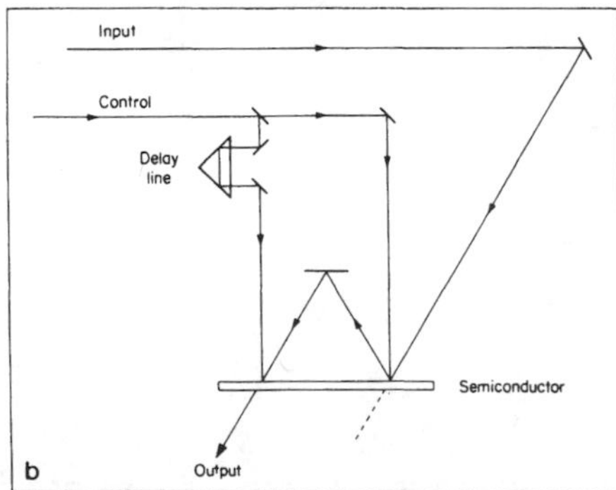

Fig. 4 a – standard semiconductor gating; b – advanced semiconductor gating arrangement producing CO_2 laser pulses which are continuously adjustable[18]

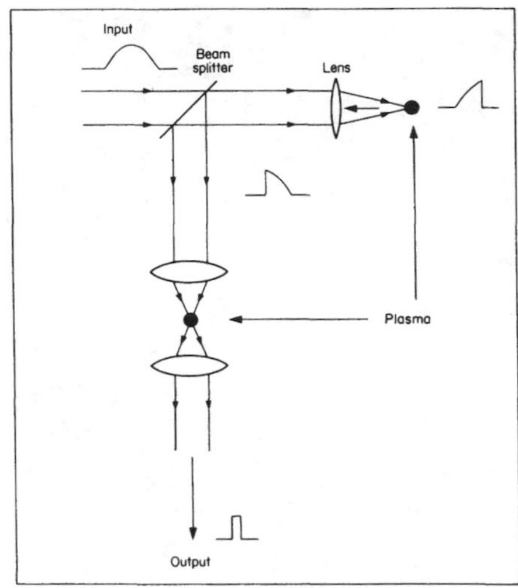

Fig. 5 Plasma gating[20]

Typical pulse widths achieved up to now are around 1 ns[17], but with a more elaborate set-up[18] (see Fig. 4b), the on/off time of the short reflected pulse can be independently controlled to yield pulses conveniently tunable between 5 and 100 ps (the shortest CO_2 pulses produced to date). This technique is only applicable to infra-red radiation, and has only been applied to the CO_2 laser. Dye (0.6 μm), ruby (0.6943 μm) and N_2 (0.337 μm) lasers and a doubled Nd laser (0.53 μm) have been used to generate the control pulse. A review of this subject has been published recently[17].

d. Plasma switching. When a laser pulse is highly focused in a gas, a plasma is created. The plasma is opaque to the laser pulse and will abruptly cut it (within a few picoseconds[19]). However, the plasma also reflects a few percent of the last part of the incident laser pulse (to which the plasma is opaque) and if we arrange another lens to focus the reflected pulse in a gas, it will again be cut, producing a more or less square pulse corresponding to the centre part of the incident laser pulse (see Fig. 5).

This scheme has been applied to CO_2 lasers[20] to yield 20 ns pulses. One drawback is that the output power is only ~ 1% of the input, which corresponds to the plasma reflectivity. The width of the output laser pulse could in principle be tuned from about 20 ps (twice the fastest possible plasma cutting time) by varying the pressure of the gases in which the plasmas are produced. A similar set-up with the lenses inside the resonator has been applied to ruby[21] and HF[22] lasers.

e. The Fabry-Perot electro-optic modulator. A Fabry-Perot interferometer (FPI) containing an electro-optical phase modulator replaces the output coupler of a cw laser (HeNe, 0.6328 μm; 20 ps)[23]. The phase modulator is driven at a frequency of around $c/2L$. This forces the resonant frequency of the FPI (maximum transmission) to vary in time as if the length of the FPI were being modulated. Under normal conditions the laser will continue lasing in a single longitudinal mode, and when the resonant frequency of the FPI corresponds to the frequency of this mode, a pulse of light is emitted. This method works only for cw lasers and has only been applied to a HeNe laser[23].

f. The velocity-matched gate. This device, not yet realized experimentally, is based on integrated optics technology[24] It is basically a directional coupler. Light propagating in a given channel can be switched to another guide if a proper index of refraction condition is chosen. The idea is quickly to switch the light from one channel to the other by rapidly varying the index of refraction using the electro-optic effect. The author claims that pulses as short as 0.1 ps (100 fs) can be achieved this way if a microwave travelling-wave, whose speed is matched to the group velocity of the light-wave propagating in the guides, induces the refraction changes.

4. Many versions of the standard electro-optic gating scheme have been devised, mostly to shorten the gated pulse or facilitate the tuning of its duration. They will be given in the following brief discussion, in which C_1 and C_2 are coaxial transmission lines of length l_1 and l_2; v is the speed of propagation of the current or voltage wave in C_1 or C_2; c the speed of light in a vacuum; T_r the rise-time of the voltage pulse $V(t)$; and T_F is the full width of the laser pulse produced.

a. Use of a voltage pulse of 2 $V_{\lambda/2}$ instead of $V_{\lambda/2}$. This allows us to produce a light pulse lasting only the steep rise (or fall) time of the voltage pulse[25]. It produces shorter laser pulses, with the same set-up, than the conventional approach of Fig. 2:

$$T_F = T_r \tag{1}$$

b. Short circuit method[26] (Fig. 6a). Electrodes E_1 and E_2 of the electro-optic element are short circuited. The short laser pulses last only the time it takes for the voltage pulse to travel from E_1 to E_2.

$$T_F = \frac{l_2}{v} + T_r \tag{2}$$

c. Method using a short-circuited transmission line[26,27] (Fig. 6b). When transmission line C_2 is short circuited, an identical voltage, although out of phase by 180°, is reflected from its end. T_F is equal to the time taken for the reflected voltage to return to E_1, cancelling $V(t)$.

$$T_F = \frac{2l_2}{v} + T_r \tag{3}$$

d. Method with two $V(t)$[26] (Fig. 7). T_F lasts until $V(t)$, propagating to E_2 along C_2, reaches E_2, resulting in zero difference of potential between E_1 and E_2.

$$T_F = \frac{l_2 - l_1}{v} + T_r \tag{4}$$

e. Method with two electro-optical elements[28] (Fig. 8). The incident pulse $I(t)$ is sliced by EO_1. The resulting short pulse then follows the dotted path to reach EO_2 where it can be gated again by the same voltage pulse propagating along C_2.

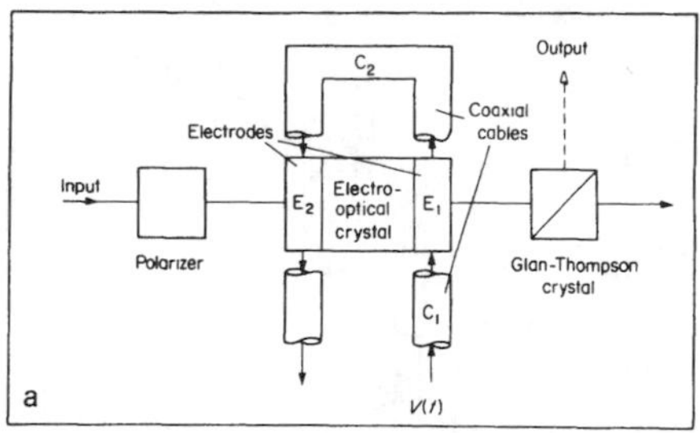

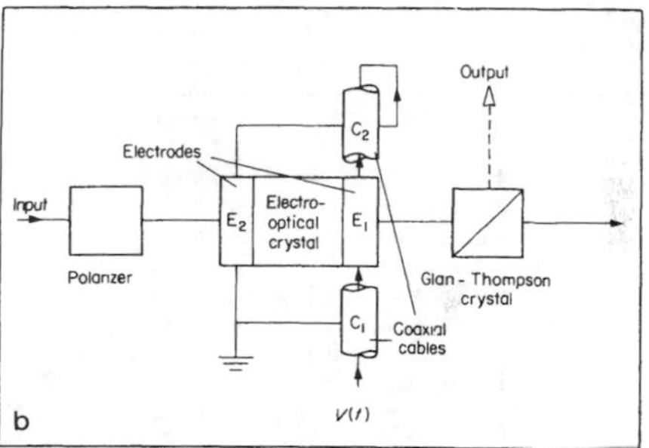

Fig. 6 a – Electro-optical gating with short circuited electrodes; b – electro-optical gating using a short circuited transmission line

The main advantage of this scheme is the ease with which T_F can be varied by changing the length (l_1) of the optical delay line AB. (In all previous techniques, the length of an electrical cable had to be changed to vary the pulse width – an inconvenient operation.)

$$T_F = \frac{l_2}{v} - \frac{l_1}{c} + T_r \tag{5}$$

f. Intra-cavity technique[29] (Fig. 9). The laser is first activated normally, creating a standing wave in the cavity. At a given time let us assume that a step voltage $V_{\lambda/2}$ is applied to the electro-optical element. The radiation contained between points C and D will go through the element twice, having

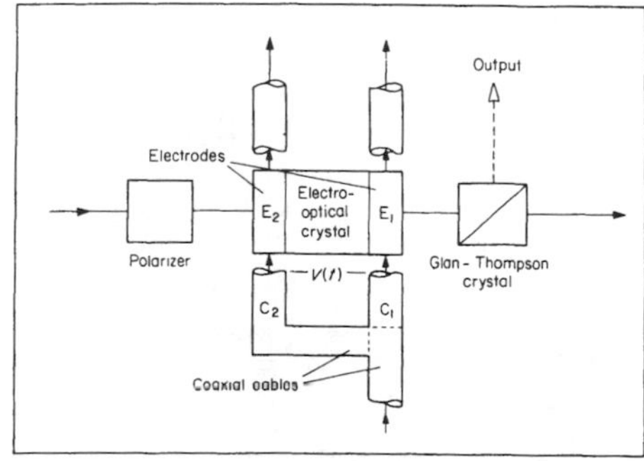

Fig. 7 Electro-optical gating using two high voltage pulses

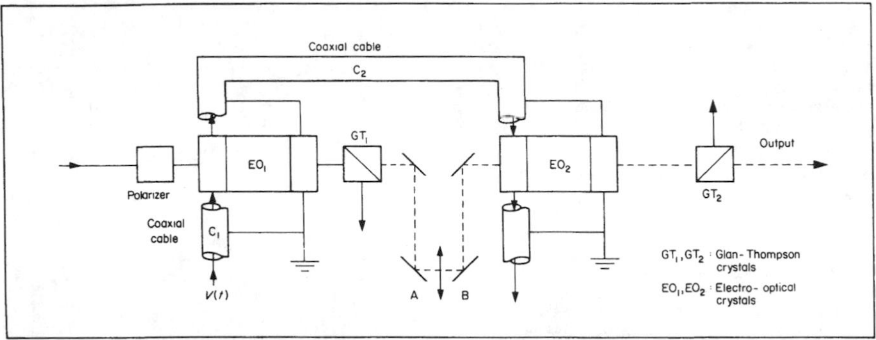

Fig. 8 Electro-optical gating using two separate Pockel cells

its polarization changed by 180°, and will not be reflected from the Glan-Thompson prism. The radiation contained between A–B will cross the electro-optical element only once, having its polarization vector changed by 90°, and will be fully reflected from the prism. Radiation contained within the element when $V(t)$ is applied will have its polarization altered by between 0° and 180° and will be partially reflected by the prism. Thus, only the radiation between A and C (of length l) will be more-or-less extracted from the cavity.

$$T_F = \frac{2l}{c} + T_r \qquad (6)$$

This scheme is very convenient when the pulse width has to be varied. The electro-optical element is mounted on a precision translation device and l or T_F can then be adjusted easily and precisely. Furthermore, this is done with a minimum of optical components. Pulses continuously tunable between 1 and 5 ns have been obtained with a ruby laser[29]. Similar arrangements, but giving a train of pulses rather than a single pulse, have also been published[30,31].

Interferometric techniques

1. We first arrange to have two laser beams interfere destructively and then break this equilibrium by either cutting off, shifting the phase, or changing the polarization of one of the beams. Interference will cease to be destructive for as long as the perturbation lasts. A laser pulse will then be produced during that time.

2. The smallest pulse width feasible is, of course, closely related to the shortest disturbance one can induce. It is most commonly produced by an electro-optical phase shifting or a plasma cutting one of the beams. In the former case the lower bound is around 50 ps, as discussed earlier in the section on 'Gating', whereas in the latter a 10 ps perturbation appears feasible[19]. The attractive feature of these schemes is that they often lend themselves to simple tuning of the pulse duration by either moving a mirror or changing the duration of the perturbation. They can also be applied to any part of the electromagnetic spectrum as long as an interferometer can be devised at that particular frequency. They yield a unique pulse whose peak power can be as large as that of the two interfering beams and is limited only by the optical damage threshold of the interferometer.

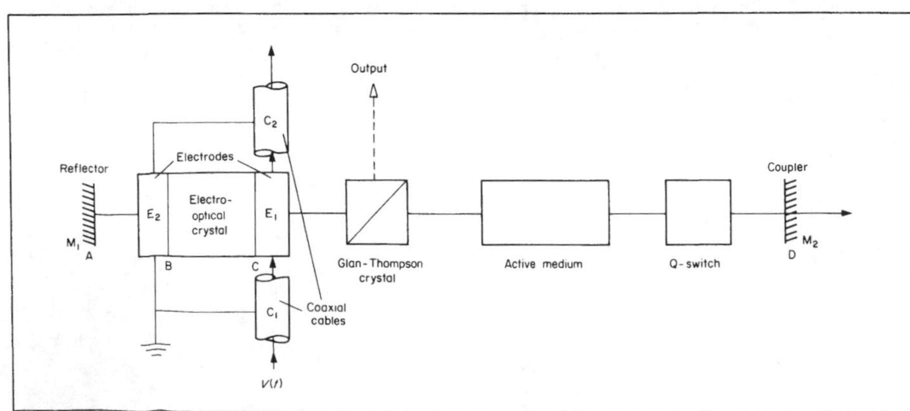

Fig. 9 Intra-cavity electro-optical gating[29]

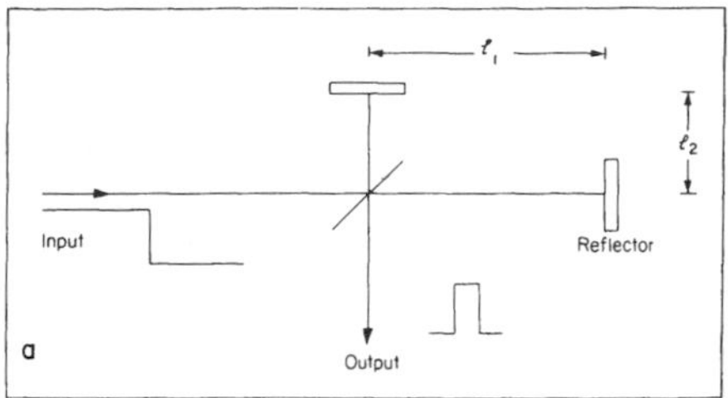

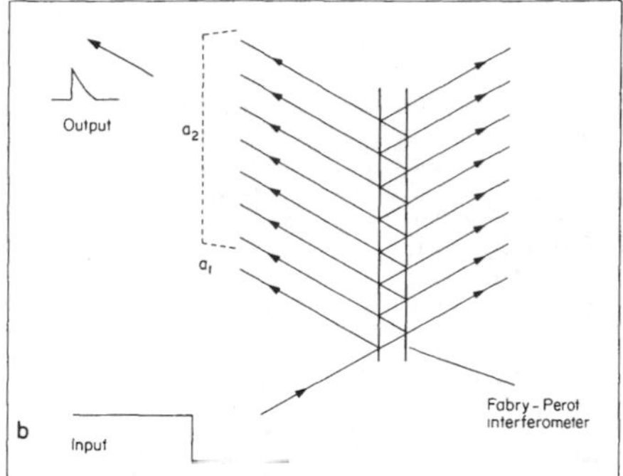

Fig. 10 Interferometric pulse generation: a – with a Michelson interferometer; b – using a Fabry-Perot interferometer

3. The different methods of setting up this technique distinguish themselves by the type of interferometer used.

a. Michelson. A single beam of light is sent into a Michelson interferometer having two arms of unequal length (l_1 and l_2) (Fig. 10). If the input beam is cut abruptly, a pulse whose duration is $(l_2 - l_1)/c$ is produced. This is because the two interfering fields do not reach the recombining mirror at the same time. The pulse width can thus be varied by varying $(l_2 - l_1)$. In practice, the speed with which the pulse is cut limits the pulse width. This method was applied to ruby[32] and CO_2[33] lasers for which pulses tunable between 500 ps and 2 ns were obtained.

b. Sagnac[34]. The same principle as used in the Michelson method applies to a Sagnac interferometer, for which two interfering beams propagate in opposite directions around the same optical path. It has been applied to a CO_2 laser (10.6 μm; 20 ns)[34].

c. Fabry-Perot[35,36]. A Fabry-Perot interferometer (FPI) is adjusted to be in resonance with an incoming laser beam (Fig. 10b). We thus have a maximum of transmission and very little light reflected. This latter condition is because the field reflected off the surface of the first mirror (a_1) is exactly of equal amplitude but 180° out of phase with the field resulting from the multiple reflections coming from inside the FPI (a_2). Fields a_1 and a_2 interfere destructively. If the incoming beam is cut abruptly, field a_1 vanishes. We are thus left with field a_2 which is our short pulse and which lasts as long as the FPI is not empty of all the energy it contains. The pulse width can be adjusted by changing the spacing between the mirrors, from around 10 ps to 1 ns[36] (theoretical). An experimental investigation[35] produced CO_2 laser pulses around 80 ps.

d. Free induction decay. An electromagnetic field in resonance with a molecular transition is completely absorbed. The zero transmittivity can be interpreted as being the result of the destructive interference between two fields: the incident field E_i and the field E_r radiated by the molecules oscillating under the influence of E_i and 180° out of phase with it. If E_i is cut abruptly, only E_r remains for as long as the molecules do not recover their equilibrium positions. Pulses as short as 30 ps for CO_2 lasers[37,38] were achieved. This technique was also applied to I_2 lasers (1.31 μm; 100 ps)[39].

The pulse width can be varied by changing the pressure of the absorber. Its power is limited by the onset of the non-linear behaviour (saturation) of the absorber. The main difficulty with this scheme is to find a suitable molecule for a given laser frequency.

The problem studied in the last three sub-sections can also be understood in the frequency domain[40]. The input electromagnetic field has a given bandwidth centred around a carrier frequency ω_0. When it is cut abruptly, the spectrum suddenly broadens. If it is cut in T_c seconds, the bandwidth of the input field then becomes approximately $1/T_c = \Delta\omega$, again centred around ω_0. If we use a spectral filter of some sort (an interferometer or a molecular absorber) this spectrum will be filtered out to a new value $\Delta\omega'$. The duration of the pulse thus obtained is $1/\Delta\omega'$. The shape of the pulse produced is determined by the spectral filter.

Pulse compression

1. Properly speaking, few of the pulse compression schemes are really compressing the total energy of the laser pulse into a shorter pulse, producing a higher peak power. Most of them reduce the duration of the input pulse at the expense of energy. The peak power remains the same or is smaller, that is, the rise-time is sharpened or the tail is cut. But it is general practice to term as compression any scheme that produces a shorter pulse out of a longer one.

2. The performances of each individual technique are discussed in the next section.

3. The different compression techniques are:

a. In laser amplifiers. It is well known that amplification of a laser pulse can shorten its duration if the intensity of the optical field reaches a value high enough to saturate the amplifier[41–43]. For this to happen, the rise-time of the incident pulse must be sufficiently short. Under that condition, the leading edge is so strongly amplified that the population inversion (and thus the gain) is considerably reduced for the trailing end. The pulse is then shorter and peaks earlier as compared to the input pulse (see Fig. 11a).

The shortest possible pulse that can be produced is $\sim 1/\Delta\nu_a$ where $\Delta\nu_a$ is the bandwidth of the amplifying transition. But this value is seldom reached and pulses are typically compressed by a factor of two to three (input divided by output duration). It should be noted that it is not easy to control the parameters of such a system to produce a pulse of predetermined shape and duration from a given input. This effect can be observed in any kind of amplifier under appropriate conditions.

b. Saturable absorber[44]. Usually, an absorbing medium can be saturated by laser radiation if the latter is strong enough

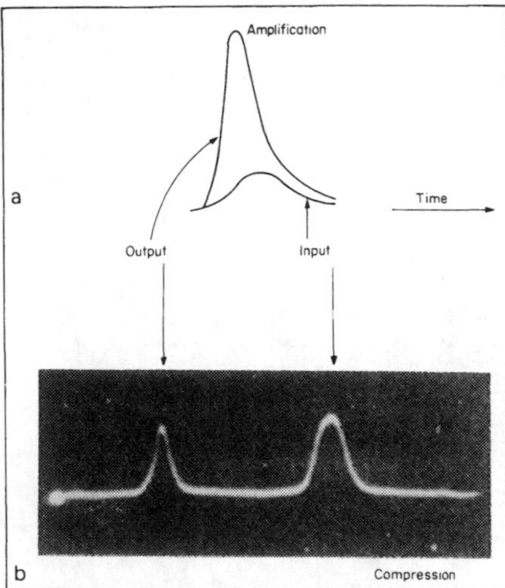

Fig. 11 a – the output of a laser amplifier superposed with the input pulse. Note the shorter pulse width and the displacement of the maximum of intensity; b – a compressed ruby laser pulse (1.1 ns; 14 kW cm^{-2}) followed by the input (2.1 ns; 1.4 MW cm^{-2}). The saturable absorber was oxazine-1-perchlorate in methanol[36]

to excite a significant proportion of the molecules out of the bottom energy level of the absorbing medium. The molecules will recover their equilibrium position in a time t_{rec}. Two regimes can then be distinguished:

(i) $T \simeq t_{rec}$ where T is the laser pulse duration: In this case, the rise-time of the pulse is used to saturate the medium before the absorber recovers to cut the trailing edge. The pulse width obtained tends to be equal to t_{rec}.

(ii) $T \gg t_{rec}$: In this case we can assume that the absorber recovers instantaneously. The pulse is shortened because the lower intensity portions of the pulse are more absorbed than the stronger, the latter saturating the absorption medium. In both regimes, typical compressions of around 2 are obtained, as shown in Fig. 11b. In the visible part of the spectrum, dyes are normally used as the non-linear molecule, but gases[45,46] or semiconductors[47] can be used with CO_2 lasers.

The output intensities obtained are usually ~ 10% of the input. This precludes the use of a chain of absorbers to attain still shorter pulses. Another severe restriction is the instability of the technique. Since significant compression is the result of a strongly non-linear effect, even small variations of the input peak power or pulse shape may result in an unpredictable output.

c. Optical Kerr Effect[48,49]. An optical Kerr cell is placed between crossed polarizers. Much like a saturable absorber, the transmission of such a device is a function of the input field intensity. Lower intensity optical fields do not significantly cross the second polarizer because the Kerr cell did not turn their polarization by a large enough amount. Pulse compression of around 2 could be obtained but has never been realized experimentally.

d. Coherent compression. This method is based on self-induced transparency experiments. It is now well known that a short pulse of duration T whose 'area' equals 2π is integrally transmitted by a two-level system if $T < T_2$, where T_2 is the transverse relaxation time[50]. But if the 'area' is not 2π, the two-level system will reshape the pulse until it reaches this value. In particular, if the 'area' of the pulse equals 3π, it is observed that the input pulse is compressed about 10 times while its peak power is increased by a factor of 3.

The greatest difficulty associated with this arrangement is certainly finding the two-level system suited to a given laser frequency. To our knowledge it has only been applied to a 10 ns $^{202}Hg_{II}$ laser pulse (0.79 μm) compressed to 1 ns in atomic rubidium[50].

e. Raman compression. This method is actually a way to transfer energy from a long to a shorter light pulse of different wavelength. For instance (see Fig. 12), a short pulse of wavelength 268 nm (I_c) and a long pulse of 249 nm (I_l) are propagating in opposite directions in a suitable Raman medium (methane in this example). The ultra-violet energy contained in I_l is transferred to I_c via the Raman transition. I_c and I_l are propagating in opposite directions so that I_c always sees a different part of I_l. About 50% of the energy contained in I_l can be transferred to I_c[51,52].

This technique is particularly interesting for excimer lasers with regard to laser fusion. Ultra-violet energy would be efficiently extracted from the active medium in long pulses and then transferred to shorter pulses more suitable for heating the targets.

f. Compression of frequency chirped pulses. Let us consider a pulse of duration T whose central frequency is ω_0. Suppose that we can linearly shift the frequency content of the pulse:

$$\omega(t) = \omega_0 + \frac{\Delta\omega}{\Delta t}\, t, \quad 0 < t < \Delta t \qquad (7)$$

where $\omega(t)$ is the instantaneous carrier frequency and $\Delta\omega$ is the maximum frequency shift we can produce in a period Δt. If $\Delta\omega > 0$, the front end of the pulse will be redder than the trailing edge (up-shifted), and vice-versa if $\Delta\omega < 0$. The

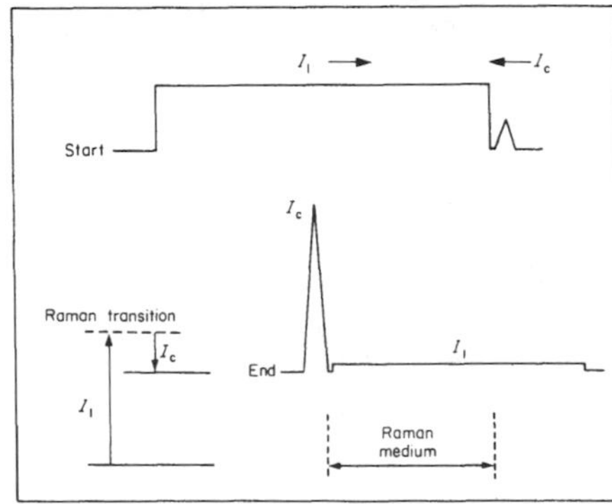

Fig. 12 Raman pulse compression scheme, before and after the interaction

spectrum is said to be chirped. Until now, the pulse shape has not been altered, only its frequency content has been modified. But if we send this light pulse through a dispersive line, that is, a system in which the blue propagates faster than the red, the blue end of the pulse will catch up with the red front. The resulting pulse is shorter and its peak power higher since energy has, at least in principle, been conserved.

This method has been applied mostly to dye lasers[53,55] but also to the Nd lasers[54]. As examples: 10 ns 0.7948 μm pulses were compressed to 1 ns[53], 100 ps to 8 ps for a Nd: YAG laser, or from cw to 500 ps, with a 14-fold increase in the peak power, for a dye laser (0.5890 μm)[55]. This method was also used to restore the shape of 3 ps dye laser pulses broadened to 13 ps by propagation through an optical fibre.

The differences in the ways this technique is actually carried out arise from the production of the chirp or the nature of the dispersive line.

(i) Generation of a chirped laser pulse: In many cases, laser pulses are already chirped[57], but they can also be produced by modulating the instantaneous phase $\phi(t)$ since

$$\omega(t) = \frac{d}{dt}\,\phi(t) \tag{8}$$

$\phi(t)$ can be varied using an electro-optic phase modulator[55], a Doppler shifter[58], an optical Kerr cell[54] (which seems to be the best approach) or by using self phase modulation in glass[56,59] or metallic vapours[60,61].

(ii) The dispersive line: To act as dispersive lines, we can use Gires-Tournois interferometers[61], a pair of gratings[62], or by using the natural dispersion in a laser amplifier[59,63] or a metallic vapour[53,55,60], or, finally, by using the interaction between a CO_2 laser frequency and phonon-polaritron propagation modes in some crystals[64].

The limiting factor of this technique is the generation of the greatest $\Delta\omega$ possible and the design of dispersive lines which are of reasonable size.

4. Many variations of these techniques have been described:

a. The combined use of compression in amplifiers and absorbers, that is, a chain of alternating amplifiers and saturable absorbers. The latter shortens the front end of the pulse and makes the compression in the amplifier more efficient. Compression of 5 has been achieved in this way[41].

b. The use of a dye for which the laser frequency falls in the overlapping region of its absorption and emission curves[65]. Part of the energy lost by the laser pulse to bleach the dye is recovered through stimulated emission. Under optimum conditions, the theory predicts that 25% amplification of the peak power is possible with a factor of 2 compression.

c. Four-wave mixing. Much like passive pulse compression in a saturable absorber, degenerate four-wave mixing behaves non-linearly and can compress a laser pulse[66].

d. The dispersive modulator[67]. A metal vapour transition almost in resonance with the pulse to be compressed, can be shifted using the Stark effect. This modulation of the transition frequency eliminates the need for a chirping mechanism since the absorbing transition shifts relative to the laser frequency, producing the same sweeping effect. Natural dispersion, close to the resonance of the transition, produces the pulse compression. This method has been applied in the 10 μm infra-red region (CO_2 and N_2O lasers) for which a 1 μs pulse was compressed in a series of 8 ns pulses, 21 ns apart[67].

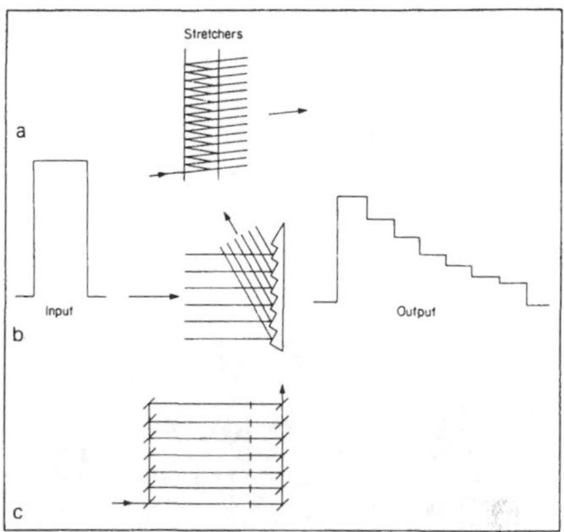

Fig. 13 Pulse stretching: a – a Fabry-Perot interferometer; b – a blazed grating; c – an array of beam splitters and attenuators

Pulse shaping

1. Many schemes have been proposed to shape or reshape laser pulses. In some experiments, such as laser thermonuclear fusion, the ability to synthesize a laser pulse of precise shape is of great importance.

The pulse compression schemes described in the previous section could be classified as pulse shaping, but in this section the aim is not necessarily to shorten the pulse. The different approaches to the present problem will be described in section **3**.

2. Most of the methods we are about to describe can be applied at any wavelength provided suitable optical components are available.

3. The main shaping systems are:

a. Interferometric. As shown in Fig. 13a, a Fabry-Perot interferometer produces a series of attenuated and delayed replicas of an incident laser pulse. When they recombine, they give rise to an elongated or stretched pulse (1.06 μm; 600 ps)[68]. Extra interferometers will increase the flexibility of the system[69]. The insertion of a non-linear medium (dye or optical Kerr cell[70]) in the arm of some types of interferometer will also change the shape of the output pulse.

b. Blazed grating. Much like a Fabry-Perot interferometer, a blazed grating will produce many replicas of an input pulse and will alter its shape, as shown in Fig. 13b[71] (theoretical).

c. Beam splitters. Again, a series of beam splitters and attenuators (Fig. 13c) will produce replicas of the input pulse that can be recombined to yield the desired pulse shape[72,73].

d. Flat lenses. Flat lenses (for example, Fresnel lenses) are used to focus short laser pulses. It can be shown that the illumination at a given radial position near the focus of the system is a function of time, that is, light from all directions

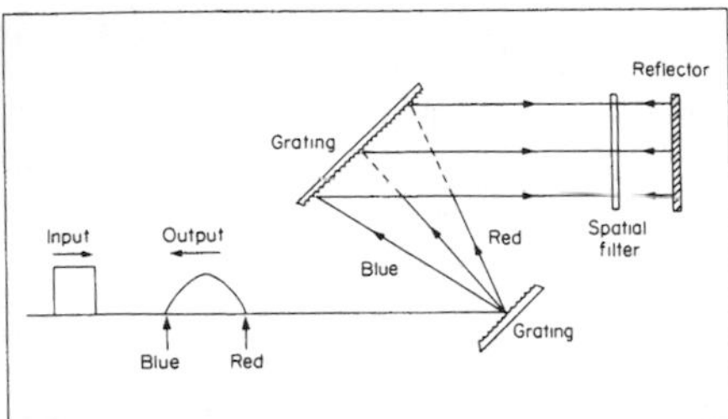

Fig. 14 Pulse stretching and shaping with gratings and spatial filtering

does not reach the focus simultaneously. Making use of properly positioned or shaped pinholes at the focal point will produce a light pulse with a shape different from the input one (theoretical[74]).

e. Grating pair. As shown in Fig. 14, a light pulse is directed onto the first grating which reflects it onto a second one. Light then travels to a mirror from which it is reflected back towards the gratings. As is evident in Fig. 14, the red portion of the spectrum travels a greater distance than the blue. The resulting output pulse is then longer, the blue being in front and the red behind[75-77]. Typically, 30 ps pulses can be stretched to 1 ns. The introduction of a filter between the second grating and the reflector will yield pulses of arbitrary shapes, as demonstrated experimentally[76]. For this technique to work properly, the input pulse must be transform limited, $T\ \Delta\omega = 2\pi k$.

In fact all the pulse shaping techniques described so far suffer from having very often high propagation losses and low beam quality because of the number of optical surfaces involved. They require very sophisticated alignment procedures and are also very dependent on the reproducibility of the light input spectrum.

f. Electro-optic sweeping. Using an electro-optical crystal, we can arrange for the beam striking it to be swept in space (Fig. 15). By placing a pinhole (and filters) at the back of the crystal, a pulse of any given shape can be synthesized[78,79]. Pulses as short as 94 ps were realized with a HeNe laser[79].

g. Electro-optic shaping. As discussed in the section on Gating, (3.i) the recent technological development of semiconductor switches makes control of high voltages possible on a picosecond time-scale[4]. The use of more than one light pulse to trigger the switch, the use of electrical circuits instead of a charged line[80], or the use of many electro-optical crystals, could enable us to 'gate' from a long input laser pulse a shorter one of a given shape.

h. Birefringent crystals. In such a crystal, the two orthogonal polarizations of an input optical field propagate at different speeds. Using a series of such crystals whose axes are correctly aligned relative to each other, we can introduce a controlled delay between the two polarization components and change the incident pulse shape in a controlled way[81].

i. Combination of many pulses. It has been suggested that to combine many pulses of programmed amplitude from independent amplifiers and delay line systems could yield a desired pulse shape[82,83]. This could alleviate the problem observed in laser fusion experiments that almost any input pulse has the same output shape at the end of a laser amplifier chain.

j. Dispersive shaping. In systems similar to those described in the section on 'Pulse compression' (3.f) (pulse compression using dispersive lines), pulse squaring effects can be achieved[84].

k. Dye shutter. It has been speculated that a laser pulse propagating in a saturable absorber could be arbitrarily shaped if other independent bleaching control pulses (Fig. 16) could alter its interaction with the absorber[85].

To conclude this section it might be worth mentioning that Bates and Henderson[86] have devised a method to evaluate the maximum possible efficiency of a passive pulse shaper.

Conclusion

Short-pulse generation techniques have been, and still are, the subject of many investigations. As a result, numerous methods have been invented. If the frequency of the short laser pulse obtained is not suitable for a given experiment, it can always be changed using harmonic generation in non-linear crystals, sum or difference frequency generation, parametric oscillation, or by pumping a dye laser which will lase at another frequency. As a result, we can say that between the vacuum ultra-violet and around 2.0 μm, the experimenter has access, in principle, to pulses of any frequency and durations of 1 ps upwards. In this region, dye and colour centre lasers, which are continuously tunable over large bandwidths, together with the frequency shifting techniques mentioned above, can completely cover the spectrum. Between 2 and 11 μm, light sources are more scarce. Frequency mixing techniques are still used but gaps in both the frequency spectrum and

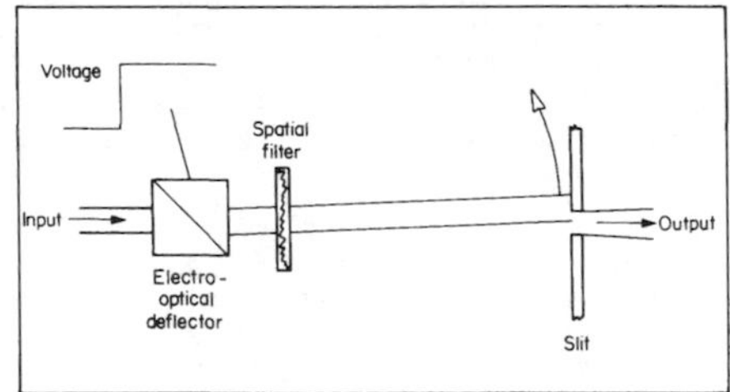

Fig. 15 Pulse shaping using an electro-optical deflector to sweep the laser beam across a slit[79]

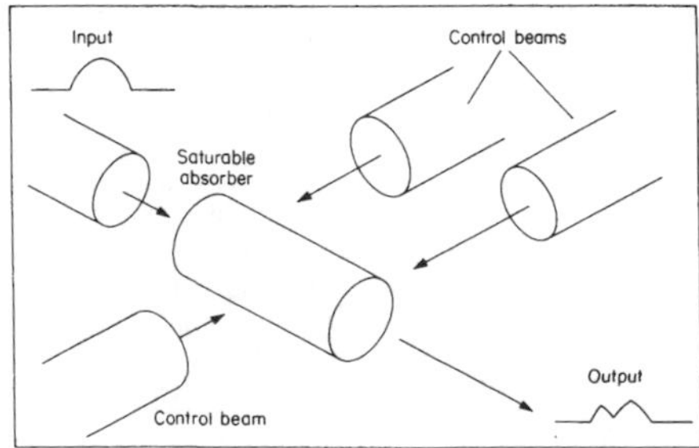

Fig. 16 Pulse tailoring using an optically gated dye cell[85]

pulse duration are still numerous. Continuously tunable, high pressure CO_2 lasers should be able to fill these gaps. Few laser types are capable of producing radiation above 11 μm.

Even if much work has already been done, there is still room for ingenious design to improve the ease with which continuously tunable light sources, both in time and frequency, could be operated.

References

1 Pacala, T.J., Laudenslager, J.B., Christensen, C.P. *Appl Phys Lett* **37** (1980) 366
2 Alcock, A.J., Richardson, M.C. *Opt Comm* **2** (1970) 65
3 Davis, S.J., Murray, J.E., Downs, D.C., Lowdermilk, W.H. *Appl Opt* **17** (1978) 3184
4 Agostinelli, J., Mourou, G., Gabel, C.W. *Appl Phys Lett* **35** (1979) 731
5 Getzinger, R.W., Ware, K.D., Carpenter, J.P., Sehott, G.L. *IEEE J Quant Electr* **QE-13** (1977) 97
6 Richardson, M.C. *Opt Comm* **10** (1974) 302
7 Figueria, J.F., Sutphin, H.D. *Appl Phys Lett* **25** (1974) 661
8 Lee, C.H. *Appl Phys Lett* **30** (1977) 84
9 Platte, W. *Opt Laser Technol* **10** (1978) 40
10 Mourou, G., Knox, W. *Appl Phys Lett* **35** (1972) 492
11 Diels, J.C. *Rev Sci Instr* **46** (1975) 1704
12 Antonetti, A., Malley, M.M., Mourou, G., Orszag, A. *Opt Comm* **23** (1977) 435
13 Plonus, M.A. Applied Electromagnetics, McGraw-Hill, New York (1978) 552
14 Hyde, R.L., Jacoby, D., Ramsden, S.A. *J Phys E: Sci Instr* **10** (1977) 1106
15 Figueira, J.F. *IEEE J Quant Electr* **QE-10** (1974) 572
16 Kruegle, H.A., Klein, L. *Appl Opt* **15** (1976) 466
17 Alcock, A.J., Corkum, P.B. *Can J Phys* **5**/ (1979) 1280
18 Jamison, S.A., Nurmiko, A.V. *Appl Phys Lett* **33** (1978) 598
19 Yablanovitch, E. *Phys Rev Lett* **32** (1974) 1101
20 McLelland, G., Smith, S.D. *Opt Comm* **27** (1978) 101
21 Hamal, K. *IEEE J Quant Electr* **QE-14** (1978) 407
22 Deke, B.K., Dyer, P.E. *IEEE J Quant Electr* **QE-15** (1979) 404
23 Kobayashi, T., Sueta, T., Cho, Y., Matsuo, Y. *Appl Phys Lett* **21** (1972) 341
24 Marcatili, E.A.J. *Appl Opt* **19** (1980) 1468
25 Morgan, P.D., Peacock, N.J. *J Phys E: Sci Instr* **4** (1971) 677
26 Scott, J.C., Palmer, A.W. *J Phys E: Sci Instr* **11** (1978) 901
27 Ireland, C.L.M. *J Phys E: Sci Instr* **8** (1975) 1007
28 Machewirth, J.P. *Laser Focus* (March 1977) 81
29 Faubert, D., Galarneau, P., Chin, S.L. *Opt Laser Technol* **13** (1981) 79
30 Liu, Y.S. *Opt Lett* **3** (1978) 168
31 Ewart, P. *Opt Comm* **28** (1979) 379
32 Milam, D., Brodbury, R.A., Hordvik, A., Schlossberg, H., Szöke, A. *IEEE J Quant Electr* **QE-10** (1974) 20
33 Szöke, A., Goldhar, J., Grieneisen, H.P., Kurnit, N.A. *Opt Comm* **6** (1972) 131
34 Trutna, R., Siegman, A.E. *IEEE J Quant Electr* **QE-13** (1977) 933
35 Fisher, R.A., Feldman, B.J. *Opt Lett* **1** (1977) 162
36 Faubert, D. PhD Thesis, Laval University (1980)
37 Kwok, H.S., Yablanovitch, E. *Appl Phys Lett* **30** (1977) 158
38 Ahrenkiel, R.K., Dunlavy, D., Sievers, A.J. *Opt Comm* **32** (1980) 503
39 Fill, E., Hohla, K., Scharppert, G.T., Volk, R. *Appl Phys Lett* **29** (1976) 805
40 Yablanovitch, E. *IEEE J Quant Electr* **QE-11** (1975) 789
41 Kryukov, P.G., Letokhov, V.S. *Sov Phys USP* **12** (1970) 641
42 Icevgi, A., Lamb Jr., W.E. *Phys Rev* **185** (1969) 517
43 Skiribanowitz, N., Kopainsky, B. *Appl Phys Lett* **27** (1975) 490
44 Penzkofer, A. *Opto-Electr* **6** (1974) 87
45 Izatt, J.R., Caudle, G.F., Bean, B.L. *Appl Phys Lett* **25** (1974) 446
46 Kluman, H. *J Appl Phys* **44** (1973) 1646
47 Feldman, B.J., Figueira, J.F. *Appl Phys Lett* **25** (1974) 30
48 Sala, K., Richardson, M.C. *J Appl Phys* **49** (1978) 2268
49 Thorne, J.M., Loree, T.R., McCall, G.H. *J Appl Phys* **45** (1974) 3072
50 Gibbs, H.M., Slusher, R.E. *Appl Phys Lett* **18** (1971) 505
51 Murray, J.R., Goldhar, J., Eimerl, D., Szöke, A. *IEEE J Quant Electr* **QE-15** (1979) 342
52 Ewing, J.J., Haas, R.A., Swingle, J.C., George, E.V., Krupke, W.F. *IEEE J Quant Electr* **QE-15** (1979) 368
53 Grischokowsky, D. *Appl Phys Lett* **25** (1974) 566
54 Lehmberg, R.H., McMahon, J.M. *Appl Phys Lett* **28** (1975) 204
55 Wigmore, J.K., Grischkowsky, O.R. *IEEE J Quant Electr* **QE-14** (1978) 310
56 Nakatsuka, H., Grisckowsky, D. *Opt Lett* **6** (1981) 13
57 Gex, J.P., Sautert, C., Vallat, P., Tourbez, H., Schelev, M. *Opt Comm* **23** (1977) 430
58 Giordamine, J.A., Duguay, M.A., Hansen, J.W. *IEEE J Quant Electr* **QE-4** (1968) 252
59 Fisher, R.A., Bischel, W.K. *Appl Phys Lett* **24** (1974) 468
60 Ya Zel'dovich, B., Sobel'man, I.I. *Sov Phys JETP-Lett* **13** (1971) 129
61 Gires, F., Tournois, P. *CR Acad Sc Paris* **T-258** (1964) 6112
62 Treacy, E.B. *IEEE J Quant Electr* **QE-5** (1969) 454
63 Konishi, S., Aoyagi, Y., Namba, S. *Opt Comm* **22** (1977) 358
64 McMullen, J.D. *J Appl Phys* **49** (1978) 16
65 Faubert, D., Chin, S.I. *Can J Phys* **57** (1979) 1359
66 Vanherzeele, H., Van Eck, J.L. *Appl Opt* **20** (1981) 524
67 Lay, M.M.T. *IEEE J Quant Electr* **QE-13** (1977) 389
68 Martin, W.E. *Opt Comm* **21** (1977) 8
69 Martin, W.E., Milam, D. *Appl Opt* **15** (1976) 3054
70 Massey, G.A., Shanmuganathan, K. *Opt Eng* **17** (1978) 247
71 Roychoudhuri, C. *Opt Eng* **16** (1977) 173
72 Hughes, J.L., Donohue, P.J. *Opt Comm* **12** (1974) 302
73 Thomas, C.E., Siebert, L.D. *Appl Opt* **15** (1976) 462
74 Hirschfeld, T., Caulfield, H.J. *J Opt Soc Am* **68** (1978) 28
75 Desbois, J., Gires, F., Tournois, P. *IEEE J Quant Electr* **QE-9** (1973) 213
76 Agostinelli, J., Harvey, G., Stone, T., Gabel, C. *Appl Opt* **18** (1979) 2500
77 Colombeau, B., Vampouille, M., Froehly, C. *Opt Comm* **19** (1976) 201
78 Massey, G.A., Elliot, R.A. *Appl Phys Lett* **29** (1976) 802
79 Kobayashi, T., Ideno, H., Sueta, T. *IEEE J Quant Electr* **QE-16** (1980) 132
80 Proud Jr, J.M., Norman, S.L. *IEEE Trans on Microwave Theory and Techniques* **MTT-26** (1978) 137
81 Bates, H.E., Alfano, R.R., Schiller, N. *Appl Opt* **18** (1979) 947
82 Krrasyuk, I.K., Lukishova, S.G., Pashinin, P.P., Prokhorov, A.M. *Sov J Quant Electr* **4** (1975) 832
83 Kryhanovskii, V.I., Mak, A.A., Sventitskaya, I.N., Serabryakov, V.A., Fligontov, Yu. A., Chertkov, A.A. *Sov J Quant Electr* **7** (1977) 190
84 Lehmberg, R.H., Reintjes, J., Eckardt, R.C. *Opt Comm* **22** (1977) 95
85 Massey, G.A., Elliot, R.A. *IEEE J Quant Electr* **QE-11** (1975) 358
86 Bates, H.E., Henderson, B.J. *J Opt Soc Am* **68** (1978) 919

An electro-optical technique to vary continuously a laser pulse

D. FAUBERT, P. GALARNEAU, S.L. CHIN

The principle of a new technique to vary continuously the duration of a laser pulse using an electro-optical modulator is described. An experiment using a ruby laser demonstrates the validity of the idea.

Many electro-optical techniques are known by now to vary the pulse length of a pulsed laser[1-4] for various applications. In all these methods, the laser pulse duration can be varied by changing the length of a cable. A more convenient way is proposed by using, in addition, an electro-optical – eo – modulator such that a laser pulse duration can be varied by simply varying the distance between the eo element and a mirror with a screw or micrometer. An experiment with a ruby laser and a numerical simulation was also performed to demonstrate the validity of the idea.

Principle of operation

A linearly polarized laser pulse of uniform intensity is transmitted completely through a correctly oriented Glan-Thompson prism – or any other appropriate polarizing or birefringent element (Fig. 1a). It then goes through an unactivated eo element (eg a Pockels cell), hits a totally reflecting mirror M and returns along the same optical path. If $T_p > 2l/c$ where T_p is the pulse duration, l the distance between the eo element and the mirror, and c the speed of light, two counter propagating electromagnetic waves will simultaneously fill the space between the eo element and the mirror when half the pulse has been reflected back onto itself. Then a high voltage 'step' function – or a very long hv 'square' pulse – of magnitude V_π is suddenly applied across the faces of the eo element, where V_π is the retardation voltage necessary to turn the polarization through an angle of $\pi/2$.

Let T be the rise time of the hv pulse. Figure 1a represents the situation of the two counter propagating em waves at time $t = 0$ when the hv is just turned on. The thickness of the eo element is assumed zero for the sake of clarity. Now, because it takes a finite time T for the hv to rise to V_π, the radiation contained in the distance $c \times T$ of region A to the left of the eo element will have its polarization effectively turned by an angle $\theta(t)$ $(0 \leqslant \theta(t) \leqslant \pi/2, 0 \leqslant t \leqslant T)$ after passing through the eo element. It will then be partially reflected by the Glan-Thompson prism. The intensity of this reflection rises from zero ($\theta(0) = 0$) to a maximum at $\theta(T) = \pi/2$ as the hv rises from zero to V_π. This is represented schematically by the rising part (region A) of the pulse in Fig. 1b. The em wave in the rest of the round trip distance between the eo element and M will follow, and its polarization will be rotated by $\pi/2$ and will be totally reflected by the prism. This is represented by the flat – because the intensity is assumed uniform – part (region B) of Fig. 1b It lasts for $(2l/c) - T$ seconds. The left propagating em wave contained in the distance $c \times T$ (region C) at the right side of the eo element will have its polarization effectively turned by an angle $\theta'(t) + \pi/2$ after two passages through the eo element $(0 \leqslant \theta'(t) \leqslant \theta/2, 0 \leqslant t \leqslant T)$, and will be partially reflected by the prism. It contributes to region C of Fig. 1b where the pulse amplitude reduces to zero at $t = T$ ($\theta'(T) = \pi/2$ at $V(T) = V_\pi$). The rest of the left propagating radiation at the right side of the last region will reach the eo element only after $V(t)$ has reached V_π. Its polarization will suffer a rotation of π in two passages through the eo element and will go through the prism without being reflected. Thus we have a reflected pulse as shown schematically in Fig. 1b. The pulse length can be varied by varying the length l.

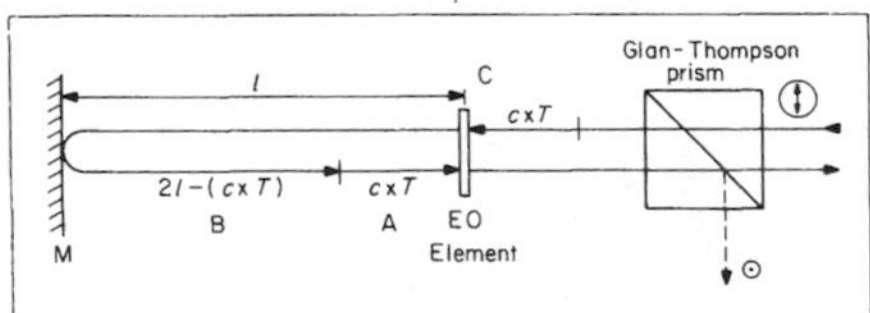

Fig. 1a Schematic of the pulse extraction arrangement at time $t = 0$ when the hv is just turned on

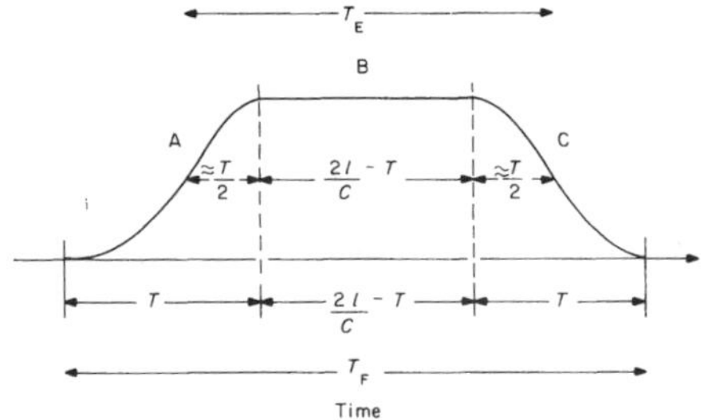

Fig. 1b The extracted pulse shape

The authors are at the Laboratoire de Recherche en Optique et Laser (LROL) Département de Physique, Université Laval Québec, G1K 7P4, PQ Canada. Received 24 November 1980

0030-3992/81/020079-04 $1981 IPC Business Press

From Fig. 1b, we see that the full width T_F is given by $T_F = T + 2l/c$ while the half width (fwhm) T_E is

$$T_E \simeq 2l/c \quad (\text{for } T \leqslant 2l/c) \tag{1}$$

As l decreases, region B of the extracted pulse shrinks and becomes zero when $2l/c = T$ at which point regions A and C fuse together. When l is reduced further, (ie $T > 2l/c$), region A starts to shrink from the right because the length of the corresponding region A in Fig. 1a is no longer $c \times T$ but $2l/c$ which is less than $c \times T$. The width of region C is unchanged but the extracted intensity doesn't start from a maximum and decrease. Instead, it will start from where the extraction of region A ends and increase through a maximum, then decrease to zero. This is because the leading part of region C (Fig. 1a) has passed through the eo element twice before $V(t)$ attains V_π. When l approaches zero, region A shrinks to zero. Region C remains and the extracted pulse increases from zero at $t = 0$ through a maximum and back to zero at $t = T$. A minimum pulse length is thus obtained at $l = 0$ (fwhm):

$$T_{\min} \cong T/2 \quad (l = 0) \tag{2}$$

The rise time T can be made as short as ~ 100 ps.[2,3,5,6] Thus in principle the minimum pulse length attainable with present day fast rise hv techniques could be as short as about 50 ps.

We note also that, in principle, the peak power of the laser pulse (Fig. 1b) will be unchanged when l is varied because a uniform laser intensity was assumed. It should also be the case for a time varying laser pulse so long as the peak region is extracted.

Experimental verification

A passively Q-switched ruby laser was used to demonstrate this idea. The eo element (a KDP crystal) and the Glan-Thompson prism were placed inside the laser cavity (Fig. 2) and the mirror M of Fig. 1a was replaced by the partially reflecting output mirror M of the laser cavity. Let us now follow the sequence of events. First the ruby was pumped above the lasing threshold by a flash lamp. When the Q-switching dye (DDI in methanol) was saturated, a linearly polarized standing wave was formed inside the Fabry-Perot laser cavity between S_1 and the totally reflecting mirror. The energy that leaked out of M triggered the spark gap and a high voltage pulse (V_π = 20.8 kV) lasting ~ 62 ns is applied across the faces S_2 and S_3 of the KDP crystal. Its rise time T was about 2 ns as shown in Fig. 3. The em energy

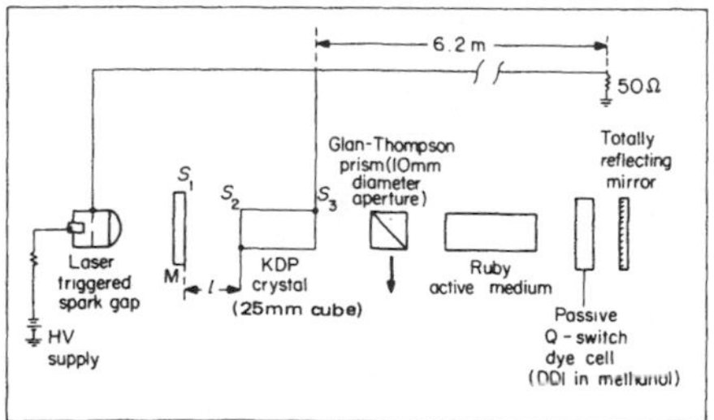

Fig. 2 The experimental setup to extract and vary a laser pulse

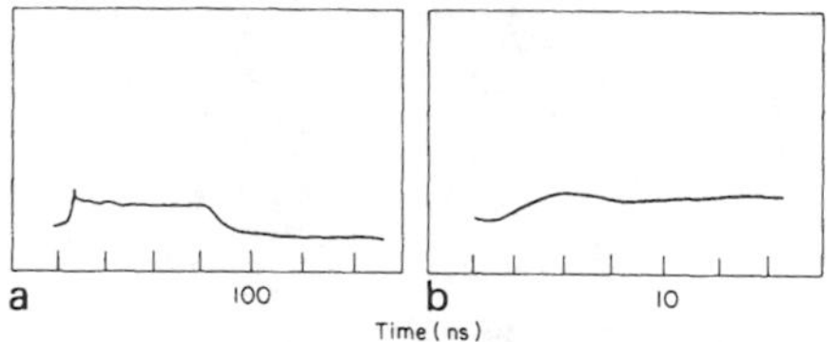

Fig. 3 Shapes of 20.8 kV pulses in the experiment measured by a hv probe of sub-nanosecond rise time (Reference 8): a – Total pulse; b – Expanded rising part of the pulse

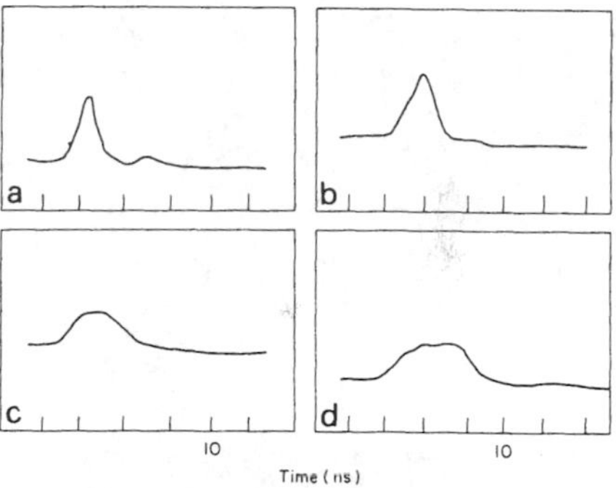

Fig. 4 Extracted laser pulses at different lengths of l; a – 1.75 cm; b – 2.25 cm; c – 30 cm; d – 62;5 cm

inside the distance l between the KDP and M was thus extracted through reflection by the prism along the direction of the thick arrow (Fig. 2). By varying l different pulse lengths were obtained. Typical pulses are shown in Fig. 4 for different l and Fig. 5 shows the quasi-continuous variation of the extracted pulse length T_E (fwhm) as a function of l (solid points) from 1 ns to 5 ns – these pulses were measured by a fast photodiode and a fast oscilloscope (Tektronix 519) with a combined rise time of about 0.5 ns. As expected, the pulse duration T_E varied with l. For example, T_E = 2 ns at l = 30 cm, as predicted by equation (1). Also, T_E reaches a minimum value of ~1 ns as predicted by equation (2) (rise $T \simeq 2$ ns). The solid line in Fig. 5 is the result of a theoretical simulation. It fits very well the experimental points below 2.5 ns. The deviation at larger T_E is attributed to the uniform intensity assumption made in the calculation (see below).

Also worth noting is the occasional appearance of a secondary pulse as shown in Figs 4a and 6. This occurred whenever there was an overshoot at the rising end of the hv pulse (Fig. 3). Qualitatively speaking (see Fig. 1) the rise time T could be considered as the time from $V = 0$ to when the overshoot descended back to $V = V_\pi$ when there was an overshoot in the hv step function. Thus, in the extracted pulse, the rising part (region A of Fig. 1b) went through a minimum after reaching the maximum value, because the polarization was turned by more than $\pi/2$ in the region of the overshoot, and consequently the extraction was reduced. Region B was unaffected and, in region C, this pulse would rise again through a secondary peak after it has descended to zero. Again, the secondary peak was caused by the extra rotation of the polarization following a rotation of π. Thus, after the pulse has descended to zero, an extra was obtained

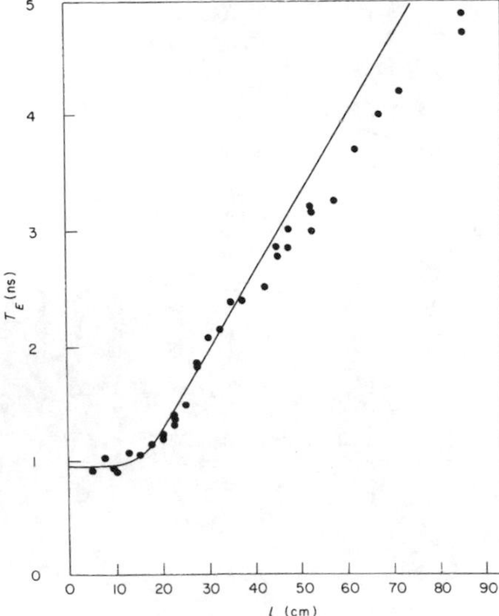

Fig. 5 The extracted pulse duration T_E (fwhm) against distance l: points, experimental; line, theoretical calculation

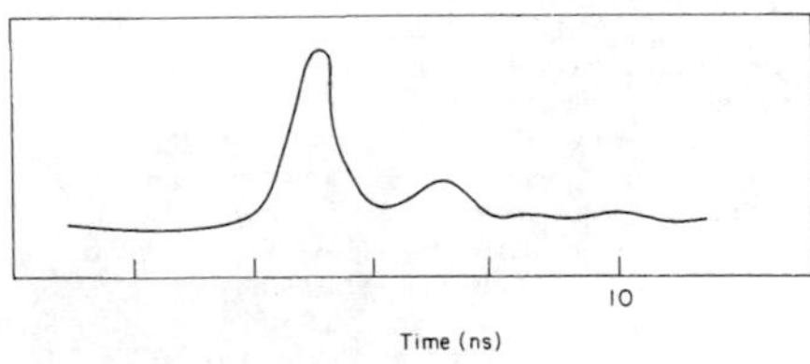

Fig. 6 Example of an extracted pulse containing a secondary peak when there was an overshoot in the hv pulse

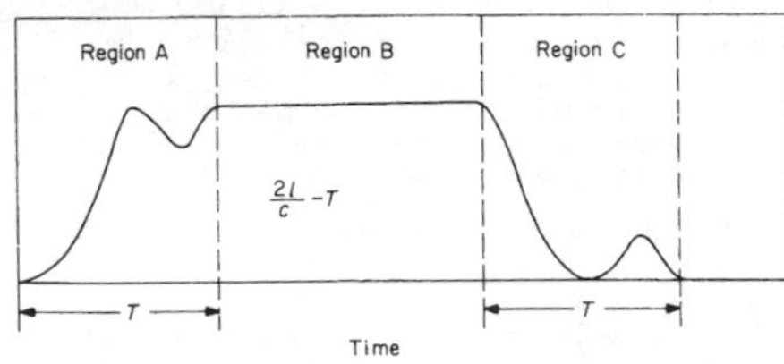

Fig. 7 Schematic of the extracted pulse shape when an overshoot occurred at the end of the rising part of the hv pulse ($2l/c > T$)

which lasted as long as the overshoot. This is shown schematically in Fig. 7. As l decreases, region B of the extracted pulse vanishes eventually at $2l/c = T$. Following the same argument given near the end of the previous section, we can see that region A (Fig. 7) shrinks from the right and that the dip disappears as $2l/c$ is reduced below T. What is left is a first pulse and a secondary peak of region C. This is the case for the results of Figs 4a and 6. (ie $2l/c < T$) and will be verified in the next section.

Numerical simulation

A simplified numerical simulation of the pulse extraction was conducted. At time $t = 0$, it was assumed that two counter propagating em waves of equal intensities and uniform in time filled the laser cavity shown in Fig. 2.

Each of those two waves was divided into an equal number of slices and the intensity of each extracted slice $I_E(t)$ was calculated using the following equations:[7]

$$I_E(t) = I_i(t) \sin^2 (\pi V(t)/2V\pi) \qquad (3)$$

$$I_i(t) = \text{constant}$$

$$V(t) = V_\pi t/T \qquad t < T$$

$$= V_\pi \qquad t \geqslant T \qquad (4)$$

where I_i is the laser intensity of the slice incident at the KDP crystal. $V(t)$, the hv across the KDP crystal, was assumed to rise linearly from $t = 0$ to $t = T$ and to flatten off at $V(T) = V_\pi$. The finite dimensions of the KDP crystal were taken into account. The potential difference that a slice of em wave sees at the KDP crystal was taken to be the average of the potential differences the slice sees when it enters and when it leaves the crystal. The extracted pulse was constructed point by point using the experimentally determined value of $T = 1.95$ ns. The fwhm (T_E) was determined for different values of l used. The variation of T_E with l is shown as the solid line in Fig. 5. The deviation from the experimental points at large l is due to the assumption that $I_i(t)$ = constant in the calculation. In the real experiment, the laser pulse has a finite duration of ~ 10 ns. Thus, if l becomes sufficiently large, the extracted pulse covers the descending (and ascending) part of the laser pulse. This renders the real width smaller than that obtained by using a uniform laser pulse.

The secondary peak (Fig. 6) can also be simulated by assuming a Gaussian overshoot in the hv step function (Fig. 8); ie a Gaussian function is superimposed at the end of the linear rise of the hv before it flattens off. From Fig. 8, we have

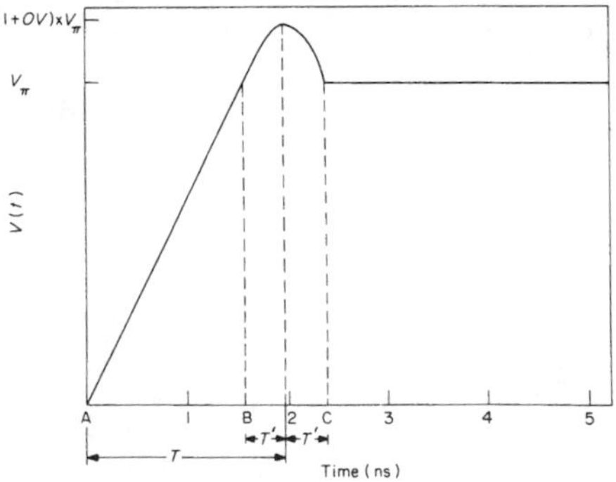

Fig. 8 Analytical hv pulse shape with a Gaussian overshoot used in the numerical calculation to verify the occurence of the secondary peak in the extracted pulse

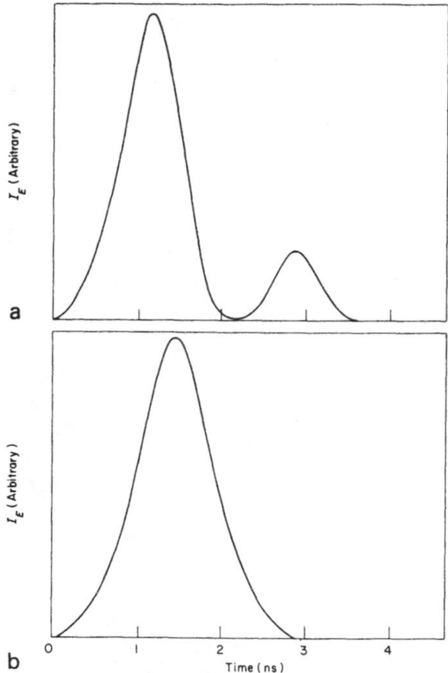

Fig. 9 Numerically simulated extracted pulses: a – in the presence of a Gaussian overshoot in the hv pulse; b – with an overshoot in the hv pulse

$$V(t) = V_\pi t/(T - T'), \quad 0 \leqslant t < T - T'$$

$$= (1 + OV)\, V_\pi \exp\left(-\left(\frac{t - T}{b}\right)^2\right) \quad T - T' < t < T + T'$$

$$= V_\pi \quad t \geqslant T + T'$$

where

$$b^2 = \frac{T'^2}{\ln(1 + OV)}.$$

Using the experimental conditions $l = 15$ cm, $T = 2.0$ ns, $OV = 0.17$, $T' = 0.4$ ns, we obtained an extracted pulse with a secondary peak as expected (Fig. 9a). For comparison, Fig. 9b shows the extracted pulse when OV was set equal to zero, ie without overshoot. Indeed, the secondary peak disappeared.

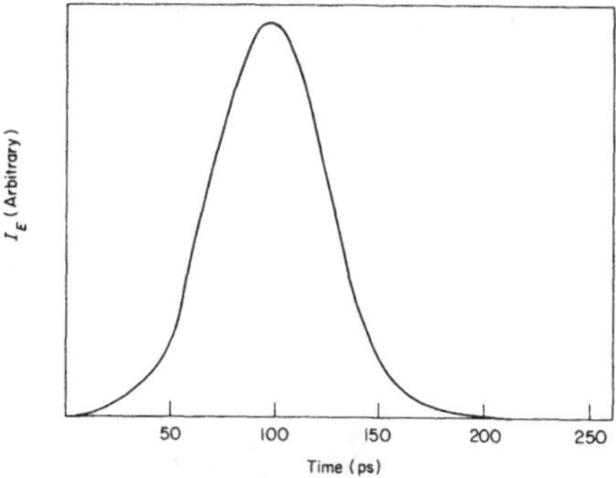

Fig. 10 Theoretical pulse shape of minimum width

Figure 10 shows a simulated shortest extracted pulse assuming that the shortest rise time of the hv was $T = 100$ ps, $l = 1$ cm. No overshoot was used in the hv. The fwhm of this minimum pulse is ~ 62 ps as expected in the qualitative discussion earlier.

Conclusion

We have proposed a new electro-optical technique for extracting the em energy of a laser pulse contained between an eo modulator and a mirror. The duration of the extracted pulse can be varied by simply varying the spacing between the eo modulator and the mirror. An experiment using a ruby laser and a numerical simulation were performed proving the validity of the principle. The minimum length (fwhm) is roughly half the rise time of the hv step function applied across the eo element. In principle, this technique can be used in any laser system so long as a proper modulator can be found for the particular laser wavelength. The element can either be inside or outside the laser cavity.

Acknowledgements

The authors appreciate the technical help of Mr. L. Turgeon during the experiment. The work was supported in part by the NSERC of Canada and the Ministry of Education of Quebec under the grant FCAC l'Action Concertée Laser CO_2.

References

1 **Alcock, A.J., Richardson, M.C.** *Opt Comm* **2** (1970) 65
2 **Scott, J.C., Palmer, A.W.** *J Phys E Sci Instr* **11** (1978) 901
3 **Ireland, C.L.M.** *J Phys E Sci Instr* **8** (1975) 1007
4 **Machewirth, J.P.** *Laser Focus* (March 1977) 81
5 **Antonetti, A., Malley, M.N., Mourou, G., Orszag, A.** *Opt Comm* **23** (1977) 435
6 **Platte, W.** *Optics and Laser Technol* **10** 1 (1978) 40
7 **Yariv, A.** 'Quantum Electronics' (J. Wiley and Sons, New York, 1975) Chapter 14
8 **Sargent, W.J., Alcock, A.J.** *Rev Sci Instr* **47** (1976) 1283

APPENDIX

Reprinted from *Laser Applications in Physical Chemistry*,
ed. D. K. Evans, 1989, pp. 39–62, by courtesy of Marcel Dekker, Inc.

2
Laser Beam Transport

S. L. CHIN Université Laval, Quebec, Canada

2.1 INTRODUCTION

Most laser applications make use of some or all of the special characteristics of a laser radiation, namely, monochromaticity, intensity, and coherence. In the case of application to chemistry, one of the very important factors is often the laser intensity. With intense laser radiation, new chemical reactions can be induced that otherwise cannot be realized using a normal monochromatic light source. These new chemical reactions are mostly due to nonlinear optical effects such as multiphoton processes, saturation, etc., which leave the chemical species in either a final product state (e.g., dissociation product) or a highly and specifically excited state followed by a final (chemical or physical) excitation to reach the product state. That is to say, very often laser chemistry relies on the nonlinear effects induced by the intense laser radiation [1,2,3,5,7]

On the other hand, intense laser radiation can produce other unnecessary nonlinear effects that will either compete with or disturb those positive effects for chemical applications. For example, when we do an experiment for efficient laser isotope separation, the laser intensity will have to be high to saturate a certain selective transition. Yet, it cannot be too high because the competing effect of laser induced breakdown (creating a spark) would occur, eliminating the selectivity. Even before breakdown occurs, self-focusing or self-defocusing of the laser beam might occur. This might aid in the chemical process, yet will deform the pulse, creating difficulties in the quantitative measurement of the laser fluence and/or intensity responsible for a particular reaction. Such inaccuracy in intensity might eventually lead to errors in absorption cross section measurement, interpretations, further studies, and designs.

In a laser chemistry experiment, the laser beam will have to be transported into the interaction region. In this chapter, some of these nonlinear effects that could exist in the region from the output end of the laser to the interaction region (inclusively) will be discussed. These effects include saturation of laser attenuators, laser-induced breakdown, self-focusing, and self-defocusing. In particular, a model describing self-focusing and/or self-defocusing of an ir (infrared) CO_2 laser in SF_6 and CDF_3 will be given. This model fits recent observations in our laboratory and elsewhere and can be used to predict the phenomena in other polyatomic absorbing gases. A consequence of these effects, as mentioned above, is the introduction of errors in the measured laser fluence and intensity. Thus, current techniques of measuring laser fluence and intensity will also be discussed. These techniques are usually assumed trivial and are seldom discussed in the literature. Yet, as can be seen in the sections that follow, they could be rather complicated.

2.2 INTENSITY CRITERION

Since laser-induced nonlinear effects occur only at high laser intensity, it is necessary to define what "high" means. Any "threshold" intensity at which a nonlinear effect starts to occur significantly is considered as a "high" intensity. Unfortunately, every nonlinear effect has a different "threshold" intensity. As such, we cannot define a universal magic intensity that takes care of all nonlinear effects.

The following are a few examples. The threshold intensity for laser-induced breakdown of gases at 1 atm (see Sec. 2.7.5 of this chapter) is of the order of a few megawatts per square centimeter using visible lasers. The threshold fluences [1,2] for ir multiphoton dissociation of SF_6 and UF_6 are about 1.2 J/cm^2 and 7 mJ/cm^2, respectively (see later chapters) at laser wavelengths of 10.6 μm and 16 μm. Although it is not strictly correct to talk about intensity in such cases, these fluences correspond to intensities of about 6 MW/cm^2 and 3.5 kW/cm^2, respectively, assuming a pulse length of 200 ns. Saturation of absorption occurs in the tens of kilowatts per square centimeter to the tens of megawatts per square centimeter range, depending on the material and laser wavelengths. For example, the saturation intensity in the visible region for chloroaluminum phthalocyanine is 100 kW/cm^2 and for cryptocyanine in methanol is 5 MW/cm^2 [3].

2.3 INTENSITY MEASUREMENT

Most laser chemistry experiments use pulsed lasers to obtain the necessary high intensity. We thus discuss only the technique of measuring pulsed laser intensity. The application of the technique to a continuous laser is straightforward.

In principle, the intensity of a laser, defined as the energy of the laser radiation that passes through a unit cross-sectional area in unit time (joules per second per square centimeter or watts per square centimeter), is trivial to measure. It is simply a matter of measuring the total pulsed energy E passing through an interaction region, the pulse duration τ_L, and the cross-sectional area A of the beam, as shown in Figs. 2.1 and 2.2. Ideally, the interaction region is empty during the measurement. The intensity I is then expressed as

$$I = \frac{E}{\tau_L A} \tag{2.1}$$

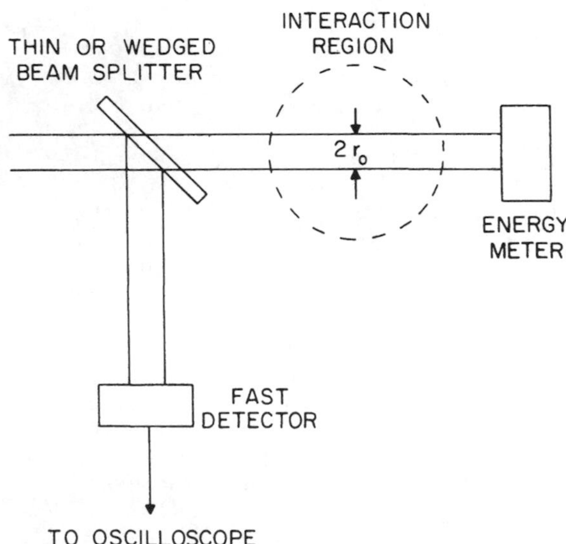

Figure 2.1 Schematic experimental set-up for a laser interaction experiment.

(a)

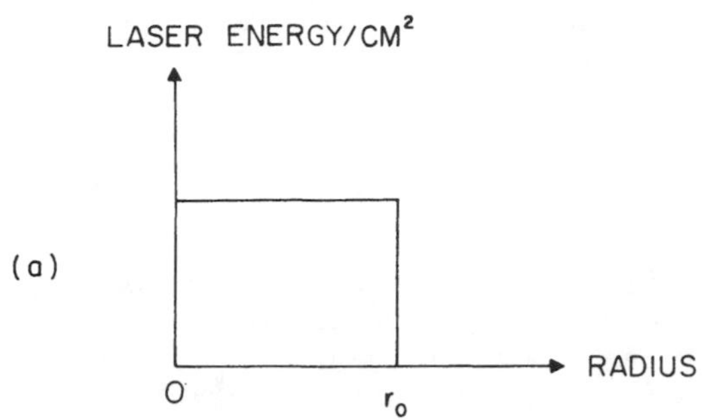

(b)

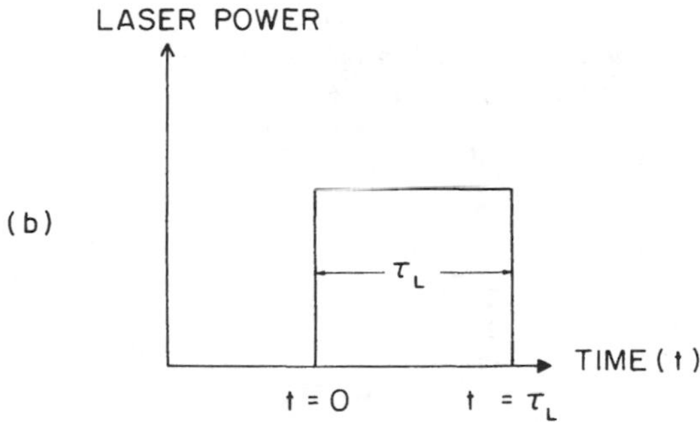

Figure 2.2 (a) Spatial and (b) temporal energy distribution across an ideal cylindrical parallel uniform laser beam of (a) radius r_0 and (b) duration τ_L.

Here, it is implicitly assumed that the laser is very stable so that a few measurements are enough, that the energy distribution across the beam is uniform up to the edge of the beam, as shown in Fig. 2.2(a) for an ideal cylindrical parallel laser beam, and that the temporal form of the pulse is a rectangle as shown in Fig. 2.2(b). Unfortunately, a real laser pulse is neither very stable, nor rectangular in time or space. One of the best compromises that is technically achievable is a cylindrical Gaussin beam, one that is Gaussian spatially and smooth temporally.

Such a laser pulse is one emitted by a laser with cylindrical symmetry oscillating spatially on the TEM_{00} mode giving a Gaussian spatial distribution [Fig. 2.3(a)] and temporally on a single longitudinal mode (SLM) giving a smooth temporal pulse form that often can be approximated as a Gaussian [Fig. 2.3(b). Thus, assuming a Gaussian pulse, the intensity at any point across the beam in the interaction region and at any time can be expressed as

$$I(\vec{r},t) = I_0 e^{-r^2/r_0^2} e^{-t^2/t_0^2} \tag{2.2}$$

where I_0 is the peak intensity one seeks to measure. Experimentally, for pulses of the order of 1 ns or longer, present electronics technology allows one to find dectectors that are fast enough to measure the laser's power profile (not intensity profile) and display it on a fast oscilloscope for most existing laser wavelengths, ranging from the ir to the uv (ultraviolet) [for example, photon-drag and fast pyroelectric detectors for the ir CO_2 laser (9 to 10 μm), vacuum photodiodes for near-ir to uv lasers, etc.]. Picosecond (10^{-12} s) and femtosecond (10^{-15} s) laser pulses require special correlation techniques to measure the pulse width [4]. Since this requires a special lengthy discussion, and is out of the scope of the present

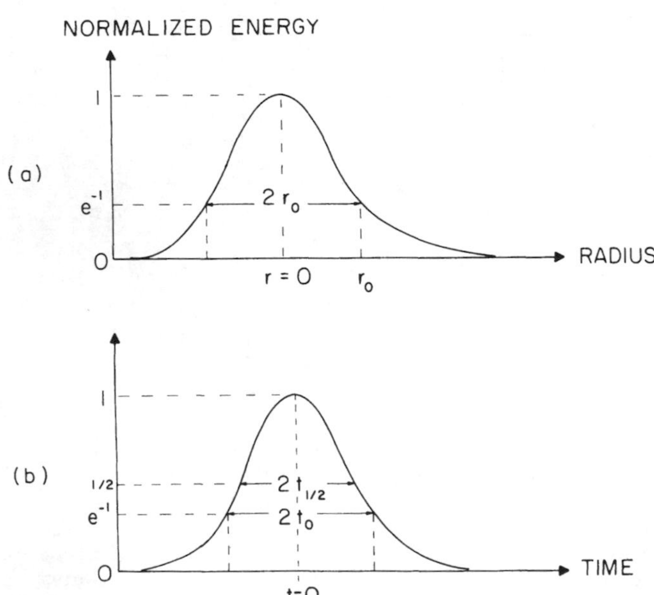

Figure 2.3 Ideal Gaussian laser pulse, (a) spatially and (b) temporally.

chapter, it will not be discussed here. Thus, using a fast detector coupled to a fast oscilloscope, one measures the temporal power profile P(t) of the laser pulse, given by

$$P(t) = P_0 e^{-t^2/t_0^2} \tag{2.3}$$

which is represented by Fig. 2.3(b). The shape of P(t) can now be measured but not yet quantified.

Also, by the definition of power (energy per second, or watt),

$$P(t) = \int_A I(\vec{r},t)\, dA \tag{2.4}$$

where the integral is over the beam cross-sectional area A in the interaction region. Substituting equation (2.2) into (2.4) for a cylindrical symmetric beam, we obtain

$$P(t) = \int_A I_0 e^{-t^2/t_0^2} e^{-r^2/r_0^2}\, r\, dr\, d\theta \tag{2.5a}$$

or

$$P(t) = I_0 e^{-t^2/t_0^2} \pi r_0^2 \tag{2.5b}$$

where r_0 is the radius of the beam at 1/e of the peak power P_0. From equations (2.3) and (2.5b), we have

$$P_0 = I_0 \pi r_0^2 \tag{2.6}$$

or

$$I_0 = \frac{P_0}{\pi r_0^2} \tag{2.7}$$

Equation (2.7) shows that thc peak intensity I_0 of a Gaussian pulse is equivalent to that of a pulse with uniform average power P_0 across the beam area πr_0^2.

After measuring the noncalibrated P_0, the noncalibrated I_0 can now be calculated, if r_0 is known. To measure r_0, many techniques can be used in principle. For example, one can shine the laser onto a photographic film or plate situated in the interaction region. This works only in the wavelength region between near ir to uv to which photographic films are sensitive. One has to make sure that the exposure is not saturated, which means linear attenuation of the beam before the measurement. A microdensitometer can then be used to measure the density distribution as shown in Fig. 2.3(a) for a Gaussian distribution from which r_0 is measured. One can also use electronic techniques such as a detector array or matrix to measure the spatial distribution directly. Another way is to use a slit coupled to an energy meter. By sliding the slit across the beam, the distribution can also be obtained point by point, assuming a stable cylindrically symmetric beam.

To calibrate the intensity quantitatively, one usually uses a calibrated energy meter as shown in Fig. 2.1. The beam splitter directs a known fraction x of the total pulsed energy E_0 into the fast detector that measures P(t). The energy that enters the energy meter is

$$E = (1 - x)E_0 \tag{2.8}$$

assuming no loss through the interaction region. Now by definition,

$$E_0 = \int P(t)\, dt \tag{2.9}$$

For a Gaussian beam [equation (2.3)],

$$E_0 = \int_{-\infty}^{\infty} P_0 e^{-t^2/t_0^2}\, dt$$

$$= P_0 \sqrt{\pi} t_0$$

Hence

$$P_0 = \frac{E_0}{\sqrt{\pi} t_0} \tag{2.10}$$

Using equation (2.8), we get

$$P_0 = \frac{E}{(1 - x)\sqrt{\pi} t_0} \tag{2.11}$$

so that equation (2.7) becomes

$$I_0 = \frac{E}{(1 - x)\sqrt{\pi} t_0 (\pi r_0^2)} \tag{2.12}$$

In equation (2.12), all the quantities at the right-hand side are calibrated measured quantities. Hence, I_0 is calibrated.

Experimentally, it is usually easier to measure $2t_{1/2}$, the full width at half maximum (FWHM) of the power profile P(t), rather than $2t_0$ [Fig. 2.3(b)]. For a Gaussian P(t), at $t = t_{1/2}$,

$$P_0 e^{-t_{1/2}^2/t_0^2} = \frac{1}{2} P_0$$

which gives

$$t_{1/2} = t_0 \sqrt{\ln 2} \tag{2.13}$$

Substituting (2.13) into (2.12) gives

$$I_0 = \frac{2\sqrt{\ln 2}}{\sqrt{\pi}} \frac{E}{(1 - x)\pi r_0^2 (2t_{1/2})} \tag{2.14}$$

The peak intensity I_0' in the interaction region in Fig. 2.1 is given by

$$I_0' = (1 - x) I_0 \tag{2.15}$$

Using equation (2.14), this becomes

$$I_0' = \frac{2\sqrt{\ln 2}}{\sqrt{\pi}} \left(\frac{E}{\pi r_0^2 \tau_L} \right) \tag{2.16}$$

where $\tau_L = 2t_{1/2}$ is the laser temporal full width at half maximum (FWHM).

Similarly, if we measure $r_{1/2}$, the spatial radius of the beam at the half maximum point, rather than r_0, we obtain, using the same argument,

$$I'_0 = \frac{2\ (\ln 2)^{3/2}}{\sqrt{\pi}} \left(\frac{E}{A\ \tau_L} \right) \tag{2.17}$$

where

$$A = \pi r^2_{1/2} \tag{2.18}$$

is the cross-sectional area at half maximum of the beam in the interaction region and

$$\frac{2\ (\ln 2)^{3/2}}{\sqrt{\pi}} \simeq 0.376 \tag{2.19}$$

All quantities on the right-hand side of equation (2.17) are measured absolute quantities. Hence I'_0 is quantified absolutely. Thus, from equations (2.17), (2.19), and (2.1), a Gaussian pulse's peak intensity can be considered as equivalent to that of a uniform cylindrical symmetric pulse whose total pulsed energy is 0.376E and whose spatial cross-sectional area and temporal width are A and τ_L, respectively.

2.4 FLUENCE

It is very often necessary to know the fluence F, defined as the laser pulse energy per unit area. Again, for the unrealistic uniform beam (Fig. 2.2), it is easy to obtain

$$F = \frac{E_0}{A} \tag{2.20}$$

For a Gaussian beam,

$$E_0 = \int_A F\ dA \tag{2.21}$$

where

$$F = F_0 e^{-r^2/r_0^2} \tag{2.22}$$

and F_0 is the peak fluence we seek to measure. (The spatial profile is assumed measured by the technique outlined in the previous section.)

From equations (2.5a), (2.9), (2.21), and (2.22), one obtains

$$F_0 = \left(\frac{E_0}{A}\right) \ln 2 \tag{2.23}$$

and

$$F'_0 = \left(\frac{E}{A}\right) \ln 2 \tag{2.24}$$

where F'_0 is the peak fluence passing through the interaction region. From equations (2.20) and (2.24), the peak fluence of a Gaussian pulse is equivalent to that of a uniform cylindrical beam whose pulsed energy is (ln 2)E and whose cross-sectional area is A. The terms E and A can be mesured and calibrated as described in the previous section. Hence, F'_0 can be measured quantitatively.

2.5 OTHER DISTRIBUTIONS

If the laser beam is not Gaussian, the same techniques still can be used to measure the temporal and spatial distribution. If the distributions are still smooth, approximations can be used. A real distribution can then be approximated to a Gaussian or a triangle, etc., and the same claculation can be carried out to obtain an approximate quantitative calibration. However, if the distribution (be it spatial or temporal) is not smooth, not much can be done to quantify the intensity or fluence. Unfortunately, most multiple-photon absorption and dissociation experiments using CO_2 or HF/DF lasers that have been reported so far in the literature (see Ch. 3) used the latter type of laser pulses. Consequently, the uncertainty of the quantities of laser fluences or intensities quoted in these work would sometimes be significant.

2.6 FOCUSED LASER BEAM

Very often, in order to obtain high intensity and/or fluence, one has to focus the laser beam down to a tiny spot so as to significantly reduce the cross-sectional area A. In such a situation, the temporal distribution of the pulse is still unchanged and can be mesured by a fast detector as shown in Fig. 2.1. If the spatial distribution is Gaussian before focusing, the spatial distribution at the focal plane is still Gaussian [20,21]. One can approximate the focal diameter D, for a TEM_{00} mode, by [20,21]

$$D \simeq f\theta$$

under the conditions $\ell >> f$ and

$$\ell >> \frac{\pi w_0^2}{\lambda}$$

Here, f is the focal length of the lens, θ the far-field divergence angle of the laser beam, ℓ the distance of the laser beam waist w_0 from the lens, and λ the laser wavelength.

In practice, due to aberrations, a direct measurement of the energy distribution at the focal plane is desirable. The laser intensity has to be attenuated linearly first. A photographic film or an electronic device etc. placed at the focus can then be exposed (visible and uv lasers) and the energy distribution can be measured by a microdensitometer [22]. With ir lasers one usually uses approximate techniques such as burning spots on different materials and measuring the spot size. This latter technique is very approximate.

An indirect photographic technique can also be used to measure the spatial distribution of a CO_2 laser. This technique uses the CO_2 laser as a heat source to increase the sensitivity of a small part of a normal photographic film that is being uniformly illuminated by an incoherent light source [23]. The technique is tedious and is seldom used by practicing laser chemists or physicists.

2.7 DIFFICULTIES

All of the above assume that all the media that the laser beam passes through before and up to reaching the interaction region are linear. Yet, when the laser intensity increases, all materials would sooner or later become nonlinear. If we were to do a quantitative laser chemistry experiment in which the laser intensity and fluence are important quantities, we have to avoid or minimize all these nonlinear effects except the one responsible for the chemical reaction. Let us use a typical experimental set up (Fig. 2.4) to illustrate this. In Fig. 2.4, a calibrated beam splitter and a calibrated energy meter are used to measure the laser pulse energy as explained in Sec. 2.3. A few calibrated attenuators are placed behind the beam splitters, after which there is a lens to focus the laser into the interaction region. Attenustors are used because one often wants to study the intensity or fluence dependence of a certain interaction. This particular set-up assumes that the laser energy is not very stable (which is often the case), so that for every shot, a measurement of the energy is necessary. It also assumes that the laser temporal and spatial profiles are Gaussian and remain unchanged from shot to shot, or a least approximately so. The temporal and spatial profile can be measured separately using the method outlined in Secs. 2.3 to 2.5. Thus, equations (2.17) and (2.24) divided by the attenustion factor of the attenustors should give the intensity and fluence reaching the interaction region.

The purpose of this set-up is to deliver a calibrated laser pulse into the interaction region so that a quantitative study of the reaction result can be made as a function of the laser fluence and intensity. Let's see what could go wrong.

2.7.1 Atmospheric Scatterers and Absorbers

The set-up is assumed to be in a normal laboratory. The empty space before the interaction region is atmospheric air. Thus, the first obvious difficulty for not being able to deliver a strong pulse into the interaction region is that there are too many scatterers

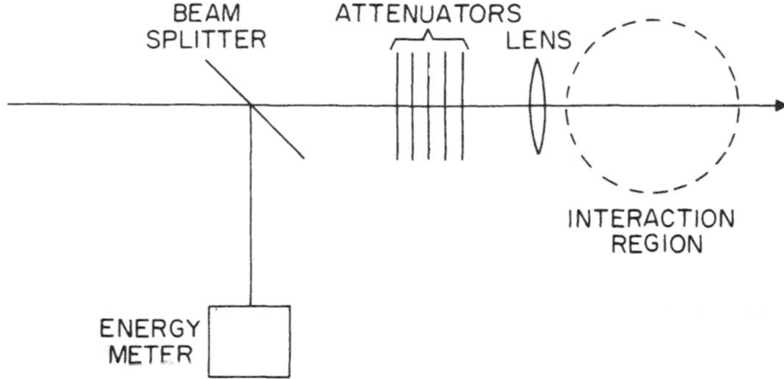

Figure 2.4 Schematic experimental set-up for a laser chemistry experiment.

and/or absorbers (dust, etc.) in the beam path, so that by the time the pulse reaches the beam splitter, the energy is already too low to be of any use. Making the laboratory "free" of dust or low in dust content is certainly the first step to solve the problem. For laser wavelength in the visible, this step may be sufficient. However, many laser chemistry experiments use also ir and uv lasers. For example, in the case of a CO_2 laser (10 μm) the presence of strongly absorbing molecular species in the laboratory's atmosphere, such as SF_6, methanol, ethanol, etc., will significantly reduce the laser energy, and in severe cases the laser might refuse to lase. This might happen especially in a chemistry laboratory where various kinds of organic vapor might often be present. If such a thing happens, one can only wait until the pollution is cleaned up through normal ventilation etc. Similar precautions have to be made in using lasers of other wavelengths, especially uv excimer lasers.

The above-mentioned loss would be very important for a big central laser facility where beam sharing requires the propagation of the laser pulse through a long distance (tens of meters or more) before reaching the target. It would be less important if the experiment takes place very near (within a meter or less) the laser.

2.7.2 Saturation of Attenuators

As the laser intensity is increased, saturation of absorption [6] in the attenuators takes place (see Fig. 2.4). Often, the attenuators are calibrated using an incoherent weak light source. With the intense laser beam, saturation of absorption means that an attenuator will suddenly become "transparent" above a certain intensity. This also affects the spatial and temporal distributions if the laser is not an ideal uniform square pulse [equation (2.1)]. The transmission of the peak intensity of the laser pulse will be more than that of the wing (or the wing is attenuated more than the peak), resulting in narrowing the pulse both spatially and temporally (a more detailed description of this effect is given in Ch. 3). One can no longer assume that the pulse arriving at the interaction region has the same profiles as measured before because E, A, and τ_L have all been changed [equation (2.17)]. Now, if we unknowingly keep the same calibration factor of attenuation, and assume that the energy delivered into the interaction zone is that measured via the energy meter divided by the calibrated attenuation, we have committed a mistake, for in reality not much has been absorbed and most of the pulse is transmitted. Such error might lead to false and inexplicable results, as we observed once using an HF laser and glass attenuators [6].

One might suggest that if the attenuators were put in front of the beam splitter, one might not have such a difficulty because the energy meter measures the pulsed energy directly. However, this will not work because the temporal and spatial profiles (i.e., A and τ_L) were changed. Also, there is a practical difficulty with this latter set-up even if the attenuation is linear. Often, the energy meter will not be sensitive enough when the attenuation is increased, thus limiting the range of energies one can measure.

2.7.3 Self-Focusing and Self-Defocusing

2.7.3.1 General Discussion

Further increase in the intensity of the laser (not fluence) may lead to self-focusing or self-defocusing of the laser beam in the optical components it passes through as well as in the interaction region. Self-focusing or self-defocusing [7] is a manifestation of the nonlinear change of the index of refraction of the material induced by the laser pulse itself. Phenomenologically, the index of a material irradiated by an intense laser is given by

$$n = n_0 + \Delta nI \qquad (2.25)$$

where n_0 is the normal index of refraction when the radiation is weak (e.g., ordinary incoherent light), Δn is the index change induced by the laser, and I is the laser intensity. The term Δn can be positive (self-focusing) or negative (self-defocusing), as demonstrated in Fig. 2.5, in which the transmission of a plane wavefront through a thin slice of nonlinear medium with $\Delta n > 0$ is shown. If the spatial intensity distribution is not uniform, as shown in Fig. 2.5, the index induced at the peak of the distribution (assuming Δn positive) will be higher than that at the wing. Since an electromagnetic wave (laser included) propagates more slowly in a higher index medium, the plane wavefront is deformed into a concave one after passing through the slice of nonlinear medium. A concave wavefront will focus by itself in space, leading to a narrower spatial intensity distribution. Thus the peak intensity and fluence as defined by equations (2.17) and (2.24) are changed, because the beam cross-sectional area A is changed.

Similarly, if Δn is negative, the transmitted wavefront becomes convex and the transmitted spatial distribution becomes wider, that is, self-defocusing.

2.7.3.2 Self-Focusing or Defocusing of a CO_2 Laser Pulse in a Polyatomic Gas

Within the interaction region, assuming a molecular gaseous medium, there is usually a strong absorption of the laser radiation (otherwise it is hard to induce much of a chemical reaction). Normally, in the

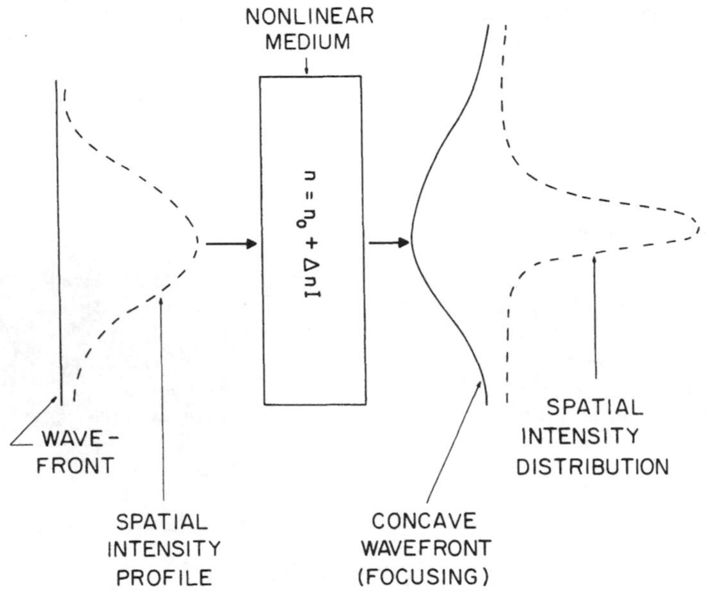

Figure 2.5 Schematic illustration of self-focusing of a Gaussian beam in a nonlinear medium.

case of a two-level system, when there is a resonant absorption, the laser pulse will neither self-focus nor self-defocus [7,8]. Thus, in an absorbing molecular system, not much attention had been paid to the possibility of self-focusing or self-defocusing effect. In the case of the interaction of a CO_2 laser pulse with absorbing gases, only a few publications have reported self-focusing and/or self-defocusing in SF_6 [9–11], NH_3 [12], CH_3F [13], and CDF_3 [14–16]. These involved also multiple photon absorption and the interaction process is not yet completely understood. However, according to our present experimental studies [14–16], the following interaction model seems to explain the existing experimental observations very well.

A. Low-Intensity Model

The key element in our explanation is the fact that the ir absorption band of the molecules is much wider than the laser linewidth. Thus, starting with low laser intensity so that only the first excited vibrational state is involved, the absorption of the laser's ir photons couples only two narrow bands in the $v = 0$ and $v = 1$ states (Fig. 2.6). The width of each band corresponds to that of the initial Boltzmann population distribution in the $v = 0$ state. Now, in the

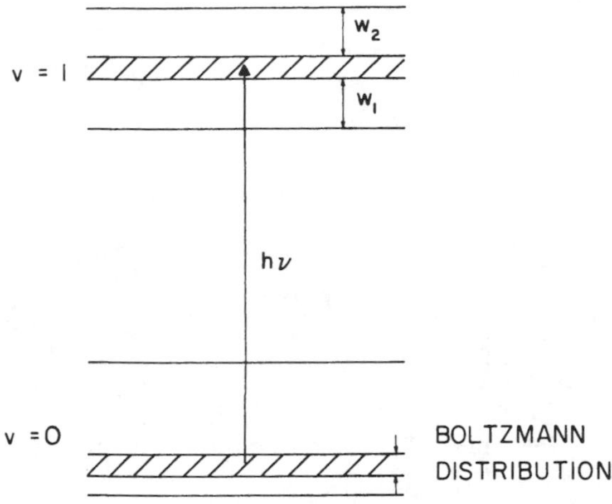

Figure 2.6 Coupling between the first two vibrational bands of a molecule by a laser radiation of frequency ν.

ideal case of a two-level system, self-focusing occurs when the transition between the two levels is at the red side of the laser line whereas self-defocusing occurs at the blue side (Fig. 2.7). In other words, if the energy difference between the two levels is less than the photon energy of the laser, self-focusing occurs and the reverse is true for self-defocusing. Thus, referring to Fig. 2.6, every individual level in the v = 1 band but at the red side of the laser line (the levels within w_1) will interact with the laser radiation, each producing a self-focusing effect of a different magnitude. In other words, since according to Fig. 2.7 the energy difference between any level in w_1 and a level within the ground-state Boltzmann distribution (the shaded area) is less than the laser photon energy, self-focusing of the laser pulse would occur during the interaction with these levels. Similarly, those at the blue side of the laser line (the levels within w_2) will produce self-defocusing.

Thus the outcome of an experiment such as the one shown in Fig. 2.1 is the result of combining simultaneously four phenomena: absorption, saturation, self-focusing, and self-defocusing. These occur to varying degrees across the temporal and spatial profiles of the pulse. Absorption reduces the transmitted amplitude of the laser pulse while saturation narrows the pulse, since the center is more highly transmitted. The latter reinforces the self-focusing effect (makes the pulse's spatial width narrower than self-focusing alone would produce) but reduces the self-defocusing effect. The combination of these effects produces a resultant effect that shows a

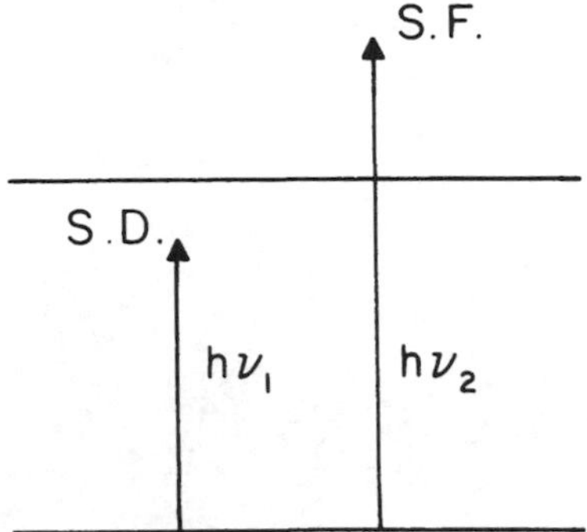

Figure 2.7 Conditions of self-focusing (S.F.) and self-defocusing (S.D.) of laser radiations in a two-level system.

universal pulse amplitude reduction (absorption) plus either spatial pulse narrowing (self-focusing) or pulse widening (self-defocusing) or nothing (no pulse widening nor narrowing). The last case corresponds to a net cancellation of the pulse narrowing and widening effects. Under this condition, the position of the laser line in the frequency/wavelength scale is defined as the central position. Since our measurement represents a combined effect, it is expected that individual rotational fine structure will not have any particular effect on our result, which is what we observed [14–16].

To summarize, at lower laser intensities involving only the ground and first vibrational states, if the laser line is to the red side of the central position, self-defocusing dominates. On the other hand, if the laser line is at the blue side of the central position, self-focusing dominates. This is in agreement with our observations in the case of CDF_3 [14–16].

B. <u>High-Intensity Model</u>

If the intensity of the laser is increased further, multiple photon absorption (or excitation of higher excited states) occurs. This introduces a similar coupling between the first and second, second and third, etc., excited states, which will lead to additional contributions to the nonlinear change of refractive index. If the anharmonicity is small, as in the case of CDF_3 [14–16], the CO_2 laser pulse can change from self-focusing to self-defocusing, depending on the selection rule $\Delta J = \pm 1$. However, if the anharmonicity of the molecule is strong, self-defocusing can change over to self-focusing independently of the selection rule [9–11]. The following paragraphs explain the proposed mechanism. More detailed experimental results are still being obtained to substantiate this model.

Figure 2.8 shows schematically the first three vibrational bands of a polyatomic molecule with a strong anharmonicity, such as SF_6. Solid vertical arrows represent absorption of photons, while S.F. or S.D. indicates that self-focusing (S.F.) or self-defocusing (S.D.) will also occur during the absorption. A dashed vertical arrow indicates only contribution to the real refractive index (self-focusing in this figure without absorption because it does not end in any real level.

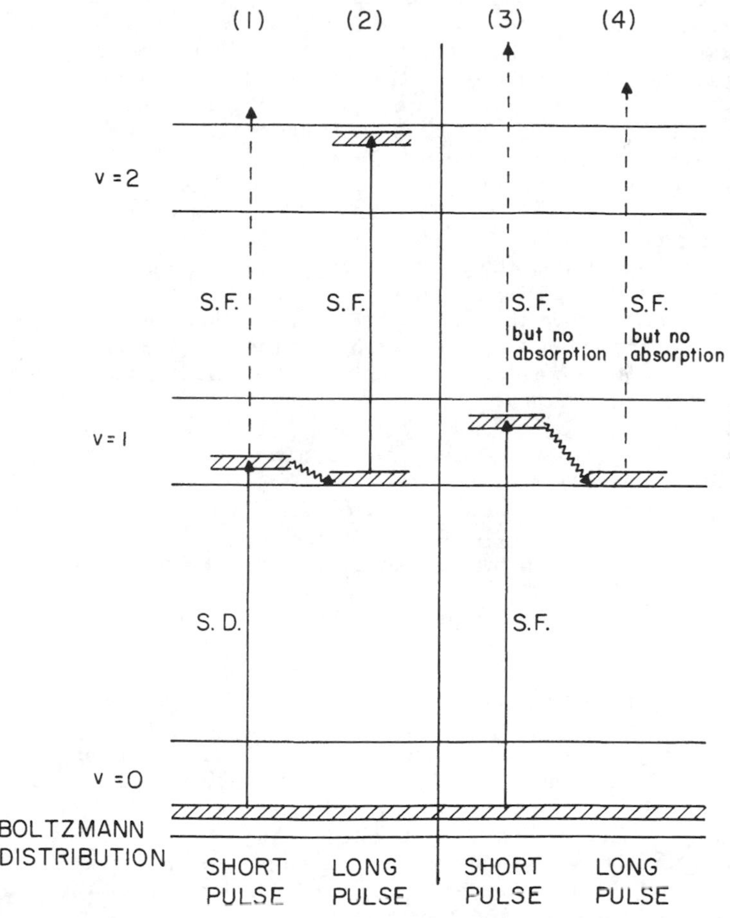

Figure 2.8 Schematic energy diagram illustrating self-focusing (S.F.) and self-defocusing (S.D.) when higher excited states with large anharmonicity are excited.

If the wavelength of the laser radiation is such that self-defocusing occurs during the transition between the v = 0 and v = 1 vibrational bands, as shown by the solid arrows indicated by S.D. under column 1, increasing the intensity of the laser leads to two possible routes of excitation into the higher excited state v = 2.

For a short laser pulse whose duration is less than the rotational relaxation time, the molecule will keep on absorbing one more photon into the next excited state, (column 1: in this particular drawing, there is no real absorption of the second photon, hence a dashed vertical line for the second photon). Because of the anharmonicity, this latter transition (interaction) leads to a self-focusing effect. Further excitation of the v = 3 band will contribute even more to self-focusing. Since the population in the excited states is increased at higher laser intensity, their S.F. contributions to the transmitted pulse become stronger while the initial S.D. contribution becomes weaker because of the depletion of population in the v = 0 band. The result is the change from S.D. to S.F. when the intensity is increased. This is what has been observed in SF_6 when an intense 2-ns CO_2 laser pulse was used [9—11].

For a longer laser pulse, the excited states in v = 1 relax first to some lower levels rotationally within v = 1 (Boltzmann distribution at a lower mean temperature than that of the distribution before relaxation) before absorbing another photon into the v = 2 band (column 2, Fig. 2.8). However, anharmonicity still permits this latter transition to contribute to S.F. Hence, the same argument in the previous paragraph applies and S.D. changes into S.F. at higher laser intensity. This has been observed in our laboratory recently using SF_6 and 200-ns CO_2 laser pulses [18].

If the laser wavelength is shorter so that self-focusing occurs during the transition between v = 0 and v = 1 bands (column 3, Fig. 2.8), further interaction with higher excited states will still lead to self-focusing but probably without any more (or with very little) absorption because of the anharmonicity. This is true whether the laser pulse is long or short (columns 3 and 4).

In the case of molecules with very small anharmonicity, such as CDF_3 (whose anharmonicity is only 0.25 cm^{-1} [14—16], increasing the laser intensity would lead to the same result as in SF_6 only if the laser pulse is very short compared to the rotational relaxation time, (see Fig. 2.9, columns 1 and 3). In our experiment, the shortest pulse used was 4 ns, but with a tail of about 10 ns. This is comparable to or larger than the rotational relaxation time, of the order of a few tens of nanoseconds [19]. As such, up until now, all our experiments can be considered to have been done under the long pulse regime (columns 2 and 4 of Fig. 2.9).

Column 2 predicts a reduction of S.F. Leading eventually to a change into self-defocusing. We observed this in CDF_3 recently

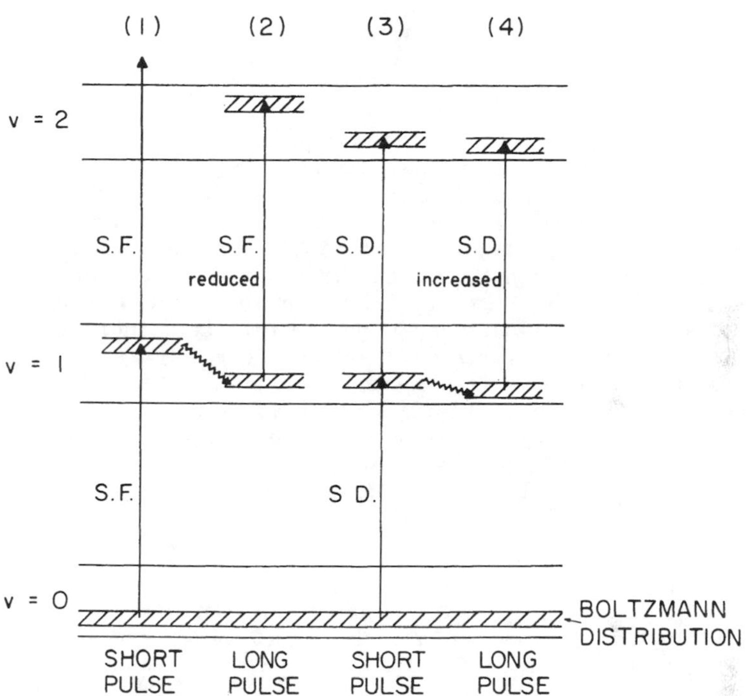

Figure 2.9 Schematic energy diagram illustrating self-focusing (S.F.) and self-defocusing (S.D.) when higher excited states with small anharmonicity are excited.

[14–16] using the 10 R (12) and 10 R (14) lines of a CO_2 laser. This is contrary to the case of SF_6, where S.D. changes into S.F. at higher laser intensity.

Finally, column 4 predicts a further increase of S.D. as the intensity increases. This experiment, in the case of CDF_3, is yet to be performed in our laboratory.

It seems that based on the above explanation, any polyatomic molecular gas will induce S.F. or S.D. of an ir laser while absorbing the laser radiation.

2.7.3.3 Other Wavelengths and Systems

The case of atomic systems using visible lasers has already been studied [7]. In the case of uv lasers whether a molecular absorbing gaseous system (suitable for laser-chemical reaction) will induce self-focusing or self-defocusing of the laser is still not known, although from what we understand at present, it seems very probable.

2.7.3.4 Consequence of S.F. and S.D.

Such self-focusing and self-defocusing effects in an interaction region have at least two consequences, one positive and one negative. The negative one is that the laser pulse is no longer well defined, while the positive one is that one might be able to make use of self-focusing to compensate for the loss of laser intensity due to absorption. A quantitative knowledge of such a compensation requires a quantitative measurement of the complex index of refraction of the gas at various laser intensities. Using Maxwell's equations and the known indices, one can then predict how the laser beam propagates in the gas. Such nonlinear indices of polyatomic molecular gases at the CO_2 laser wavelengths are still unknown. In our laboratory, we have recently devised a simple but powerful technique to measure the nonlinear indices of polyatomic absorbing gases using the CO_2 laser. Preliminary results using SF_6 were very satisfactory [18].

2.7.4 Thermal Blooming

Absorption of the laser energy in the interaction region will lead to heating of the system. If the time of relaxation of the excited energy into translational energy (e.g., V-T relaxation when excited by an ir laser) is shorter than the laser pulse duration, the heating of the system will lead to a further divergence of the laser pulse (thermal blooming [17]) because the index of refraction in the heated region is lower. The consequence is similar to that due to self-focusing.

2.7.5 Optical Breakdown and Damage [7]

Further increase in laser intensity will lead to ionization of the atoms and molecules in the path of the laser beam. If the atomic or molecular density is high (e.g., solids, liquids, or gases at pressures greater than 1 torr), a spark will be created and the material will be damaged.

Such a spark is the result of a chain of interactions. Consider an atomic gas. The spark is started with the creation of a few free electrons through multiphoton ionization of some impurity molecules whose ionization potential is low. These free electrons are accelerated in the strong laser field in collision with the surrounding atoms (inverse bremsstrahlung). The kinetic energy of these electrons will increase to the point that in colliding with the next atoms, they ionize them, creating more electrons, which are in turn accelerated in the laser field through collisions. They will then acquire enough energy to ionize more atoms, and so on. An avalanche ionization soon takes place, resulting in a plasma. The sudden expansion of the plasma leads to an acoustic wave (a snapping sound), while the recombination of the free electrons and ions results in the emission of a broadband bright light (spark).

In the case of a transparent solid, the free electrons can be those excited from the valence band into the conduction band of the solid. Once the free electrons are created, the same chain reaction takes place, creating a breakdown in the solid.

Sometimes, before reaching the interaction zone, the laser beam has already self-focused in passing through the optical components. Such self-focusing increases the laser intensity in the material, leading to a breakdown. Optical windows or lenses are often damaged internally through such a mechanism.

When a lens is used, multiple reflections may lead to accidental focusing of the laser radiation back onto the lens itself or on some other components (Fig. 2.10). This often leads to damage also and should be avoided. A way to do so is to use antireflection coating. Even this might cause damage if the laser is very intense.

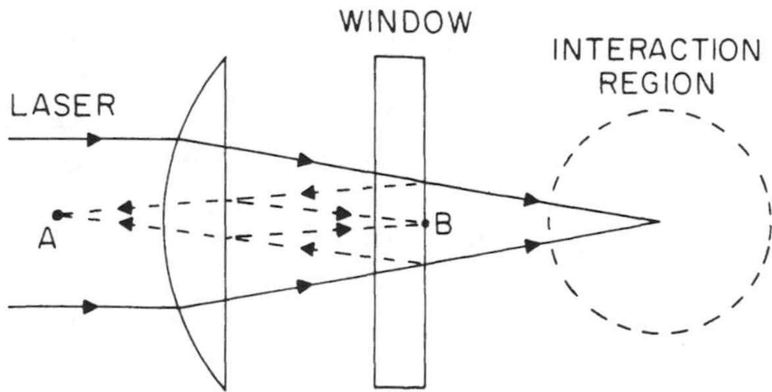

Figure 2.10 Schematic diagram showing two of the many secondary foci after a laser beam passes through a focusing lens and a window. Refraction in the optical components is neglected for simplicity. If the laser is very intense, focus A will create a spark. If the lens is near this focus, it will be damaged. Similarly, focus B will damage the window. Similar consideration of multiple reflections at other surfaces or any other combination of lens and windows must be given to minimize damage.

2.8 SUMMARY

In this chapter, present techniques of measuring laser intensity and fluence have been described. Different phenomena and effects that could lead to errors in the measurements and in the laser-chemistry process were discussed. Some emphasis was given to some of our recent work on self-focusing and/or self-defocusing of a CO_2 laser pulse in polyatomic gases. Other phenomena included optical breakdown, thermal blooming, and scattering.

REFERENCES

1. P. Kolodner, C. Winterfeld, and E. Yablonovitch, Opt. Commun., 20:119 (1977).
2. A. Kaldor, P. Rabinowitch, D. M. Cox, J. A. Horsley, and R. Brickman, "Laser Chemistry Experiments with UF_6," paper presented at the International Quantum Electronics Conference, Atlanta, June (1978).
3. W. Koechner, Solid State Laser Engineering, Springer-Verlag, Berlin, p. 438 (1976).
4. S. L. Chapiro, ed., Ultrashort Light Pulses, Springer-Verlag, Berlin (1977).
5. M. D. Levenson, Introduction to Nonlinear Laser Spectroscopy, Academic Press, New York (1982).
6. S. L. Chin, D. K. Evans, and M. D. McAlpine, Possible source of anomalous results in i.r. laser photochemistry, Appl. Opt., 22:963 (1983).
7. Y. R. Shen, The Principles of Nonlinear Optics, Wiley, New York (1984).
8. L. Allen and J. H. Eberly, Optical Resonance and Two-Level Atoms, Wiley, New York (1975).
9. A. Nowak and D. Ham, Self-focusing of 10μ laser pulses in SF_6, Opt. Lett., 6:185 (1981).
10. P. Bernard, P. Galarneau, and S. L. Chin, Self-focusing of CO_2 laser pulses in low pressure SF_6, Opt. Lett., 6:139 (1981).
11. J. Ackerhalt, H. Galbraith, and J. C. Goldstein, Opt. Lett., 6:377 (1981).
12. I. A. Al-Saidi, O. J. Biswas, C. A. Emshary, and R. G. Harrison, Opt. Commun., 52:336 (1985).
13. M. R. Siegrist, P. D. Morgan, and M. R. Green, J. Appl. Phys., 49:3699 (1978).

14. P. Galarneau, Y. Beaudoin, and S. L. Chin, "Self-focusing and self-defocusing of a TEA-CO_2 laser pulse in CDF_3 vapor," Proc. SPIE Quebec International Symposium, 2–6 June (1986).

15. Y. Beaudoin, P. Galarneau, and S. L. Chin, An experimental study of self-focusing and self-defocusing of a TEA-CO_2 laser pulse in CDF_3, Appl. Phys. B, 42:225 (1987).

16. P. Galarneau, Z. Y. Niu, F. Yergeau, S. L. Chin, D. K. Evans, and R. D. McAlpine, Simultaneous nonlinear absorption and index effects in the propagation of intense TEA-CO_2 laser pulses through CDF_3, Appl. Opt., 24:2804 (1985).

17. W. W. Duley, CO_2 Lasers, Effects and Applications, Academic Press, New York (1976).

18. Y. Beaudoin, I. Golub, and S. L. Chin, Direct measurement of the nonlinear dispersion of SF_6 with a TEA-CO_2 laser, Opt. Commun., 63:325 (1987).

19. D. Harradine, B. Foy, L. Laux, M. Dubs, and J. I. Steinfeld, Infrered double resonance of fluoroform-d with a tunable diode laser, J. Chem. Phys., 81:4267–4280 (1984).

20. A. Siegman, Lasers, University Science Books, Mill Valley, California (1986).

21. D. Marcuse, Light Transmission Optics, 2nd ed., Van Nostrand Reinhold, New York (1982).

22. L. A. Lompre, G. Mainfray, and J. Thibault, Instantaneous measurement of the intensity distributing of a focused high power laser pulse, Rev. Phys. Appl., 17:21 (1982).

23. G. R. Mitchel, B. Grek, T. W. Johnston, F. Martin, and H. Pepin, Nanosecond photography at 10.6μ using silver halide film, Appl. Opt., 18:2422 (1979).

INDEX